T0137402

Springer Theses

Recognizing Outstanding Ph.D. Research

Aims and Scope

The series “Springer Theses” brings together a selection of the very best Ph.D. theses from around the world and across the physical sciences. Nominated and endorsed by two recognized specialists, each published volume has been selected for its scientific excellence and the high impact of its contents for the pertinent field of research. For greater accessibility to non-specialists, the published versions include an extended introduction, as well as a foreword by the student’s supervisor explaining the special relevance of the work for the field. As a whole, the series will provide a valuable resource both for newcomers to the research fields described, and for other scientists seeking detailed background information on special questions. Finally, it provides an accredited documentation of the valuable contributions made by today’s younger generation of scientists.

Theses are accepted into the series by invited nomination only and must fulfill all of the following criteria

- They must be written in good English.
- The topic should fall within the confines of Chemistry, Physics, Earth Sciences, Engineering and related interdisciplinary fields such as Materials, Nanoscience, Chemical Engineering, Complex Systems and Biophysics.
- The work reported in the thesis must represent a significant scientific advance.
- If the thesis includes previously published material, permission to reproduce this must be gained from the respective copyright holder.
- They must have been examined and passed during the 12 months prior to nomination.
- Each thesis should include a foreword by the supervisor outlining the significance of its content.
- The theses should have a clearly defined structure including an introduction accessible to scientists not expert in that particular field.

More information about this series at http://www.springer.com/series/8790

Silvia Celli

Gamma-ray and Neutrino Signatures of Galactic Cosmic-ray Accelerators

Doctoral Thesis accepted by
the Gran Sasso Science Institute, L'Aquila, Italy

Springer

Author
Dr. Silvia Celli
Department of Physics
Gran Sasso Science Institute
L'Aquila, Italy

Supervisor
Prof. Felix Aharonian
High Energy Astrophysics
Max Planck Institute for Nuclear Physics
Heidelberg, Germany

ISSN 2190-5053 ISSN 2190-5061 (electronic)
Springer Theses
ISBN 978-3-030-33126-9 ISBN 978-3-030-33124-5 (eBook)
https://doi.org/10.1007/978-3-030-33124-5

This Springer imprint is published by the registered company Springer Nature Switzerland AG
The registered company address is: Gewerbestrasse 11, 6330 Cham, Switzerland

To my parents,
whom I admire
for the education, nurture and support
they devoted to me.

To Francesco,
for his persistent dedication and love.

Supervisor's Foreword

The current paradigm of Galactic cosmic rays assumes that supernova remnants (SNRs)—the results of gigantic explosions of stars—are responsible for the locally measured fluxes of cosmic rays at energies below 10^{15} eV. For decades, this belief has been based on phenomenological arguments and theoretical meditations. The kinetic energy contained in shock waves of young SNRs is one of the strongest arguments supporting SNRs as the major factories of Galactic cosmic rays. Also, the so-called diffusive shock acceleration has been established as a viable mechanism for the effective acceleration of particles in young SNRs.

Over the last decade, we have seen remarkable progress towards the understanding of the origin of cosmic rays in two directions. The first one concerns the precision and high quality of measurements of primary and secondary cosmic rays. Although these results can be interpreted in the framework of the 'standard model', the detected 'excess' of antiparticles (positrons and antiprotons) requires non-negligible modification of the current paradigm of Galactic cosmic rays. Still, several key issues regarding, in particular, the sites and sources of cosmic-ray production are not fully resolved and identified. A real breakthrough in this regard is expected from gamma-ray observations. The recent discoveries of gamma-ray emission at high and very high energies from many young and middle-aged SNRs confirm, in general, the early theoretical predictions, but at the same time raise new questions and challenges. Advanced and new detailed calculations concerning the acceleration and escape of particles in SNRs and their interactions leading to the gamma-ray and neutrino production become 'hot' topics in the field.

Some of these topics constituted the basis of the Ph.D. thesis of Dr. Silvia Celli. She has completed a solid Ph.D. thesis, performed on a high professional level. The problems discussed and studied in the thesis are related to (1) the acceleration, propagation, and radiation of particles in supernova remnants; (2) very high-energy neutrinos from the Galactic Center; (3) the potential of the next-generation gamma-ray and neutrino detectors, CTA and KM3NeT, for the study of extended non-thermal astrophysical sources in the Galaxy. Here, she has demonstrated excellent computational skills and a deep understanding of the underlying physics. The obtained results and conclusions are based on extensive analytical and numerical simulations. They are formulated and presented in transparent and convenient forms that can be readily used in the interpretations of gamma-ray and neutrino observations as well as in confident predictions for future measurements. This concerns especially the identification of nature ('hadronic or leptonic?') of gamma-ray emission from young and middle-aged supernova remnants and the search for cosmic-ray PeVatrons.

Heidelberg, Germany
July 2019

Prof. Felix Aharonian

Preface

Supernova remnants are believed to be the major contributors to the observed Galactic cosmic-ray flux, though indisputable observational pieces of evidence of such statement are still missing. A crucial aspect of the supernova remnant paradigm for the origin of Galactic cosmic rays is that particle acceleration, as due to diffusive shock acceleration, requires effective confinement of particles in the shock region to let them achieve energies up to the so-called *knee*, around $\sim 10^{15} - 10^{16}$ eV. However, the current theoretical description of cosmic-ray acceleration and propagation within and around supernova remnants suffers from certain limitations, which also affect the predictions on the shape of the energy spectra of secondary gamma rays and neutrinos. In particular, in this thesis, two relevant aspects of this theory are investigated: the particle acceleration at shocks propagating in clumpy non-homogeneous environments and the particle escaping process from the acceleration site. The standard diffusive shock acceleration model usually assumes that shocks expand into ideally uniform environments, while a more realistic picture should consider an inhomogeneous gas distribution where supernova remnants develop. In this work, I conducted a detailed study on the particle acceleration and propagation through non-homogenous structures and its effect on the resulting secondary radiation. Regarding the particle escape from the acceleration site, I developed a phenomenological model to investigate this process and its impact on the gamma-ray emission from middle-aged supernova remnants, where particle escape is expected to be effective. I will show that spectroscopic and morphological studies of the gamma rays coming from both inside and immediately outside of those remnants can provide insight into the escaping process in general, and in particular, will shed light on their ability to act as *cosmic-ray PeVatrons*. So far, the only hint of the presence of a PeVatron has been found in the Galactic Center region, whose nature is, however, unclear. Under the assumption that the observed gamma-ray flux originates from hadronic interactions, I calculated the expected flux of multi-TeV neutrinos in order to investigate its detectability with future km^3-scale neutrino telescopes. Finally, a comparative analysis of the performances of the two major upcoming detectors, namely CTA and KM3NeT, is presented in the context of future studies on the origin of Galactic cosmic rays through respectively gamma-ray and neutrino observations.

The thesis is organized as follows:

- In Chap. 1, the supernova remnant paradigm for the origin of cosmic rays is introduced and a discussion concerning possible Galactic PeV accelerators is presented. As gamma rays and neutrinos constitute observational signatures of particle acceleration and propagation, a review of their properties and detection techniques is provided.
- In Chap. 2, the propagation of accelerated particles within supernova remnants is investigated in the presence of strong shocks evolving through non-homogeneous media. A numerical approach to the particle transport under these conditions is here provided for the first time, conditions that represent realistic situations for the environments where sources as supernova remnants usually expand. Since dense molecular clumps constitute ideal targets for accelerated protons, enhanced gamma-ray and neutrino emissions are expected. The model is shown to provide an adequate description of the broadband gamma-ray emission of the Galactic supernova remnant RX J1713.7-3946 both in terms of total flux and spectral shape.
- In Chap. 3, a phenomenological description of particle escape from middle-aged supernova remnants is presented, and it represents the first attempt of studying this process within the context of extended sources. A proper description of this phenomenon is extremely relevant for the correct interpretation of the radiation spectrum observed in these sources, which reflects not only the acceleration mechanism and the interaction processes, but also the particle escape from the acceleration site. The model is applied to three interesting middle-aged Galactic supernova remnants, namely IC 443, W 51C, and W 28N. A major implication of the presence of particle escape is represented by the possible production of high-energy radiation also outside of the remnant shock, characterized by a very peculiar bump-like energy spectrum. This feature is interesting from the point of view of both gamma-ray and neutrino emissions, being experimentally connected to potentially background-free regions. Moreover, the escaping process is particularly relevant for a correct understanding of the cosmic-ray spectrum observed at Earth and to disentangle the propagation effects through the Galaxy.
- In Chap. 4, a candidate source of PeV cosmic rays located at the center of the Galaxy is discussed. The Galactic Center, as recently observed in multi-TeV gamma rays, shows a central emission with spectral cut-off energy at an energy of about 10 TeV. Nonetheless, a diffuse emission surrounding the central source shows no visible cut-off up to the energies currently probed by H.E.S.S.: the possibility of an intense infrared radiation field absorbing gamma rays from the central source is investigated for the first time. The detection of very high-energy neutrinos in angular correlation with the electromagnetic radiation would confirm the hadronic hypothesis for the origin of gamma rays. Hence, expectations from current and next-generation neutrino instruments are

provided, indicating the relevance of a Northern Hemisphere detector for the observation of this region with a clean event sample.
- In Chap. 5, the performances of the next-generation gamma-ray and neutrino detectors are investigated, and differential sensitivities of CTA and KM3NeT for extended sources are derived. This study represents one of the first attempts towards the understanding of instrumental performances for extended sources related to spectroscopic detection of gamma rays through the imaging technique and capabilities of neutrino telescopes. Sensitivity analyses are, hence, applied to some interesting PeV cosmic-ray candidate sources, as the Galactic Center Ridge and the aforementioned supernova remnant RX J1713.7-3946.
- The main results of the work are summarized and discussed in Chap. 6.

The thesis contains three appendices, specifically separated from the text in order to facilitate its reading. In Appendix A, an overview of the equations regulating the magnetohydrodynamical properties of astrophysical plasma is presented, together with an insight into the numerical code adopted for the solution of this system of equations. In Appendix B, a detailed description of the numerical methods adopted for the solution of the particle transport equation in the presence of molecular clumps is provided. It is a technical appendix, intended to support the interested reader in reproducing the physical results discussed in Chap. 2. Its content represents an original work developed by the author. Finally, Appendix C provides the mathematical framework developed in order to derive the analytical solution of the diffusive transport equation, satisfied by the escaping particle density function and presented in Chap. 3.

L'Aquila, Italy | Dr. Silvia Celli

Part of this thesis has been published in the following journal articles

This thesis is based on the original results obtained by the author. Part of it has already been published in the following peer-reviewed journal papers:

- S. Celli, A. Palladino & F. Vissani, *Neutrinos and γ rays from the Galactic Center Region after H.E.S.S. multi-TeV measurements*, European Physics Journal C **77** (2017) 3;
- L. Ambrogi, S. Celli & F. Aharonian, *On the potential of Cherenkov Telescope Arrays and KM3 Neutrino Telescopes for the detection of extended sources*, Astroparticle Physics Journal **100** (2018) 69A;
- S. Celli, G. Morlino, S. Gabici & F. Aharonian, *Supernova remnants in clumpy media: particle propagation and gamma-ray emission*, Monthly Notices of the Royal Astronomical Society **487** (2019) 3;
- S. Celli, G. Morlino, S. Gabici & F. Aharonian, *Exploring particle escape in middle-aged supernova remnants through gamma rays*, Monthly Notices of the Royal Astronomical Society **490** (2019) 3, https://doi.org/10.1093/mnras/stz2897.

Acknowledgements

I am undoubtedly grateful to the High-Energy Astrophysics Group in GSSI, for the very stimulating environment where I spent the years of my Ph.D. In particular, a heartfelt thank is addressed to my advisors Felix, Giovanni, Stefano, and Francesco, for their firm and careful guidance through this work, which would not have been possible without their precious knowledge and expertise. Many thanks also go to the technical staff of GSSI, for their help through many different situations, and to the High-Energy Astrophysics Group at APC, for their hospitality during my visit to Paris. A warm embrace is directed to my friends in L'Aquila and in Rome, for all the encouraging conversations we had. Finally, all my gratitude is devoted to my family, for the great motivation and support they provided to me, and to Francesco, for his extraordinary patience and love.

Acknowledgments

Syadvada

(Theory of Sevenfold Predications)

Once upon a time, there lived six blind men in a village.
One day the villagers told them: 'Hey, there is an elephant in the village today'.
They had no idea what an elephant was.
They decided: 'Even though we would not be able to see it, let us go and feel it anyway'.
All of them went where the elephant was.
Every one of them touched the elephant.
'Hey, the elephant is a pillar', said the first man who touched his leg.
'Oh, no! It is like a rope', said the second man who touched the tail.
'Oh, no! It is like a thick branch of a tree', said the third man who touched the trunk of the elephant.
'It is like a big hand fan', said the fourth man who touched the ear of the elephant.
'It is like a huge wall', said the fifth man who touched the belly of the elephant.
'It is like a solid pipe', said the sixth man who touched the tusk of the elephant.
They began to argue about the elephant and every one of them insisted that he was right. It looked like they were getting agitated.
A wise man was passing by and he saw this.
He stopped and asked them: 'What is the matter?'
They said: 'We cannot agree on what the elephant is like'.
Each one of them told what he thought the elephant was like.
The wise man calmly explained to them:
'All of you are right. The reason every one of you is telling it differently is because each one of you touched the different part of the elephant.
So, actually the elephant has all those features what you all said'.
'Oh!' everyone said.
There was no more fight.
They felt happy that they were all right.

Contents

Chapter 1
Introduction

This thesis deals with topics from both the theoretical and the experimental fields. Therefore, in this introductive chapter, both aspects are covered in order to provide the reader with a complete picture concerning the current understanding of the physical processes responsible for the formation of the cosmic-ray (CR) flux observed at Earth and the detection principles of their secondary emission.

1.1 Galactic Cosmic Rays

CRs are charged particles constantly impacting the Earth atmosphere. They are characterized by two components: a hadronic one, mainly composed by protons ($\sim$87%), Helium ($\sim$12%) and heavier nuclei ($\sim$1%), and a leptonic one, which is about 100 times less abundant. The observed all-particle differential energy spectrum of CRs is shown in Fig. 1.1: it extends without any distinct feature as a power law $\propto E^{-2.7}$ up to the so called *knee*, located at an energy of about $E_{\rm knee} \simeq 3\,\mathrm{PeV}$. Beyond this energy, the spectrum softens to $E^{-3.1}$: thus the bulk of CRs basically resides in the GeV domain. There is evidence that around the knee, the composition changes [141], suggesting a rigidity[1]-dependent maximum CR energy at the source, with more massive particles having their break at higher energies. Up to an energy of about 10^{17} eV, CRs are believed to originate in our own Galaxy. On the contrary, particles with energy larger than $\sim 10^{18}$ eV, usually referred to as Ultra High Energy Cosmic Rays (UHECRs), cannot be confined in the Galaxy, because their Larmor radius in the typical Galactic magnetic field is of the same order of the Galactic Disc thickness, or even larger. Hence, if they were produced in the Galaxy, the particle arrival direction should trace the source position in the sky, given the contained angular deflection. On the contrary, as the incoming spatial distribution of UHECRs is nearly isotropic

[1]The rigidity of a particle is $R \equiv p/q$, with p its momentum and $q = Ze$ its charge (e being the elementary charge).

S. Celli, *Gamma-ray and Neutrino Signatures of Galactic Cosmic-ray Accelerators*, Springer Theses, https://doi.org/10.1007/978-3-030-33124-5_1

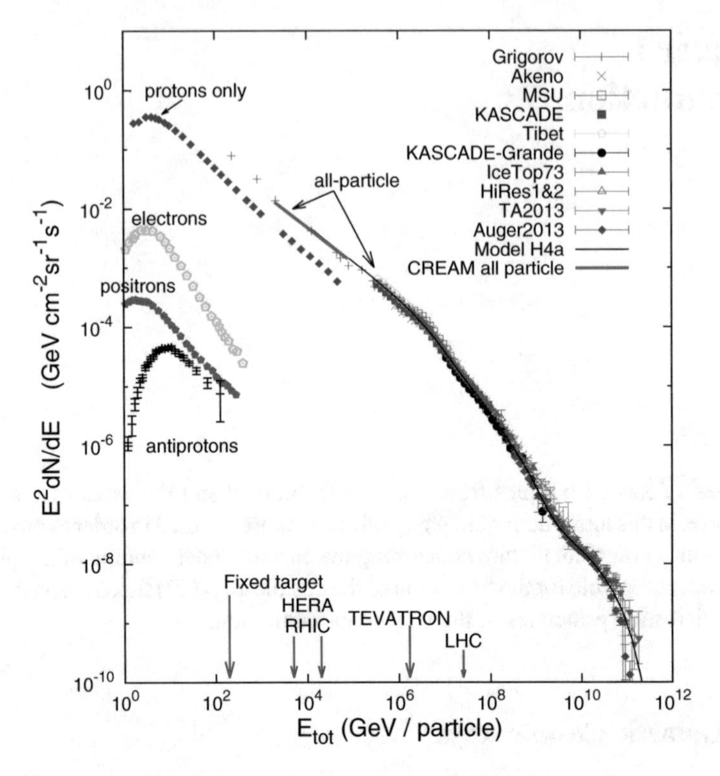

Fig. 1.1 Energy flux of CRs observed at Earth. Figure reproduced from Gaisser et al. [125] with permission of the Cambridge University Press through PLSclear

[1], the general opinion is that these particles come from extra-galactic sources. A second feature, called the *second knee*, is visible around 2×10^{17} eV: interpreting the knee as connected to the maximum energy reached in the accelerator and assuming a rigidity-dependent maximum energy, then the second-knee would naturally arise as a consequence of the acceleration of the heaviest nucleus, namely iron. Indeed, if the maximum energy of protons could reach 3×10^{15} eV, then heavier nuclei with charge Z would reach Z times larger energies: in this scenario, Fe would have an energy of $26 \times E_{\rm knee}$. Hence, the region between the knee and the second knee would result from the superposition of cut-offs in the spectra of different chemical elements. Finally, another break, the so-called *ankle*, is located around $E_{\rm ankle} \simeq 3 \times 10^{18}$ eV, where the CR spectrum flattens back to $E^{-2.7}$: as discussed above, the ankle might be connected with the introduction of an extra-galactic component of CRs, whose origin is unknown [65], though this interpretation is still subject of active debate [235].

Baade and Zwicky [57] first proposed that supernovae can provide the adequate energetics to explain the observed flux of CRs.[2] Later, it was recognized [56, 58, 70, 155] that relativistic particles can be effectively accelerated via Fermi mechanism [107, 108] at shock waves that form during the expansion of supernova remnants (SNRs) in the interstellar medium (ISM). The present formulation of this idea is often referred to as the *Supernova Remnant paradigm for the origin of CRs*, as it will be discussed in Sect. 1.2. The paradigm relies on the following facts, namely:

- SNRs can provide the power needed to sustain the CR flux at the observed level, if a fraction of about 10% of the kinetic energy released at the supernova explosion is converted into relativistic particles (see e.g. [71]);
- Diffusive Shock Acceleration (DSA) can operate at SNR shocks, providing a possible mechanism to accelerate particles [56, 58, 70, 155]. However, the simplest model of diffusive shock acceleration, namely the test-particle approach, fails to achieve PeV energies [158], as required to explain the CR knee.
- A possible solution is represented by the fact that the accelerated particles themselves can amplify the magnetic field at shocks during the acceleration process via various plasma instabilities ([59], see also Sect. 1.2.4), and consequently the magnetic field strength at SNR shocks might increase as to allow the acceleration of particles up to PeV energies and beyond;
- Signatures of amplified magnetic fields are inferred from the observed narrow filaments of non-thermal X-ray emission in several young SNRs [240], the most striking evidence being in Tycho [247].

All of these ideas support the conviction that SNRs are indeed the sources of CRs, though an unambiguous and conclusive proof of such a statement is still missing. Particle acceleration in SNRs is accompanied by the production of gamma rays and neutrinos due to interactions of the accelerated CR protons and nuclei with the ambient medium (the so-called *pp* interaction). During the latest years, numerous detections of SNR shells in TeV gamma rays have confirmed the theoretical predictions that SNRs can operate as powerful CR accelerators. If these objects are responsible for the bulk of Galactic CRs (GCRs), then they should accelerate protons and nuclei to 10^{15} eV and beyond, i.e. act as cosmic *PeVatrons*. The model of DSA allows, under certain conditions, acceleration of particles to such high energies and their gradual injection into the ISM. However, the details of how particles propagate within the SNR, escape the shock and are released into the ISM are poorly understood, though these processes strongly influence the VHE emissions observed.

This thesis is centered on the investigations of the particle propagation both inside and outside of the acceleration site and on the evaluation of the effects produced on the radiation spectra. Though the theory of DSA is often developed for homogeneous media, SNR shocks usually expand in highly turbulent and inhomogeneous media, with a clumpy structure shaped by the interaction among the progenitor wind and

[2] Note that Baade and Zwicky did not believe that CRs were originated in the Galaxy, as they opened their 1934 paper by writing: 'Two important facts support the view that CRs are of extra-galactic origin'. See [228] for an early formulation of the supernova hypothesis.

the circumstellar medium (CSM). In fact, massive stars explode during the red giant phase, and more massive ones also enter the Wolf-Rayet phase, during which they lose all of their hydrogen envelope because of high mass loss rates ($0.2 - 10\,M_{\odot}\,\mathrm{yr}^{-1}$). For instance, massive O-stars spend about 90% of their life in the main sequence, during which a fast wind is blown at a speed of about $v_w \sim 1500$ to 3000 km/s: this wind carves a low density region into the CSM, where the SNR shock will expand later on. Thus, it is not unusual for shocks to expand within regions of different densities and temperatures, particularly in the case of core-collapse SNe [248]. In a simplified picture, the ISM can be schematized as a multi-phase medium, whose density is regulated by SN explosions. These phases are expected to be kept in rough pressure equilibrium. In particular, most of the space would be filled by a hot low-density medium, the HIM, with typical values of density and temperature respectively of $10^{-2.5}\,\mathrm{cm}^{-3}$ and $10^{5.7}$ K and a filling factor of about 0.8. This component embeds cold, neutral and dense clouds (CNM), with density and temperature respectively of $10^{1.6}\,\mathrm{cm}^{-3}$ and $10^{1.9}$ K and a filling factor of about 0.02. Lastly, as the clouds are immersed in the UV and soft X-ray radiation fields emitted by stars and SNRs, a warm ionized medium (WIM) with temperature of 10^4 K and filling factor of 0.2 is expected to surround the clouds [171]. The particle acceleration in shock fronts propagating through the inhomogeneities of the ISM has been first studied by [69], with the aim of explaining the synchrotron emission observed in the radio spectrum of evolved SNRs. However, these first estimates assumed that particles freely propagate inside the coldest and densest structures of the ISM, without accounting for the dynamical interaction among the shock and the dense clouds, which allows for magnetohydrodynamic (MHD) instabilities to develop in case of large density contrasts among the clumps and the HIM. Such instabilities have in fact shown to be able to amplify the magnetic field all around the densest clouds [143] and possibly delay the particle propagation inside them. A detailed modeling of this scenario will be presented in Chap. 2, where particle propagation within SNRs expanding in non-homogenous media is considered in view of the remarkable effects produced on the spectra of secondaries. In Chap. 3, instead, particle propagation is considered during the escape process, which introduces recognizable features in the spectra of secondaries. Such effects have to be accounted for when attempting a phenomenological interpretation of the origin of the observed radiation. In fact, it is necessary to consider the mechanism of CR escape from their sources in a complete acceleration theory in order to determine the spectrum observed at Earth. Hence, a fully consistent description of the origin of CRs should include the particle acceleration, the particle escape and the particle propagation both inside the source and in the ISM traversed towards the Earth. In particular, within this scenario, it is worth to note that the number of SNRs acting as PeVatrons, and consequently expected to be bright in >10 TeV gamma rays, should be rather limited, because multi-PeV protons can be accelerated only during a relatively short period of the SNR evolution, namely at the time of the transition between the free-expansion and the Sedov phase, when the shock velocity is large enough to allow a sufficiently high acceleration rate. The recent detection of a PeVatron located in the Galactic Centre as claimed by the H.E.S.S. Collaboration [19] has successfully demonstrated the feasibility of conducting PeVatron searches in the TeV gamma-ray sky, naturally arising the question concerning analogous searches in the neutrino sky. For this reason, The Galactic Center region is investigated in Chap. 4, where precise predictions for neutrinos are derived for current and future generation of neutrino telescopes. At the same time, the H.E.S.S. claim has also suggested that sources other than SNRs might accelerate PeV CRs in the Galaxy. Resolving morphological and spectral features is thus extremely relevant to shed light on CR acceleration and propagation, and possibly to unveil their sources. Hence, Chap. 5 is devoted to the analysis of the performances of the next-generation gamma-ray and neutrino instruments.

The remaining part of this chapter is devoted to provide the background information which will be used in the rest of the thesis. In Sect. 1.2 the basic picture of the SNR paradigm is summarized, emphasizing the open problems which are going to be analyzed in the following. In Sect. 1.3 the PeVatron case is presented not only in terms of SNRs, but also considering other sources as possible contributors to the GCR flux: in particular, the Galactic Center and stellar clusters with OB associations are investigated. The electromagnetic and weak emissions arising from CR interactions with radiation, matter and magnetic fields are described in Sect. 1.4, while Sect. 1.5 is dedicated to the detection principles of very-high-energy (VHE) gamma rays and neutrinos.

1.2 The SNR Paradigm for the Origin of Galactic Cosmic Rays

Based on energetic considerations, [57] have been the first to propose that supernovae constitute viable sources for addressing the origin of CRs and, even nowadays, this idea remains the most popular scenario to explain the origin of GCRs. Their consideration about the source energetics can be summarized as follows. The power needed to maintain the GCRs at the observed flux level with respect to the energy losses connected with their escape from the Galaxy approximately amounts to

$$P_{\rm CR} \sim \frac{U_{\rm CR} V_{\rm CR}}{\tau_{\rm res}} \sim (3-10) \times 10^{40}\ \text{erg/s} \tag{1.1}$$

where $U_{\rm CR} \sim 0.5\,\text{eV/cm}^3$ is the CR energy density measured at Earth, $V_{\rm CR} \sim 400\,\text{kpc}^3$ is the volume of the Galactic halo where CRs are effectively confined and $\tau_{\rm res} \sim 5 \times 10^6$ yr is the typical residence time in such a volume [178]. The energy released by a SN explosion in the form of bulk kinetic energy of the expanding shell is estimated to be $E_{\rm SN} \sim 10^{51}$ erg, with an explosion rate of $R_{\rm SN} \sim 3$ SNe per century in the Galaxy: therefore, the total power injected into the Milky Way by SNe is

$$P_{\rm SNR} \sim R_{\rm SN} E_{\rm SN} \sim 3 \times 10^{41}\ \text{erg/s} \tag{1.2}$$

Assuming that a fraction of about 10% of the total SN mechanical energy is converted into non-thermal particles, then SNe can actually account for the flux of CRs observed at Earth [216]. The idea that such a mechanism might effectively work follows from the development of a consistent theory, namely the stochastic acceleration occurring at the SNR shocks, as suggested by [56, 155] and quantitatively formulated by [58, 70]. It represents a particular case of the acceleration process proposed by [107, 108], with particles scattering back and forth across a shock wave due to the presence of magnetic inhomogeneities: Sect. 1.2.1 will focus on the description of this mechanism.

The particle energy that marks the transition between Galactic and extra-galactic CRs is most likely located between the knee and the ankle. As a consequence, the acceleration mechanism connected to supernovae must be able to accelerate particles up to the PeV energy range and beyond. The standard linear treatment of diffusive shock acceleration does not allow acceleration of protons at energies larger than $\sim 10^{14}(B/3\mu\text{G})$ eV [158], thus failing to reproduce the position of the knee, unless a strong amplification of the magnetic field in the upstream region is assumed. From the observational point of view, the main evidence for large magnetic fields in the shock region is represented by the detection of narrow X-ray synchrotron filaments surrounding young SNRs [245]. Recently, an advanced non-linear approach was adopted in [166] for the treatment of the shock and accelerated particles as a symbiotic self-organizing system: however, even accounting for non-linear and non-resonant magnetic field amplification, it is not guaranteed that the system is able to achieve PeV energies [59, 75]: in particular, type II SNe appear the best candidate, though the PeVatron phase most probably lasts few hundred years. The acceleration of CRs of these energies must be accompanied by the production of gamma rays and very-high-energy neutrinos. Note that the detection of several SNRs in TeV gamma rays is compatible with this scenario, but it cannot be considered a proof of the fact that SNRs can accelerate CR protons. This is because electrons can as well be accelerated at shocks, and their inverse Compton (IC) scattering on the Cosmic Microwave Background (CMB) as well as on infrared radiation fields can also account for the observed TeV radiation [138]. Thus the ambiguity between the hadronic or leptonic origin of the gamma-ray emission observed from SNRs is the main obstacle in proving (or falsifying) the SNR paradigm for the origin of CRs. In general, the two scenarios mainly differ in the required magnetic field strength: the IC scenario usually requires a very low magnetic field (of the order of 10 μG) in order to simultaneously account for radio, X-ray and gamma-ray emission, while the hadronic scenario requires a much larger value, of the order of tens to hundreds of μG. Such a large magnetic field cannot result from the simple compression of interstellar magnetic field, but requires some amplification, which is, in turn, a possible signature of efficient CR acceleration itself.

1.2.1 Diffusive Shock Acceleration

In the late seventies, a mechanism for the acceleration of particles that operates at SNR shocks was proposed, as soon as it was realized that particles can be accelerated at shock waves via a first-order Fermi mechanism. A characteristic prediction of this model is a $\propto E^{-2}$ differential energy spectrum for the accelerated particles. This is achieved when particles are scattered back and forth across the shock numerous times through *collisionless* scattering, as represented in Fig. 1.2. The fact that the shock is collisionless is due to the very low density that characterizes the ISM, with typical values of $n \sim 1\,\text{cm}^{-3}$. In such a low density, in fact, the mean free path for particle-particle interaction due to Coulomb scattering is $\lambda_{pp} \sim 10^{20} n^{-1}$ cm, which is much larger than the typical size of an SNR. As a consequence, the formation of a shock should result from a collective interaction among thermal particles in the plasma and turbulent electric and magnetic fields. In this case heating takes place over distances of c/ω_{pe}, where $\omega_{pe} = 4\pi e^2 n_e (m_e)^{1/2}$ is the plasma frequency (m_e and n_e being respectively the electron mass and density) and c is the speed of light: this corresponds to a shock thickness of about $\Delta x \sim 10^7 n_e^{-1/2}$ cm, which is thirteen orders of magnitude smaller than the range over which Coulomb collisions operate. For this reason, shocks in low density environment are called collisionless shocks.

The diffusion process results from particle scattering on random MHD waves [68, 218], which arise in magnetized plasma in response to perturbations. It is likely that these perturbations are provided by CRs themselves, rendering the entire system self-regulating. In principle, MHD fluctuations can be either Alfvén waves, namely magnetic waves moving parallel to the background magnetic field at the Alfvén speed $v_A = B_0/\sqrt{4\pi n_i m_i}$ (where B_0 is the background ordered field, n_i is the number density of the ions composing the plasma and m_i is the ion rest mass), or fast magnetosonic waves, i.e. magnetic waves propagating perpendicular to the ordered magnetic field with phase velocity $\gg v_A$. However, the Alfvén waves are expected to dominate over the magnetosonic ones because the latter are more efficiently damped [217].

In the following, the energy spectrum of shock-accelerated particles is derived. The computations are performed in the shock reference frame, as depicted in Fig. 1.2, and a non-relativistic shock is considered. In addition, the background ordered magnetic field B_0 is assumed to be oriented in the direction of motion of the shock. In the shock frame, the upstream fluid is moving at speed $v_1 = -v_s$ (v_s being the shock speed) towards the shock surface, with a relative velocity with respect to the downstream fluid of $\beta = v_1 - v_2 > 0$ (in units of c), v_2 being the speed of the downstream fluid.

Starting for simplicity with a particle that is already relativistic with an initial energy E, as measured in the upstream frame, we will consider a full cycle across the shock. When the particle gets advected in the downstream, it will have an energy equal to

$$E_d = \gamma E(1 + \beta\mu) > 0 \tag{1.3}$$

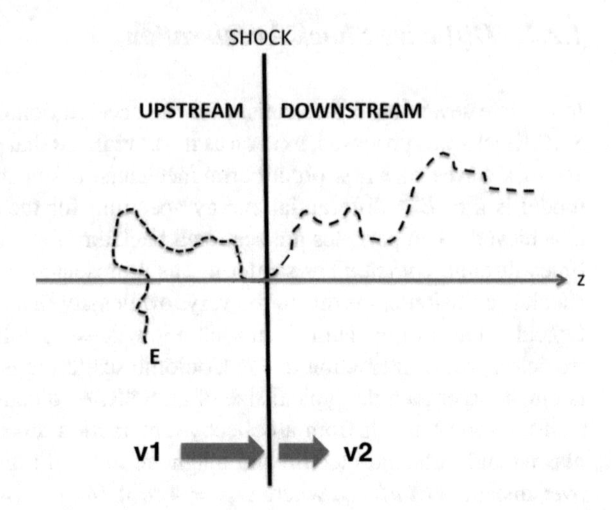

Fig. 1.2 A particle of initial energy E is diffused back and forth of the shock by Alfvén waves and gains energy at each complete cycle across the shock. This figure is depicted in the shock frame: all quantities in the upstream are denoted by the subscript 1, while those in the downstream are denoted with the subscript 2. Note that the ordered background magnetic field is $\mathbf{B_0} = B_0\hat{z}$

because of the energy-momentum Lorentz transformation, where μ is cosine of the pitch angle, namely the angle between the ordered background field B_0 and the trajectory of the particle. In this crossing, μ is constrained to be $0 \leq \mu \leq 1$. Possibly, the particle is able to re-enter the upstream, thanks to the scattering on magnetic inhomogeneities which reflect the particle back towards the shock. This is equivalent to require that $-1 \leq \mu' \leq 0$. After such a reflection, the particle in the upstream reference frame will have an energy equal to

$$E_u = \gamma E_d(1 - \beta\mu') > 0 \tag{1.4}$$

At each crossing, the energy gain is bounded to be positive. The final energy of the particle will be

$$E_u = \gamma^2 E(1 + \beta\mu)(1 - \beta\mu') \tag{1.5}$$

In the assumption that the particle density has been isotropized in the downstream, thanks to the scattering on magnetic perturbations, it is possible to compute the particle average energy gain per crossing. The flux of particles moving from the downstream into the upstream will be

$$J_{\mathrm{d}\to\mathrm{u}} = \int d\Omega \frac{N}{4\pi} c\mu = \frac{Nc}{4} \tag{1.6}$$

where N is the density of particles accelerated at the shock, namely the particles that have been extracted from the thermal bath and injected into the acceleration process, and the integral is performed with $d\Omega = -2\pi d\mu$ and $-1 \leq \mu \leq 0$. On the other hand, the flux of particles moving into the downstream will be

$$J_{\text{u}\to\text{d}} = -\frac{Nc}{4} \tag{1.7}$$

so that, in general, the flux from one surface to the other reads as

$$J = \pm\frac{Nc}{4} \tag{1.8}$$

with the convention of symbols defined above. The probability that a particle is crossing the shock in a given direction μ is (normalizing to the total flux J)

$$P(\mu)d\mu = \frac{\frac{N}{4\pi}c\mu d\Omega}{\pm\frac{Nc}{4}} = \frac{-\frac{N}{4\pi}c\mu 2\pi d\mu}{\pm\frac{Nc}{4}} = \mp 2\mu d\mu \tag{1.9}$$

and hence the average energy gain per cycle is

$$\left\langle\frac{E_u - E}{E}\right\rangle_{\mu\mu'} = -\int_0^1 d\mu 2\mu\int_{-1}^0 d\mu' 2\mu'[\gamma^2(1+\beta\mu)(1-\beta\mu') - 1] = \frac{4}{3}\beta \tag{1.10}$$

Since the energy gain is proportional to β (and since $v_2 \ll v_1$), the efficiency of such a process is first-order in the shock speed.

1.2.2 Particle Spectrum

The most remarkable property of the first-order Fermi mechanism consists in the production of a particle spectrum which is a featureless power law, regardless of the details of the acceleration process. Such universality comes as a consequence of the balance between the energy gained by particles and the probability that they escape the accelerator. In order to derive such a property, the distribution function of accelerated particles $f(z, p)$ has to be introduced: it corresponds to the particle density in the phase space and it is normalized in such a way that the number of particles located between z and $z + dz$ is $N(z) = 4\pi\int p^2 f(z, p)dp$. In the reference frame of the downstream, the distribution function at the shock position ($z = 0$) f_0 results isotropic because of the multiple scattering on the turbulent magnetic waves. Hence, the flux of particles crossing the shock with pitch angle between μ and $\mu + d\mu$ reads as

$$\phi = f_0(v_2 + \mu)d\mu \tag{1.11}$$

where v_2 is the downstream speed (see Fig. 1.2), expressed in units of the speed of light. The flux entering the downstream is characterized by $0 \le v_2 + \mu \le 1$, thus

$$\phi_{\text{in}} = \int_{-v_2}^1 d\mu f_0(v_2 + \mu) = \frac{f_0}{2}v_2(1 + v_2)^2 \tag{1.12}$$

On the other hand, particles moving towards upstream have $-1 \leq v_2 + \mu \leq 0$ and the flux exiting the downstream is

$$\phi_{\rm out} = \int_{-1}^{-v_2} d\mu f_0(v_2 + \mu) = \frac{f_0}{2} v_2 (1 - v_2)^2 \tag{1.13}$$

Thus, the probability that a particle can return to upstream, in the limit $v_2 \ll 1$, is

$$P_{\rm d \to u} = \frac{\phi_{\rm out}}{\phi_{\rm in}} = \frac{(1 - v_2)^2}{(1 + v_2)^2} \simeq 1 - 4v_2 \tag{1.14}$$

Since $v_2 = v_s/4c \simeq 10^{-3}$, then $P_{\rm d \to u} \simeq 1$: therefore it is highly probable that particles are scattered back in the upstream. On the other hand, in the case of relativistic shocks, where the energy gain per cycle is huge (as well as v_2), the probability to come back to the upstream is much more reduced.

The energy spectrum of particles can be derived in the assumption of a Markovian stochastic process. Moving back to physical units, after a whole cycle (upstream, downstream and upstream again or viceversa) the particle energy gain is (see Eq. (1.10))

$$\left\langle \frac{\Delta E}{E} \right\rangle = \frac{4}{3}(v_1 - v_2) \equiv \frac{4}{3} V \tag{1.15}$$

If the seed in the plasma consists of N_0 particles at energy E_0, part of these particles will have an energy E_1 after the first cycle such that

$$\frac{E_1 - E_0}{E_0} = \frac{4}{3} V \rightarrow E_1 = E_0 \left(1 + \frac{4}{3} V \right) \tag{1.16}$$

After the second cycle, particles will have achieved an energy E_2 such that

$$E_2 = E_1 \left(1 + \frac{4}{3} V \right) = E_0 \left(1 + \frac{4}{3} V \right)^2 \tag{1.17}$$

and so on. In correspondence of the k-th cycle, the particle energy has become

$$E_k = E_0 \left(1 + \frac{4}{3} V \right)^k \rightarrow \ln \left(\frac{E_k}{E_0} \right) = k \ln \left(1 + \frac{4}{3} V \right) \tag{1.18}$$

In order to asses the relevance of this process, it is necessary to evaluate the number of particles which undergo a full cycle across the shock. This can be evaluated as (see Eq. (1.14))

$$N_1 = N_0 (1 - 4v_2) \tag{1.19}$$

at the first cycle, and after k cycles they will be

$$N_k = N_0(1 - 4v_2)^k \rightarrow \ln\left(\frac{N_k}{N_0}\right) = k \ln(1 - 4v_2) \quad (1.20)$$

Using Eqs. (1.18) and (1.20), the number of cycles required to achieve an energy E_k amounts to

$$k = \frac{\ln\left(\frac{E_k}{E_0}\right)}{\ln\left(1 + \frac{4}{3}V\right)} = \frac{\ln\left(\frac{N_k}{N_0}\right)}{\ln(1 - 4v_2)} \quad (1.21)$$

and consequently

$$\ln\left(\frac{N_k}{N_0}\right) = \frac{\ln(1 - 4v_2)}{\ln\left(1 + \frac{4}{3}V\right)} \ln\left(\frac{E_k}{E_0}\right) \equiv -s \ln\left(\frac{E_k}{E_0}\right) \quad (1.22)$$

Thus, the stochastic process of acceleration naturally produces a particle energy spectrum distributed as a power-law

$$N_k = N_0 \left(\frac{E_k}{E_0}\right)^{-s} \quad (1.23)$$

The value of the spectral index s can be derived through a series expansion of

$$\begin{aligned} &\ln(1 - 4v_2) \simeq -4v_2 \qquad (v_2 \ll 1) \\ &\ln\left(1 + \frac{4}{3}V\right) \simeq \frac{4}{3}V \qquad (V \ll 1) \end{aligned} \quad (1.24)$$

so that

$$s = -\frac{\ln(1 - 4v_2)}{\ln\left(1 + \frac{4}{3}V\right)} \simeq \frac{3v_2}{v_1 - v_2} \quad (1.25)$$

Defining the *compression ratio* as

$$r = \frac{v_1}{v_2} \quad (1.26)$$

then the number of particles with energy larger than E is

$$N(> E) = N_0 \left(\frac{E}{E_0}\right)^{-s} \qquad s = \frac{3}{r - 1} \quad (1.27)$$

For a strong shock in a monoatomic gas, where $r \simeq 4$ (Appendix A), it holds $s \simeq 1$. Finally, the differential energy spectrum, namely the density of particles with energy between E and $E + dE$, is obtained as

$$\frac{dN}{dE} \equiv \frac{N_0}{sE_0}\left(\frac{E}{E_0}\right)^{-\gamma} \qquad \gamma = s + 1 = \frac{r+2}{r-1} \tag{1.28}$$

If $s \simeq 1$, then $\gamma \simeq 2$. It is worth noting here how the shape of the spectrum of the accelerated particles does not depend upon the microphysics of the process. In order to give an idea about how many shock crossings are needed to reach 10^{15} eV, one can assume that a proton is accelerated by a shock at $v_s = 5000$ km/s, starting with an initial kinetic energy of 100 keV. As the gain per crossing amounts to 1.7% (see Eq. (1.15)), the number of scattering required is $k = \ln(10^{15}/10^5)/\ln(1.017) \sim 1400$ (see Eq. (1.21)) and the probability for the full process to be completed is about 0.003.

An alternative and more complete derivation of the differential energy spectrum of accelerated particles can be obtained by solving the transport equation [128], namely the equation regulating the evolution of the particle distribution function in time, space and momentum. This equation reads as

$$\frac{\partial f}{\partial t} + v\frac{\partial f}{\partial z} = \frac{\partial}{\partial z}\left[D\frac{\partial f}{\partial z}\right] + \frac{1}{3}p\frac{\partial f}{\partial p}\frac{dv}{dz} + Q(z, p) \tag{1.29}$$

for unidimensional propagation in the configuration space, namely a shock speed entirely directed along z. The particle distribution function will evolve affected respectively by the advection in the plasma velocity field, the diffusion in the plasma turbulent magnetic field, the adiabatic compression/expansion of the plasma and the injection at the shock position (in $z = 0$, so that $Q(z, p) \propto q(p)\delta(z)$). In order to derive the particle spectrum, we are going to solve Eq. (1.29) in two regions of space. Note that all quantities labeled with the subscript 0 will be computed in the shock position, 1 in the upstream and 2 in the downstream. Starting with integrating Eq. (1.29) across the shock surface, namely for $0^- \leq z \leq 0^+$, and assuming both the stationarity ($\partial f/\partial t = 0$) and the regularity of the distribution function (i.e. its continuity across the shock so that $\int_{0^-}^{0^+} dz\, v\partial f/\partial z = 0$), we obtain

$$0 = D\frac{\partial f}{\partial z}\bigg|_2 - D\frac{\partial f}{\partial z}\bigg|_1 - \frac{1}{3}(v_1 - v_2)p\frac{\partial f_0}{\partial p} + q_0(p) \tag{1.30}$$

as

$$v(z) = \begin{cases} v_1 & z < 0 \\ v_2 & z > 0 \end{cases} \longrightarrow \frac{dv}{dz} = -\delta(z)(v_1 - v_2) \tag{1.31}$$

On the other hand, by integrating Eq. (1.29) between the shock upstream and the infinite upstream, namely for $-\infty < z \leq 0^+$, we obtain

$$v_1 f_0 = D\frac{\partial f}{\partial z}\bigg|_1 \tag{1.32}$$

Introducing Eq. (1.32) into Eq. (1.30), and by considering that the distribution function was isotropized in the downstream ($\partial f/\partial z|_2 = 0$), we derive

$$- v_1 f_0 - \frac{1}{3}(v_1 - v_2) p \frac{\partial f_0}{\partial p} + q_0(p) = 0 \quad (1.33)$$

where any connection with the microphysics of the acceleration process, previously contained in the diffusion coefficient, has now disappeared. By solving the homogeneous equation associated with Eq. (1.33), one finally obtains

$$f_0(p) \propto p^{-\frac{3r}{r-1}} \quad (1.34)$$

that is the analogous of Eq. (1.28) when computing the differential energy spectrum, as it follows from the relation $N(p)dp = N(E)dE = 4\pi p^2 f_0(p)dp$. However, describing the acceleration spectrum in terms of the particle distribution function is preferable with respect to describing it in terms of the integral spectrum, in that the former naturally embeds the description of both relativistic and non-relativistic particles, while Eq. (1.28) was derived for relativistic particles only.

Currently, one of the longstanding problems of the DSA model is represented by the injection problem, namely how charged particles are injected into the acceleration process and how do they reach sufficient energy to come back to the upstream. In fact, the injection results from the microphysics of the particle motions at the shock: it is often assumed that these particles belong to the tail of the thermal Maxwellian distribution, which makes up the plasma. Furthermore, a second concern regards the maximum energy which particles would be able to achieve during the acceleration process. In fact, the power-law spectrum derived so far extends in principle up to an infinite particle energy, as the concept of maximum energy of accelerated particles is not naturally embedded in the test-particle theory of DSA. This is clearly connected to the fact that a stationary acceleration process was considered, which is a safe assumption if particles are at some point able to escape the accelerator. In other words, a realistic description of the process should include in Eq. (1.29) an escape term able to balance the source term. The last issue is discussed in the next Section.

1.2.3 Maximum Energy of Particles

As shown until now, the theory of DSA in the test-particle approach, where particles are considered with no dynamical role, predicts an energy spectrum for the accelerated particles which is a featureless power law. The spectral slope in momentum reads as $\alpha = 3r/(r-1)$ (see Eq. (1.34)), depending only on the compression ratio r of the shock, regardless of any details of the acceleration mechanism. In the presence of an effective particle escape process, it is reasonable to assume that a maximum energy exists, where the spectrum drops. For strong non-relativistic shocks, where the mechanism is believed to operate very efficiently, the slope of the spectrum below the cut-off coincides with the canonical slope $\alpha = 4$, independently of the details of the acceleration. On the other hand, the spectral shape of the cut-off depends dramatically on both the microphysics of the diffusive acceleration process and the physical mechanism regulating the particle maximum energy [236]. A qualitative estimate for the maximum energy attained by particles is given by the so-called *Hillas criterion*, proposed by [137], which defines the condition for the size R of an accelerator to be able to magnetically confine a particle with gyroradius equal to its Larmor radius r_L. For SNRs this is equivalent to saying that the diffusion length of particles ahead of the shock should not exceed a given fraction $\xi \sim 0.1$ of the shock radius, or in other terms $D(E_{\rm max})/v_s \leq \xi R_s$, where $D(E_{\rm max})$ is the diffusion coefficient for protons of energy $E_{\rm max}$. Such a condition translates into

$$E_{\rm max} \simeq 1 \left(\frac{v_s}{10^3\ {\rm km/s}}\right)\left(\frac{R_s}{\rm pc}\right)\left(\frac{B}{1\ \mu{\rm G}}\right)\ {\rm TeV} \tag{1.35}$$

where Bohm diffusion was considered. Characteristic values for a young SNR, in which a strong shock is expanding during the ejecta-dominated phase, are $v_s \simeq 10^4$ km/s, shock size $R_s \simeq 1$ pc in a typical interstellar magnetic field of $B \simeq 3\,\mu$G. Hence, the maximum energy expected for confined particles is $E_{\rm max} \simeq 30$ TeV. As was first pointed out in [158], in order to be able to get up to PeV energies, one should assume extremely large shock speeds, which however might be reasonable only at the very initial stages of a remnant evolution, lasting for few tens of years. Alternatively, some mechanisms in the amplification of the magnetic field should be operating upstream of the shock.

The Hillas condition is a necessary but not sufficient condition for an efficient acceleration. A more detailed description of the acceleration process allows to identify the value of the maximum energy in the competition between the acceleration time $t_{\rm acc}$ and the shortest among these three timescales: (i) the age of the accelerator $t_{\rm age}$, (ii) the particle escape time from the accelerator $t_{\rm esc}$, and (iii) the particle energy loss time $t_{\rm loss}$. Consequently,

$$t_{\rm acc} \leq \min(t_{\rm age}, t_{\rm esc}, t_{\rm loss}) \tag{1.36}$$

For age-limited systems, the maximum achievable energy is determined by the condition $t_{\rm acc}(E_{\rm max}) = t_{\rm age}$. The case of the particle acceleration limited by escape requires a careful modeling, which will be discussed in detail in Chap. 3. Finally, the case when the acceleration is limited by the energy losses is particularly relevant for electrons, for instance, where the maximum energy is in most cases limited by radiative synchrotron and IC losses in the ambient magnetic and photon fields, while for hadrons such a limitation is usually negligible.

A theoretical estimate of the acceleration time can be derived considering the time taken by a particle to perform a complete cycle, $t_{\rm cycle}$, across the shock that is

$$t_{\rm acc} = \frac{t_{\rm cycle}}{\Delta E/E} = \frac{3}{v_1 - v_2}\left(\frac{D_1}{v_1} + \frac{D_2}{v_2}\right) \simeq 8\frac{D_1}{v_s^2} \tag{1.37}$$

where the last passage was obtained by assuming that the downstream turbulence is compressed at the shock by the same compression factor of the plasma, $\delta B_2 = 4\delta B_1$.

Note that, since the acceleration time is dominated by the upstream conditions, magnetic field amplification is required *upstream*, where only the accelerated particles can arrive. It is therefore natural to expect that the magnetic field turbulence may be excited by the same particles that are being accelerated. Hence, a self-consistent description of DSA requires the treatment of the dynamical reaction of CRs on the turbulent magnetic field and therefore the introduction of a non-linear theory of DSA, which is briefly discussed in the next Section.

1.2.4 Non-linear Theory of Diffusive Shock Acceleration

The development of a non-linear theory of DSA (NLDSA) can be ascribed to the need for considering three main aspects of the theory, i.e.: (i) the dynamical reaction of the accelerated particles, namely the pressure that they exert on the plasma around the shock, that affects both the shock dynamics and the acceleration process; (ii) the plasma instabilities induced by the same accelerated particles and connected with the amplification of the magnetic field, a necessary ingredient to achieve efficient particle acceleration up to the knee energy; and (iii) the dynamical reaction of the amplified magnetic field, namely the fact that, during the process of field amplification, the magnetic pressure might become larger than the upstream thermal pressure of the incoming plasma, affecting the compression factor at the shock. These motivations clearly show the intrinsic non-linearity of DSA. In fact, the first and the last process act by modifying the compression factor of the plasma at the shock, changing as a consequence also the spectrum of accelerated particles (as shown in Eq. (1.28)). Concerning the second mechanism, the non linearity of DSA appears in that likely the same accelerated particles are responsible for the process of magnetic field amplification, since this has necessary to take place upstream of the shock in order to reduce the acceleration time (see Eq. (1.37)). As the magnetic field determines the diffusion coefficient that describes the particle motion, once the field amplification comes to be affected by the distribution of particles in the acceleration region, then the diffusion coefficient itself (that tells the particles how to evolve) becomes dependent upon the distribution function of accelerated particles. These aspects show that the accelerated particles and the magnetic fields are a self regulating system, and it is their interplay that determines the high non-linearity of the problem. The remaining part of this Section provides a more detailed description of the three non-linear processes involved in the CR acceleration at shocks. The reader is referred to [71, 166] for a comprehensive review of NLDSA.

The dynamical effect on the shock due to the presence of accelerated particles is twofolds: (i) the pressure in accelerated particles slows down the incoming upstream plasma as seen in the shock reference frame, thereby creating a precursor and thus leading to a plasma compression factor that depends on the location upstream of the shock, and (ii) the escape of the highest energy particles from the shock region makes the shock radiative, thereby inducing an increase of the compression factor between upstream infinity and downstream. Both these effects result into a modification of the spectrum of accelerated particles, which turns out to be no longer a pure power law. In fact, particles with momentum p diffuse upstream within a distance that is proportional to the diffusion coefficient $D(p)$, that is usually a growing function of momentum. This implies that particles with low momentum experience a compression factor $r < 4$, as they mostly diffuse close to the shock, while higher momentum particles trace a compression factor $r > 4$. As a consequence, the spectrum is expected to be steeper than p^{-4} at low momenta and harder than p^{-4} at high momenta, with the transition typically occurring around a few GeV/c. The change of slope in the spectrum is directly related to the formation of a precursor upstream of the shock [166].

The role of the plasma instability in the process of particle acceleration in SNR shocks is connected with the super-Alfvénic streaming of charged particles in a plasma [256]. The collective effect of the streaming of CRs induces the self-generation of the same waves responsible for particle diffusion, through the growth of resonant waves with wavenumber $k = 1/r_L$, where r_L is the Larmor radius of the particles generating the instability. As the waves can be resonantly absorbed by individual particles, the net effect is the particle pitch angle diffusion. The resonance condition would lead to expect that the growth stops when the turbulent magnetic field becomes of the same order as the pre-existing ordered magnetic field $\delta B \sim B_0$, so that the saturation level of this instability has often been assumed to occur when $\delta B/B_0 \sim 1$. Therefore, because of the intrinsic resonant nature of the instability, the possibility to reach sufficiently high energy seems inhibited, not because of the timescale, but rather because of the resonant nature that limits δB to be $\sim B_0$ at most. Note that in the non-resonant case, the saturation condition would not hold anymore.

The third aspect of the non-linearity of CR acceleration at shocks consists of the dynamical reaction of magnetic fields produced on the shock by CRs upstream. As the magnetic field acts by reducing the plasma compressibility when the magnetic pressure becomes comparable with the thermal pressure of the upstream gas, it acts in the opposite direction with respect to the action of CRs, creating larger values of the compression ratio. This is the reason why taking into account the effect of magnetic fields on the shock dynamics leads to predictions of less modified shocks, and correspondingly less concave spectra of accelerated particles [74].

1.2.5 The Problem of Particle Escape

DSA is the most successful and widely accepted mechanism for explaining the acceleration of CRs in several astrophysical environments. Despite the basic physics of this process being very robust, some fundamental issues still need to be addressed. In fact, in the context of the SNR paradigm for the origin of CRs, one of the least understood mechanisms is represented by particle escape from the source. The flux of escaping particles plays an essential role within the acceleration scenario up to the PeV, as well as in the formation of the Galactic CR flux detected at Earth. If there were no escape from upstream, all particles accelerated in an SNR would be advected downstream and undergo severe adiabatic energy losses before being injected into the ISM: in this case, the requirements to reach PeV energies would be even more severe than they already are. Another important consideration is that the non-resonant instability requires a flux of escaping particles, which are those providing the current able to excite the instability itself. Without escape, this instability would not be triggered and therefore magnetic field amplification would be limited to the resonant channel, that however is constrained by the saturation condition. Note that, the flux of escaping particles is tightly connected to the CR flux observed at Earth, by means of all those processes involved during particle propagation towards the Earth. Unfortunately, the details of how CRs escape SNRs still remain unclear. Nonetheless, the escape represents an essential process in the CR production mechanism, closely bounded to the acceleration process. The lack of 100 TeV gamma-ray observations in astrophysical sources is then either connected with the intrinsic properties of the accelerators or alternatively might be due to the escape scenario, namely to the fact that PeV protons have already escaped the acceleration site. A complete understanding of the physical processes operating in cosmic sources is still missing and consequently the correct phenomenological interpretation of SNR spectra, which is especially relevant in two contexts, namely both for young and for middle-aged SNRs. In fact, a crucial issue arising from the VHE gamma-ray observations in SNRs concerns the spectral shape of their energy flux: all of the measurements show power-law behavior different from the expected $dN/dE = \phi(E) \propto E^{-2}$. From the observed sample of TeV SNRs, two classes are emerging

(i) SNRs mainly emitting in the GeV band, as W 44, IC 443 and W 51C;
(ii) SNRs whose emission extends up to the TeV band, as Cas A, Tycho, SN 1006, RX J1713.7 – 3946 and RX J0852.0 – 4622 (also called Vela Junior).

Among TeV SNRs, two different varieties arise: (i) very young remnants (Cas A, Tycho, SN 1006) with a typical age of about 300 yr and steep spectra, and (ii) young SNRs (RX J1713.7 – 3946 and RX J0852.0 – 4622), with typical age of few 10^3 yr and hard spectra. The spectral discrepancy could possibly be ascribed to different environmental conditions where the remnants are expanding, which are commonly not accounted for in the context of DSA theory. The presence of dense inhomogeneities in the shock environments, in fact, strongly modifies the spectrum of particles emitting radiation, with respect to the acceleration spectrum, as will be extensively discussed in Chap. 2. GeV SNRs, on the other hand, show typical steep spectra $\phi(E) = E^{-\gamma}$ with slopes $\gamma \simeq 2.3 - 2.5$. Such steep spectra could represent an intrinsic feature of the acceleration process. Though this conclusion has not yet been proven, [91] showed that, by adopting a statistical model based on a population study of universal SNR spectra, only spectral indices steeper than 2 are consistent with the number of detections realized by the current generation of gamma-ray instruments. Alternatively, the steepness might be interpreted as the signature of particle escape from the shock region, as will be further discussed in Chap. 3.

Among young SNRs, an interesting and debated case is represented by RX J1713.7 – 3946: its non-thermal emission has challenged theorists, given that both the one-zone leptonic and hadronic interpretations failed to reproduce the data.

A pronounced hardening was observed in the GeV domain, and it was considered as a strong argument in favor of the leptonic scenario. However, the highly non-uniform environment in which the remnant is expanding adds complexity to the problem, as will be explained in Sect. 2.6, being potentially able to alter the gamma-ray spectrum expected by the neutral pion decay in a uniform target medium. Note that such a situation is very relevant for neutrino emissions as well and that neutrinos are smoking guns for the identification of hadronic sources.

It should be kept in mind that, within the context of hadronic production processes, the spectral index for gamma rays closely resemble the one of CRs accelerated in situ. Now, linear shock acceleration theory predicts a $\phi(E) \propto E^{-2}$ energy spectrum for accelerated particles, while the introduction of non-linear effects due to the CR pressure in the upstream plasma would favor an even harder spectrum, up to $\phi(E) \propto E^{-1.5}$. However, the spectrum of particles released into the ISM does not coincide with the accelerated spectrum, since particle escape can modify the final spectrum injected in the ISM. Therefore, a detailed modeling of CR escape from the shock region is required, accounting for the temporal, the spatial and the energy dependence of the phenomenon.

One would naively expect that CRs are released as the shock slows down [204], and that they are released gradually, the ones having the highest energy first, and the ones with lower energies at later times. Qualitatively, in order to understand when CRs with different energies leave the SNR, one should equate the particle diffusion length with a fraction of the SNR radius, so as to get $E_{\max} \sim R_s v_s B \propto t^{-1/5}$ (within the hypothesis that escape is effective at the Sedov-Taylor stage). This zero-order estimate provides a qualitative description of the process of CR escape from SNRs, indicating that the decrease with time of the maximum energy of confined particles is quite mild. Note however that this conclusion is uniquely connected with the fact that a diffusion coefficient constant in time was assumed in the derivation. Indeed, if the magnetic field at the shock is amplified by CR streaming instability, a faster decrease of $E_{\max}$ with time may be expected, since as the shock speed decreases the amplified magnetic field decreases too [59]. A dedicated discussion on particle escape will be approached in Chap. 3: this study will be limited to hadrons, as leptons suffer from severe energy losses due to synchrotron radiation, which needs to be taken into account for a correct description of the escape. Investigations of the energy spectrum of gamma rays in the cut-off region might shed light on this poorly understood process, in that very sharp cut-offs are expected if particles are suddenly released as soon as they reach the maximum energy.

1.3 Galactic PeVatrons

One of the key objectives of CR studies remains the identification of the principal contributors to the Galactic component of CRs. Most likely, hundreds or thousands of objects contribute to it. If so, the identification of these objects on a source-by-source basis would be rather challenging, especially given that many of these sources may also not be active anymore. A more viable approach is connected with the search for a source population, the best-studied representatives of which should (i) collectively provide the production rate of CRs in our Galaxy, and (ii) explain the basic characteristics of CRs up to the knee, including their composition. The current paradigm of the GCR origin is based on the fact that SNRs satisfy both requirements. However, because of the large uncertainties on the gas density in the gamma-ray production regions, the level of contribution of SNRs to the CR production in the Galaxy is not yet observationally established. More importantly, so far gamma-ray observations have failed to demonstrate that SNRs can accelerate particles beyond 100 TeV. The term 'PeVatron' refers to an object accelerating protons with a hard ($\propto E^{-2}$) energy spectrum without any break up to $E_p \sim 1$ PeV. In hadronic interactions, the spectrum of secondary gamma rays and neutrinos almost mimics the spectrum of the parent protons, being shifted towards lower energies by a factor of ~10–20. Thus, measurement of secondary particles with a hard power-law energy spectrum extending up to several tens of TeV would imply an unambiguous detection of a PeVatron. So far, the observations of young SNRs didn't show such hard multi-TeV energy spectra. Only a few SNRs have been detected above 10 TeV, but in all cases with steep spectra. The large power-law indices and the early cut-offs in the photon spectra (typically less than tens of TeV) imply that the spectra of parent protons do not extend much beyond 100 TeV. Such an unexpected result is a hint that either young SNRs do not accelerate CR protons to PeV energies or the production of PeV protons takes place in a sub-class of SNRs, which so far have not been detected in gamma rays, or alternatively the production occurs at the very beginning of the SNR life ($t \leq 100$ yr). The first option would imply the inability of SNRs to play a major role in the production of Galactic CRs. The second option leaves room for a marginal contribution of SNRs among the possible sources of the GCR flux. However, in this case, only a minor fraction of SNRs would contribute to CRs, at least at highest energies. Consequently, the efficiency of energy conversion in these objects should significantly exceed the standard 10% value. The third option would strongly reduce the number of detectable PeVatrons, as the very energetic phase would last for a limited amount of time and correspondingly it would be less likely to catch it observationally.

As a final confirmation (or falsification) about the SNR paradigm for the origin of Galactic CRs is still missing, one should consider possible alternative explanations to the energy and amount of flux required to match the CR observations. Recently, the H.E.S.S. observations of the Galactic Center [19] have revealed the presence of a VHE gamma-ray emitter around Sgr A*, which shows a remarkably featureless spectrum up to many tens of TeV. The hypothesis of a PeVatron in the Galactic Center arises naturally, if this radiation is attributed to CR collisions as likely in a region populated by massive clouds. This scenario is further discussed in Sect. 1.3.1. On the other hand, the VHE gamma rays point towards a region located within 60 pc from the central radio emitters: given the large uncertainties that affect the very inner parsecs of our Galaxy (in both the gas and source distribution), the hypothesis that there might be sources other than the supermassive black-hole itself is realistic. In fact, several classes of gamma-ray emitters populate the region and, potentially, some of these might provide non-negligible contribution to the observed CR flux.

In this regard, the clusters of young massive stars are of special interest, given the collective effects provided by the multiple SNRs and stellar wind shocks present inside these regions. The acceleration could take place close to the stars or in the so-called superbubbles, multi-parsec structures caused by the collective activity of massive OB star winds around the compact stellar associations [78, 82], where the acceleration on multiple shocks may potentially increase the maximum energy of CR protons to 10^{15} eV [152], making clusters of massive stars PeVatron candidates, to be considered as either complimentary or alternative candidates for the origin of GCRs to the Galactic Center itself (see discussion in Sect. 1.3.2).

1.3.1 The Galactic Center

The VHE radiation observed by H.E.S.S. in the Galactic Center (GC) region comes from a diffuse region, extended for few tens of parsecs in latitude and about 200 pc in longitude from the central radio emitter. In particular, considering the radial distribution of gamma rays, in the annular region contained between about 20–60 pc from the radio source the spectrum of gamma rays shows remarkably no sign of energy break or cut-off, up to energies of about 30 TeV [19]. In a hadronic scenario, these photons would derive from very energetic protons, possibly showing evidence for the first PeVatron source. A crucial aspect of the observations performed in this region is constituted by the inferred radial distribution of CRs, that appears fundamental in the PeVatron scenario. In fact, the CR radial profile and the distribution of the target gas in the Central Molecular Zone (CMZ) shape the morphology of the VHE gamma-ray emission. Figure 1.3 shows the radial profile of the $E \geq 10$ TeV cosmic-ray energy density w_{CR} up to $r \sim 200$ pc, as derived through the gamma-ray luminosity and the amount of target gas. Such a CR density in the CMZ is found to be one order of magnitude larger than that of the 'sea' of CRs, namely the diffuse flux of particles that fills the Galaxy, while the energy density of low-energy (GeV) CRs in this region does not sizably differ from the average Galactic value [250]. This scenario requires the presence of one or more accelerators of multi-TeV particles operating in the CMZ. Moreover, the radial distribution is compatible with a $w(r) \sim 1/r$ profile, as shown in Fig. 1.3: this is expected in the case of a continuous injection of particles from a central source, which is then followed by diffusion across the Galaxy. Indeed, considering three dimensional diffusion in spherical symmetry, the CR density function at equilibrium conditions (namely $\partial f/\partial t = 0$) satisfies the following equation

$$Q\delta(r) = \frac{1}{r^2}\frac{\partial}{\partial r}\left(r^2 D \frac{\partial f}{\partial r}\right) \tag{1.38}$$

where D represents the diffusion coefficient and Q the source flux, here assumed to originate at the very center of the system, namely in $r = 0$. Therefore, if $r \neq 0$, then

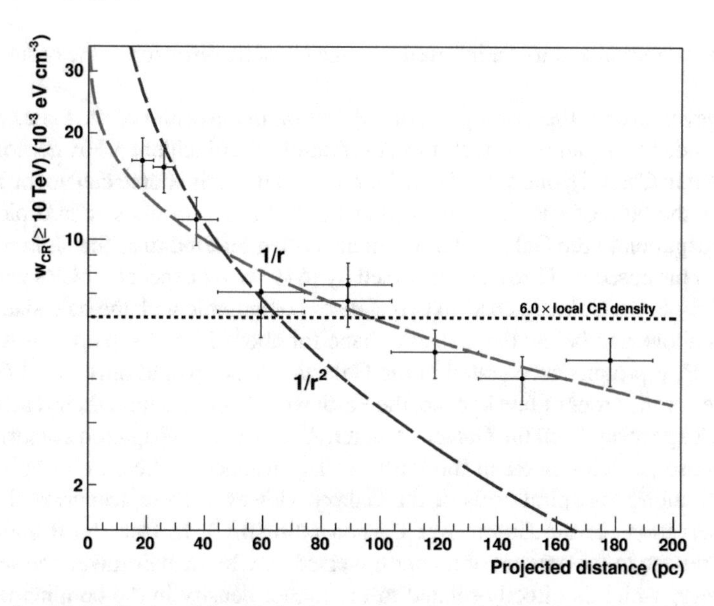

Fig. 1.3 Spatial distribution of the CR density versus projected distance from Sgr A*. The vertical and horizontal error bars show the 1σ statistical plus systematical errors and the bin size, respectively. A fit to the data of a $1/r$ (red short-dashed line), $1/r^2$ (blue long-dashed line) and a homogeneous (black dotted line) CR density radial profiles integrated along the line of sight are shown. The best fit of a $1/r^\alpha$ profile to the data is found for $\alpha = 1.10 \pm 0.12\,(1\sigma)$. Figure from Abramowski et al. [19], reprinted by permission from Springer Nature

$$r^2 D \frac{\partial f}{\partial r} = \text{const} \tag{1.39}$$

Furthermore, assuming the diffusion coefficient to be independent of the radial position (at least within the spatial scale where the gamma rays are observed), then one derives

$$f(r) \propto \frac{1}{r} \tag{1.40}$$

On the other hand, if CRs were advected into a wind one would expect $f(r) \propto 1/r^2$, while the case of a burst-like CR injection would result into a constant radial profile up to the typical diffusion length $\sim\sqrt{D t_{\rm burst}}$. Therefore, the $1/r$ profile of the CR density up to 200 pc in longitude indicates a continuous (or quasi) injection of protons into the CMZ from a centrally located accelerator on a characteristic timescale exceeding the time of diffusive escape of particles from the CMZ, i.e.

$$\tau_{\rm diff} \simeq \frac{R_{\rm CMZ}^2}{6D} \simeq 2 \times 10^3 \left(\frac{D}{10^{30}\,{\rm cm^2\,s^{-1}}} \right)^{-1} \text{yr} \tag{1.41}$$

where D is normalized to the inferred average Galactic diffusion value of multi-TeV CRs.

However, even if the assumption of a PeVatron in the center of the Galaxy would be verified, for instance through the observation of coincident VHE neutrinos (as discussed in Chap. 4), one should consider the fact that this source can not be responsible for the bulk of CR flux observed at Earth. In fact, in the standard picture of CR propagation in the Galactic halo, which is often referred to as the *Galactic Halo model* as proposed by [128] and endorsed by [64], we do expect that CRs propagate diffusively in the Galactic Disc up to distances comparable with the halo size, which extends above and below the Galactic Plane for about $H \simeq 3 - 4\,\text{kpc}$. As a consequence, PeV protons accelerated in the Galactic Center would only travel Galactic distances of the order of few kpc, i.e. the Earth would not be within their reach, being located 8 kpc away from the Galactic Center. Alternative propagation scenarios may allow those particles to reach the Earth, as for instance in the case of anisotropic diffusion along the spiral arms of the Galaxy. However, these scenarios challenge the observations of the Boron over Carbon ratio (B/C). In fact, the B and C flux ratio is related to the amount of matter traversed by CRs in their travel, the so-called *grammage*, which is directly related to the matter density in the confinement volume. If CRs would diffuse mainly along the arms, they would in fact accumulate a grammage much larger than observed [71].

1.3.2 Stellar Winds and OB Associations

The role of stellar winds in the acceleration of Galactic CRs was suggested already in the 1980s by [94] and [78], and later on extensively discussed by [176], as a consequence of the discovery in the same years of a spatial coincidence between SNRs, OB associations, and a gamma-ray *hot spot* in COS B data [175] that soon lead to consider O and B-type stars as relevant candidates for particle acceleration. These are massive stars ($M > M_{\odot}$) which lose a substantial amount of mass in the form of stellar winds, blowing at supersonic speed v_w (2000 to 3500 km/s), thereby shining at kinetic luminosities of the order of 10^{36} to 10^{37} erg/s. Integrating such luminosity values over the star lifetime (a few million years), one can derive an energy release which is comparable to that of supernovae explosions. Another important feature of OB association, which is particularly relevant for the confinement of CRs, is their ability to produce large HII regions around them (typically extending few tens of parsecs) [82]. Moreover, a spatial correlation between OB associations, SNRs and molecular clouds (MCs) is expected, since massive stars are the progenitors of core-collapse supernovae. The hypothesis that COS B hot spot could be explained by the interactions among the uniform 'sea' of Galactic CRs and the dense gas of the MCs was soon discarded, as the emission was as bright that an excess of about one order of magnitude in the flux of CRs was necessary to explain the data [120]. On the other hand, the gamma-ray emission from such associations, named SNOBs, was interpreted as a result of the decay of neutral pions produced in the interactions of CRs accelerated at SNRs with the dense gas of the parent MC. This example shows the potential of SNR/MC associations in providing insight into the process of CR acceleration at shocks: thus, gamma-ray observations of such associations are fundamental in the context of testing the SNR hypothesis for the origin of GCRs.

Recently, a remarkable similarity was discovered in both the energy and radial distribution of CRs around massive Galactic stellar clusters [36], that can be summarized into $w_{\rm CR}(E,r) \propto E^{-2.3} r^{-1}$, as shown in Fig. 1.4. Such a trend, inferred from gamma-ray observables around Westerlund 1, Westerlund 2 and the Cygnus Cocoon, soon brought back the interest on the possibility that clusters of massive stars could power the GCR flux. As for the case of the GC, the radial profile $\propto r^{-1}$ suggests a continuous injection over $\sim$ Myr timescale, which is difficult to explain through SN explosions only. It is possible that the CR radial profile observed around the GC (see Sect. 1.3.1) might be ascribed to the three compact clusters rather than to the central black-hole, respectively the Arches, the Quintuplet and the Young Nuclear Cluster, which are located within 50 pc from the GC. The strong similarity might be considered as an evidence that the same phenomenon is acting in these sources: the simplest interpretation is that CRs have been continuously injected and diffused

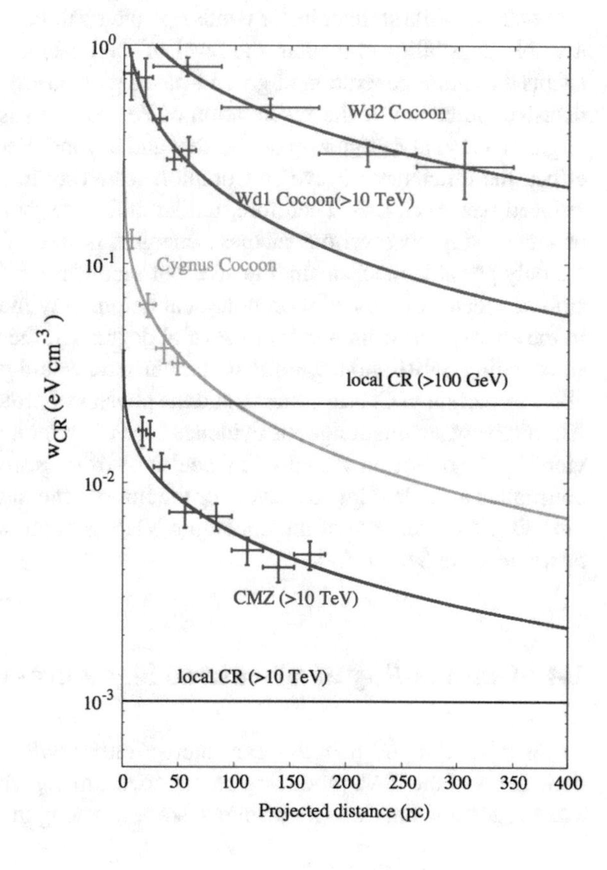

Fig. 1.4 CR proton radial distributions in the Cygnus Cocoon and Westerlund 2 above 100 GeV, and in Westerlund 1 and the CMZ above 10 TeV. For comparison, the energy densities of CR protons above 100 GeV and 10 TeV based on the measurements by AMS are also shown. Figure from Aharonian et al. [36], reprinted by permission from Springer Nature

Table 1.1 Physical parameters of four extended gamma-ray structures and the related stellar clusters. Table from Aharonian et al. [36], reprinted by permission from Springer Nature

Source	Westerlund 2	Cygnus Cocoon	CMZ	Westerlund 1
Extension (pc)	300	50	175	60
Age of cluster (Myr)	1.5–5	3–6	2–7	4–6
L_{kin} (erg/s)	2×10^{38}	2×10^{38}	1×10^{39}	1×10^{39}
Distance (kpc)	5	1.4	8.5	4
$w_{CR}(>100\,\text{GeV})$ (eV/cm^3)	6	0.2	0.26	4.8

away through the ISM. The characteristic timescales are determined by the age of the cluster, $T \sim$ few 10^6 years, while the distance scales span from ten to hundreds parsecs in the ISM. GeV and TeV emission all along such a length scale could be hardly due to electrons, because of the strong energy losses they undergo. Moreover, the luminosity budget of these clusters would be enough to explain the CR power of around $P_{CR} = (0.3 - 1) \times 10^{40}$ erg/s, as visible from Table 1.1 where the main parameters of the aforementioned massive clusters are reported.

A scenario in which SNRs provide efficient particle acceleration up to several tens of TeV, while stellar cluster winds operate up to PeV energies, seems to provide a viable possibility to explain the level of the observed CR flux. Its confirmation awaits the future generation of ground-based gamma-ray and neutrino telescopes. A decisive indication of the acceleration of PeV protons is provided by observations of gamma rays at energies up to 100 TeV and beyond. Because of the Klein–Nishina effect, the efficiency of inverse Compton scattering in this energy band is largely reduced (see Sect. 1.4). Therefore, unlike other energy intervals, the interpretation of gamma-ray observations at these energies is free of confusion and reduces to the only possible mechanism, the decay of secondary π^0-mesons. Hence, the extension of spectrometric and morphological gamma-ray measurements up to 100 TeV in the energy spectrum and up to several degrees in the angular size, from regions surrounding SNRs and powerful stellar clusters, would provide crucial information about the origin of CRs in general, and the physics of proton PeVatrons, in particular. Alternatively, an unambiguous evidence for PeV proton acceleration would be provided by the detection of multi-TeV neutrinos from specific sources, which however constitutes a challenging task even for the future generation of instruments of multi-km^3 size. An overview of the upcoming VHE gamma-ray and neutrino instrument is presented in Sect. 1.5.

1.4 Gamma-Ray and Neutrino Signatures of Cosmic Rays

Accelerated charged particles can interact either with matter, radiation and magnetic fields of the ISM, producing secondaries, among which photons and neutrinos. The suggestion that these messengers, since traveling in straight line, can reveal the sites of particle acceleration in the Universe, constitutes the basis of gamma-ray and neutrino astronomy fields. Two main channels of productions are identified into: (i) leptonic processes, involving primary electrons and positrons, from which photons emerge, and (ii) hadronic processes, involving mainly primary protons and helium nuclei, from which both photons and neutrinos arise. While the flux of secondaries resulting from hadronic processes is governed by the CR density and the target gas density, the flux resulting from the leptonic ones is traced by the electron density and the radiation/magnetic fields. Among the leptonic processes, pair production, Compton scattering, Bremsstrahlung, synchrotron and ionization losses are included. On the other hand, proton-proton and proton-photon interactions constitute the hadronic channels. A complete description of these processes can be found in [206]. To distinguish the sources of CR protons from the sources of CR electrons, gamma-ray observations are often not sufficient, so that multi-wavelength observations have to be taken into account. These measurements—mostly in the radio and X-ray bands—indicate that the best candidate sources for the origin of GCR electrons are Pulsar Wind Nebulae (PWNe), while those for GCR protons are SNRs. As neutrinos are only produced in hadronic channels, their detection appears extremely relevant in the search for CR-proton sources. Once produced, secondary particles can be absorbed in their travel towards the Earth, for instance in pair production via photon-photon interactions, due to the radiation fields present within the source or the cosmic radiation background fields: this phenomenon is mainly affecting gamma rays at the highest energies, as discussed in Sect. 4.3, while it does not affects neutrinos. For this reason, neutrinos constitute the almost unique astronomical messengers in the energy range from about 100 TeV to 10 EeV [43]. Moreover, being stable and weakly interacting, they are able to reach Earth after traveling over cosmic distances, allowing for directly pointing to their production site. In addition, they allow to probe the deepest interiors of sources, which might result opaque to photons. In the following, a summary of the relevant gamma-ray and neutrino production mechanisms is presented.

1.4.1 High-Energy Radiation Processes in Cosmic-Ray Accelerators

Any interpretation of an astronomical observation requires the unambiguous identification of the radiation mechanism responsible for the observation itself. Therefore, in this Section, the principal features produced by different radiation processes on the spectrum of secondaries are highlighted. The gamma-ray radiation is the only component of the electromagnetic spectrum that cannot be produced by the thermal motion of charged particles, as the required minimum temperature would be unreasonably high. Indeed, considering a black body, the temperature required to emit MeV gamma rays is $T = 10^9$ K (for comparison, the nuclear fusion reactions inside the Sun require 10^7 K and consequently result in keV photons). Generally, low-energy gamma rays are mainly produced by Bremsstrahlung and proton-proton interactions, while in the high-energy domain IC scattering accompanies the proton-proton channel. However, interesting considerations concerning the magnetic field can be derived through the multi-wavelength observations of the source spectra, as synchrotron losses strongly affect the electromagnetic spectrum, from the radio to the X-ray energy bands.

The computation of secondary emission spectra requires the knowledge of the cross-section of the relevant processes and of the energy distribution of the parent particles $N(E)$. The cross-sections of electromagnetic processes are well know from classical electrodynamics theory, while the cross-sections of hadronic processes are measured in accelerators and extrapolated to higher energies for astrophysical applications. The particle distribution function $N(E)$ depends on time, and its calculation requires the knowledge of (i) the injection spectrum $Q(E)$, which depends on both the acceleration mechanism acting at the source and the escape process from the source, and (ii) the diffusion and advection effects on the particle transport. A simplified equation describing the particle distribution evolution is [128]

$$\frac{\partial N}{\partial t} = \frac{\partial}{\partial E}[PN] - \frac{N}{\tau_{\rm esc}} + Q \tag{1.42}$$

being $P(E) = -dE/dt$ the energy loss rate and $\tau_{\rm esc}$ the escape time of particles due to diffusion and advection, namely $\tau_{\rm esc} = (1/\tau_{\rm diff} + 1/\tau_{\rm adv})^{-1}$. To solve this equation, stationarity is assumed in the following, i.e. $\partial N/\partial t = 0$: the solution derived is called *steady state particle distribution*. Moreover, assuming that the escape time is much longer than the energy loss rate, the term $N/\tau_{\rm esc}$ can be neglected and Eq. (1.42) is solved by

$$N(E) = \frac{1}{dE/dt}\int_E^\infty Q(E')dE' \tag{1.43}$$

where the energy loss rate dE/dt accounts for all the processes relevant in the source. In the remaining part of the Section, several processes are investigated and their effect on the spectrum of parent particles is highlighted.

Bremsstrahlung

Radiation due to the acceleration of a charged particle in the Coulomb field of another charge is called *Bremsstrahlung* or *free-free* emission. In the case when a population of relativistic particles interacts with a thermal particle population, the emission mechanism is called *non-thermal Bremsstrahlung*. The interaction with the electric field of an ion (or nucleus) produces a change in the particle trajectory. Since the emission probability scales as the inverse square of the mass of the emitting particle m, namely $\sigma_{\rm Br} \propto (e^2/m)^2$, then electron Bremsstrahlung is strongly favored with respect to proton (with mass $m_{\rm p}$) Bremsstrahlung, its interaction probability being a factor $(m_{\rm p}/m_{\rm e})^2 \simeq 4 \times 10^6$ times larger. Bremsstrahlung losses are catastrophic, in the sense that high-energy electrons radiate almost all of their energy in one or two photons: in other words, at every radiative phenomenon, half of the electron energy is given to the photon, which mostly contributes to the gamma-ray radiation in the sub-TeV domain. A relevant parameter for the electron emission in a medium of density n is the so-called *radiation length*

$$X_0 = \frac{7}{9(n\sigma_0)} \tag{1.44}$$

namely the average distance over which the relativistic electron reduces its energy to 1/e of its initial value (here σ_0 is the cross-section asymptotic value). For instance, in gaseous hydrogen $X_0 \simeq 60$ g/cm^2. A second important parameter is the so-called *critical energy*, below which ionization losses dominate over Bremsstrahlung losses: in a gaseous hydrogen $\epsilon_{\rm cr} = 700 m_e c^2 \simeq 350\,\text{MeV}$, m_e being the electron mass. At higher energies, effective particle multiplication in a cascade is achieved. Introducing the average energy loss rate for electrons

$$-\left(\frac{dE_e}{dt}\right)_{\rm Br} = \left(\frac{cm_{\rm p}n}{X_0}\right)E_{\rm e} \tag{1.45}$$

one can derive the electron lifetime due to Bremsstrahlung losses as

$$\tau_{\rm Br} = \frac{E_{\rm e}}{dE_{\rm e}/dt} \simeq 4\times10^7\left(\frac{n}{1\,\text{cm}^{-3}}\right)^{-1}\ \text{yr} \tag{1.46}$$

Since the electron loss rate is proportional to the electron energy, the electron lifetime is energy independent. This implies that for an initial power-law spectrum of electrons $Q(E_{\rm e}) \propto E_{\rm e}^{-\Gamma}$, Bremsstrahlung losses do not change the original electron spectrum $N(E_{\rm e}) \propto E_{\rm e}^{-\Gamma}$. As a consequence, the Bremsstrahlung gamma-ray spectrum would simply reproduce the shape of the electron spectrum. However, at low energies, when ionization losses dominate, the electron spectrum becomes flatter, following a distribution like $N(E_{\rm e}) \propto E_{\rm e}^{-\Gamma+1}$, and one generally expects to see a quite hard gamma-ray spectrum with slope $\Gamma - 1$.

Synchrotron Radiation

Particles accelerated in a magnetic field B emit synchrotron radiation. For non-relativistic speeds, the nature of the radiation is rather simple and it is called *cyclotron radiation*: the frequency of emission coincides with the frequency of gyration in the magnetic field. However, for ultra-relativistic particles, the frequency spectrum is much more complex and can extend to many times the gyration frequency. This radiation is known as *synchrotron radiation*. Synchrotron radiation is more efficiently produced by electrons: indeed gyrating protons produce synchrotron radiation with a characteristic energy which is reduced by a factor $(m_{\rm p}/m_{\rm e})^3 \simeq 7\times10^9$ with respect to that produced by electrons, while the proton cooling time is $(m_{\rm p}/m_{\rm e})^4 \simeq 10^{13}$ times longer. Thus, in the following, a focus on electron emitted radiation is provided.

The classical treatment of this mechanism provides an accurate description of the process as long as the condition

$$\frac{E}{m_e c^2}\frac{B}{B_{cr}} \ll 1 \tag{1.47}$$

is satisfied, $B_{cr} = m_e^2 c^3/e\hbar \simeq 4.4 \times 10^{13}$ G being the critical value of the magnetic field relevant to quantum effects, while E is the energy of the gyrating particle in the magnetic field B. Under this assumption, the total emitted power from an individual electron amounts to

$$P_{sync} = \frac{4}{3}\sigma_T c \beta^2 \gamma^2 U_B \tag{1.48}$$

where $\sigma_T = (8\pi/3)^2(e^2/m_e c^2)^2 \simeq 6.65 \times 10^{-25}$ cm^2 is the Thomson cross-section for the elastic scattering of a photon off a free charged particle (as described by classical electromagnetism). In the previous equation, $U_B = B^2/8\pi$ represents the energy density of the magnetic field, while $\beta = v/c$ and $\gamma = E_e/m_e c^2$ are respectively the particle speed and its kinetic energy. The emitted power defines the energy loss rate of electrons by synchrotron losses, since $(dE/dt)_{sync} = -P_{sync}$. For instance, the cooling time for electrons in a magnetic field is [124]

$$\tau_{sync} = 6\pi \frac{m_e^2 c^4}{c\sigma_T B^2 E_e} = 1.3 \times 10^{10} \left(\frac{B}{1\,\mu\text{G}}\right)^{-2} \left(\frac{E_e}{1\,\text{GeV}}\right)^{-1} \text{yr} \tag{1.49}$$

and the characteristic energy of the emitted photon is

$$\epsilon_c = h\nu_c = \frac{3}{4\pi}\frac{\hbar e B \sin\alpha E_e^2}{m_e^3 c^5} \tag{1.50}$$

where α is the pitch angle. Hence, the stronger the magnetic field (and/or the more energetic is the emitting particle) and the higher is the frequency of emitted radiation. The spectral energy distribution of radiated photons by an individual electron in an isotropic magnetic field is (ϵ being the energy of the emitted photon)

$$P_{syn}(E_e, \epsilon) = \frac{\sqrt{2}}{h}\frac{e^3 B}{m_e c^2} F(x) \tag{1.51}$$

where $F(x) = x \int_x^\infty K_{5/3}(\xi) d\xi$, $K_{5/3}$ is the modified Bessel function of order 5/3 and $x = \epsilon/\epsilon_c$. Such an emissivity peaks in correspondence of $\epsilon \simeq 0.29\epsilon_c$, with a sharp drop at higher energies. Generally, the synchrotron spectrum will derive from the radiation emitted by a population of CR electrons: assuming a power-law distribution for the initial electron population with slope Γ, the steady state distribution of cooled electrons will show a steepening to $\Gamma + 1$, while the emitted radiation spectrum will have a slope $(\Gamma + 1)/2$. Furthermore, the shape of the cut-off appears shallower in the emitted radiation: indeed, if the cooled electrons are distributed like

$$\frac{dN}{dE} \propto E^{-(\Gamma+1)} \exp[-(E/E_0)^{\beta}] \quad (1.52)$$

then emitted synchrotron photons will be distributed as

$$\frac{dN}{d\epsilon} \propto \epsilon^{-(\Gamma+1)/2} \exp[-(\epsilon/\epsilon_0)^{\beta'}] \quad (1.53)$$

with $\beta' = \beta/(\beta+2)$ [254]. Thus, a simple exponential cut-off in electron spectrum ($\beta = 1$) would produce a shallow feature in the emitted radiation, yielding $\beta' = 1$. This is likely the case of an age-limited system, where the maximum energy in the electron spectrum is connected with the time spent in the acceleration process. In the case of a loss-limited system, where the maximum energy in the electron spectrum is connected with the cooling process, one would expect $\beta = 2$ and consequently $\beta' = 1/2$ [254]. Note that an erroneous conclusion would be derived within the δ-functional approximation, which assumes that the synchrotron spectrum is concentrated around $\epsilon \simeq 0.29\epsilon_c$ thus predicting a drop as sharp as $\beta' = \beta/2$. On the other hand, in the presence of a spatially distributed magnetic field, the spectrum of the synchrotron radiation is produced with $\beta' = \beta/3$ [255].

The synchrotron emission has a continuous spectrum. While its intensity in the radio domain is proportional to $dN/d\epsilon \propto N_e B^2$, thus showing a degeneracy among electron density and magnetic field, the intensity in the X-ray domain is proportional to $dN/d\epsilon \propto N_e$. Hence, observations in the X-ray band provide a powerful tool to infer the electron density distribution, while joint radio observations allow to derive the amplitude of the magnetic field. Furthermore, in young SNRs, the cut-off photon energy $\epsilon_{\text{cut-off}}$ of synchrotron X-ray photons is independent of the magnetic field and it is directly connected to the shock speed as [32]

$$\epsilon_{\text{cut-off}} = 0.55\eta^{-1} \left(\frac{v_s}{3000\,\text{km/s}}\right)^2 \text{keV} \quad (1.54)$$

where $\eta \leq 1$ is the particle gyrofactor, namely the ratio among the particle mean free path and its gyroradius ($\eta = 1$ in the case of Bohm diffusion). Thus, X-ray observations are particularly relevant in that any deviation from the Bohm regime produces a shift in the cut-off photon energy. Alternatively, in highly turbulent regions where Bohm diffusion is more likely achieved, X-ray observations allow to constrain the shock speed.

Inverse Compton Scattering

The inverse Compton scattering is a fundamental electromagnetic process, which frequently takes place in cosmic environment, thanks to the uniform presence of the 2.7 K CMB radiation field. Since IC scattering of protons is suppressed by a factor $(m_e/m_p)^4$ with respect to that of electrons [31], only high-energy electrons scattering off low energy photons will be discussed in the following. In this interaction, the electrons transfer part of their energy to photons, producing a VHE emission up to tens of TeV. The total cross-section of the process depends only on the product k of the interacting electron energy ϵ_e and the target photon energy ω, i.e. $k = \epsilon_e \omega$, with all the energies given in units of $m_e c^2$. For $k \ll 1$, the process occurs in the non-relativistic regime, or *Thomson regime*. In this case, the cooling time of electrons amounts to [31]

$$\tau_{\mathrm{IC}} = 3 \times 10^8 \left(\frac{U_{\mathrm{ph}}}{1\,\mathrm{eV/cm^3}} \right)^{-1} \left(\frac{E_{\mathrm{e}}}{1\,\mathrm{GeV}} \right)^{-1} \mathrm{yr} \tag{1.55}$$

and the interaction cross-section scales as

$$\sigma_{\mathrm{IC}} \simeq \sigma_{\mathrm{T}}(1-k) \qquad k \ll 1 \tag{1.56}$$

For $k \gg 1$ the ultra-relativistic regime, or *Klein–Nishina regime*, takes place and the cross-section is modified into

$$\sigma_{\mathrm{IC}} \simeq \frac{8}{3} \sigma_{\mathrm{T}} \frac{\ln(4k)}{k} \qquad k \gg 1 \tag{1.57}$$

The total power emitted by an individual electron is

$$P_{\mathrm{IC}} = \frac{4}{3} \sigma_{\mathrm{T}} c \beta^2 \gamma^2 U_{\mathrm{ph}} \tag{1.58}$$

where $U_{\mathrm{ph}} = n_{\mathrm{ph}} \omega$ is the energy density of the seed photons. The ratio of the radiative losses caused by synchrotron emission to those generated by inverse Compton scattering is equal to the ratio between the magnetic field energy density and the photon energy density (see Eqs. (1.48) and (1.58)). Therefore, in the case of a complete leptonic origin of gamma rays, one can determine the strength of the magnetic fields directly by comparing P_{sync} with P_{IC}.

Note that in the Klein–Nishina regime, a single interaction is sufficient to transfer a significant fraction of the electron energy to the photon. For a power-law energy spectrum of electrons with slope Γ and an isotropic target photon distribution, the corresponding spectrum of the emitted gamma rays turns out to be steeper, i.e. $dN/d\epsilon \propto \epsilon^{-\alpha}(\ln(4\omega\epsilon) + \mathrm{const})$ with $\alpha = \Gamma + 1$. On the other hand, in the Thomson regime the average energy of the up-scattered photon is $\epsilon = \omega E_{\mathrm{e}}^2$, thus only a fraction $\epsilon / E_{\mathrm{e}} \simeq k \ll 1$ of the primary electron is converted in photon energy: assuming an electron power-law spectrum and an isotropic target radiation field, the resulting gamma-ray spectrum is still a power-law with slope $\alpha = (\Gamma + 1)/2$ [30]. Therefore, in the Klein–Nishina regime, the IC gamma-ray spectrum will be significantly steeper. Thus, even a power-law distribution of electrons will produce a gamma-ray spectrum with a break, due to the onset of the Klein–Nishina regime.

Proton-Proton Interaction

Inelastic collisions of protons and nuclei with the ambient gas produce mostly pions, with a smaller contribution from kaons and etas, as

$$p + p \longrightarrow \begin{cases} p + p + \pi^0 \\ p + n + \pi^+ \end{cases} \tag{1.59}$$

The minimum kinetic energy of the proton for the production of a neutral pion amounts to $E_{\rm th} = 2m_{\pi^0}c^2(1 + m_{\pi^0}/4m_{\rm p}) \simeq 280\,{\rm MeV}$. The cross-section for π production is almost energy independent and amounts to about $\sigma_{pp} = 2 \times 10^{-26}\,{\rm cm}^2$, as reported in Fig. 1.5a. The cooling timescale of protons via the pp process can be described as [31]

$$\tau_{\rm pp} = \frac{1}{nf\sigma_{\rm pp}c} \simeq 5.3 \times 10^7 \left(\frac{n}{1\,{\rm cm}^{-3}}\right)^{-1}\ {\rm yr} \tag{1.60}$$

where $f \sim 0.5$ represents the inelasticity coefficient. At high energies, where multiple pion production is achieved, π^+, π^- and π^0 are produced in very similar amounts due to isospin symmetry. Their full decay chain reads as

$$\begin{cases} \pi^0 \longrightarrow \gamma + \gamma \\ \pi^+ \longrightarrow \mu^+ + \nu_\mu \longrightarrow e^+ + \nu_e + \overline{\nu}_\mu + \nu_\mu \\ \pi^- \longrightarrow \mu^- + \overline{\nu}_\mu \longrightarrow e^- + \overline{\nu}_e + \nu_\mu + \overline{\nu}_\mu \end{cases} \tag{1.61}$$

The mean energy of the leading pion is about 20% of the proton initial energy. In the case of π^0 meson production, the two photons will equally share this amount, leading to individual photon energies of $E_\gamma \simeq 0.1E_p$. A characteristic feature of the gamma-ray differential energy spectrum, as a consequence of hadronic production, is the so-called 'pion bump' [222]: this is a distinct bell-type feature, which appears between 100 MeV and a few GeV. Indeed, because of the fast decay of π^0 into two gamma-ray photons (decay time $\tau_{\pi^0} \simeq 8.5 \times 10^{-17}\,{\rm s}$), each with an energy of $E_\gamma = m_{\pi^0}c^2/2 = 67.5\,{\rm MeV/c}^2$ in the rest frame of the neutral pion, the resulting gamma-ray number spectrum dN/dE is symmetric around such value in a log-log representation. Thus, the spectrum of gamma rays always presents a spectral maximum which is independent of the distribution of parent pions [251]. The gamma-ray emissivity $q_\gamma(E_\gamma)$ is directly defined through the neutral pion emissivity $q_\pi(E_\pi)$ as

$$q_\gamma(E_\gamma) = 2\int_0^1 q_\pi\left(\frac{E_\gamma}{x}\right)\frac{dx}{x} \tag{1.62}$$

where $x = E_\gamma/E_\pi$. The factor 2 accounts for the two photons produced in the final state, and the emissivity of neutral pions is

$$q_\pi(E_\pi) = cn_\mathrm{H} \int \sigma_\mathrm{pp}(E_\pi, E_\mathrm{p}) n_\mathrm{p}(E_\mathrm{p}) dE_\mathrm{p} \tag{1.63}$$

Here, the inclusive interaction cross-section σ_pp was introduced, as well as the differential energy distribution of the parent protons $n_\mathrm{p}(E_\mathrm{p})$ and the target density n_H. It is worth to recall that the AGILE and Fermi-LAT Collaborations have claimed the detection of the pion bump towards two middle-aged SNRs, respectively IC 443 and W 44 [23], and interpreted such a feature as an evidence for acceleration of CR protons and nuclei in SNRs.

In the case of charged pion production, the four leptons in the final state of the pion-muon decay chain will equally share the energy of the leading pion, resulting into neutrinos and electrons of about $E \simeq 0.05 E_p$. The spectra of secondaries will resemble the spectra of the projectile protons, in case of simple power-law behaviors of the primaries: the shape of a cut-off, instead, gets modified, as extensively discussed by [150]. Analogously to Eq. (1.62), the neutrino emissivity $q_\nu(E_\nu)$ is directly connected to the charged pion emissivity: for muon neutrinos it holds that

$$q_{\nu_\mu}(E_{\nu_\mu}) = 2 \int_0^1 (f_{\nu_\mu^{(1)}}(x) + f_{\nu_\mu^{(2)}}(x)) q_\pi \left(\frac{E_{\nu_\mu}}{x} \right) \frac{dx}{x} \tag{1.64}$$

where $x = E_{\nu_\mu}/E_\pi$. The factor 2 accounts for both the contributions from π^+ and π^- mesons, while $f_{\nu_\mu^{(1)}}$ is the kernel function describing muon neutrinos produced in the two-body pion decay and $f_{\nu_\mu^{(2)}}$ refers to those produced in the three-body muon decay (see [150] for further details). On the other hand, for electron neutrinos (with $x = E_{\nu_\mathrm{e}}/E_\pi$)

$$q_{\nu_\mathrm{e}}(E_{\nu_\mathrm{e}}) = 2 \int_0^1 f_{\nu_\mathrm{e}}(x) q_\pi \left(\frac{E_{\nu_\mathrm{e}}}{x} \right) \frac{dx}{x} \tag{1.65}$$

since ν_e only comes from the muon decay. According to the environment where the pp (and analogously the $p\gamma$) interaction happens, a different neutrino flavor ratio is produced at the source and consequently expected at Earth. For instance, in the case of full pion decay chain, as in Eq. (1.61), the flavor ratio produced at the source is $(\nu_\mathrm{e} : \nu_\mu : \nu_\tau) = (1 : 2 : 0)$. However, in the case of very dense sources, also called *damped-muon sources*, the muon can interact before it decays and consequently the production ratio modifies into $(\nu_\mathrm{e} : \nu_\mu : \nu_\tau) = (0 : 1 : 0)$. Alternatively, in the case of neutron sources, the flavor ratio becomes $(\nu_\mathrm{e} : \nu_\mu : \nu_\tau) = (1 : 0 : 0)$. Applying neutrino oscillations over cosmic distances, the flavor ratio at Earth is expected to be $(\nu_\mathrm{e} : \nu_\mu : \nu_\tau) = (1 : 1 : 1)$ in the full pion decay case, while a reduced amount of ν_e and ν_τ is expected respectively in the case of muon-damped sources and neutron sources [196].

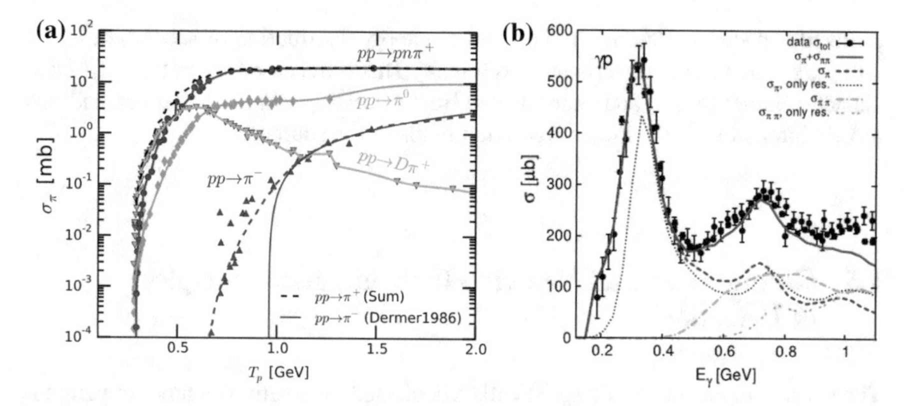

Fig. 1.5 *Left:* proton-proton interaction cross-sections for π-meson production as a function of incident proton kinetic energy $T_p < 2\,\mathrm{GeV}$. Credit: Yang et al. [251], reproduced with permission ©ESO. *Right:* Photo-absorption cross-section on proton targets as a function of the photon energy. Reprinted from Lalakulich & Mosel [160], with the permission of AIP Publishing

Proton-Photon Interaction

The hadronic interaction of protons with radiation fields proceeds through the resonant production of the Δ^+ hadron ($m_{\Delta^+} = 1232\,\mathrm{MeV/c^2}$), i.e. the interaction cross-section peaks in correspondence of the hadron rest energy, assuming a value as high as $\sigma_\Delta \simeq 5 \times 10^{-28}\,\mathrm{cm}^2$, as reported in Fig. 1.5b. The resonance subsequently decays into mesons (mostly pions) according to

$$p + \gamma \xrightarrow{\Delta^+} \begin{cases} p + \pi^0 & \mathrm{BR} = 2/3 \\ n + \pi^+ & \mathrm{BR} = 1/3 \end{cases} \tag{1.66}$$

where the branching ratios (BR) are computed assuming isospin conservation. For the center-of-mass energy of a proton-photon interaction to exceed the threshold energy for the Δ-resonance production, the proton energy must meet the condition:

$$E_\mathrm{p} \geq \frac{(m_\Delta^2 - m_\mathrm{p}^2)}{4E_\gamma} \simeq 1.6 \times 10^8 \left(\frac{1\,\mathrm{eV}}{E_\gamma}\right) \mathrm{GeV} \tag{1.67}$$

Therefore, high-energy protons interact resonantly with low-energy radiation fields. Considering for instance the CMB as a target field, the threshold for the production of a neutron and a pion in the final state amounts to $\sim 10^{20}$ eV. Such a process, which only affects UHECRs, is also called GZK process [130, 252]. Given the uniform distribution of the CMB radiation field through the Universe, such an energy value also defines the maximum energy for protons to propagate over cosmic distances without being absorbed and therefore it sets an end to the CR spectrum.

Because of the resonant nature of the process, the spectrum of projectile protons is not strictly reproduced by secondaries: it is rather the convolution of the target and projectile spectra to determine the spectral energy distribution of secondaries, that will peak around a characteristic energy scale. This is the case for instance of Active Galactic Nuclei (AGN) and Gamma-Ray Bursts (GRBs), where accelerated hadrons mostly interact with the intense radiation fields of the source jet.

1.5 Gamma-Ray and Neutrino Instruments: Principles of Detection

The last part of this thesis (Chap. 5) will be dedicated to sensitivity studies of gamma-ray and neutrino telescopes, with a focus on their detection capability towards extended sources that possibly populate the VHE Galactic sky. For this reason, in this Section, the reader is provided with a brief introduction concerning some key aspects of the detection technique implemented in current instruments. Open issues are also discussed, as these constitute the scientific goals of future instruments.

Gamma rays are particularly interesting since they allow to study two important aspects about CRs, namely (i) particle propagation within and around sources, providing hints concerning the diffusion properties of turbulent environments as in the case of shocks, and (ii) particle escape from their acceleration site. Precision measurements have been performed during the last decades, improving our understanding of the gamma-ray sky, thanks to the combination of GeV instruments (as Fermi-LAT and AGILE), ground-based Imaging Atmospheric Cherenkov Telescopes (IACTs) (as H.E.S.S. and MAGIC), and water Cherenkov detectors (as HAWC). An era of significant increase in the number of sources has started with these instruments, leading to an improved understanding of the different source classes that populate the high-energy sky. As a result, today, we see three main components in the gamma-ray sky:

(i) the Galactic diffuse emission, generated by the interaction of the GCRs with the interstellar gas and radiation fields;
(ii) individual sources, both located in the Galactic Plane and at higher latitudes;
(iii) the weak isotropic diffuse emission, detectable at all Galactic latitudes.

The relative contribution of these emissions to the total gamma-ray flux changes with energy, as expected considering that the source energy spectrum is typically harder than the diffuse Galactic component. In fact, while about 80% of photons comes from the Galactic diffuse emission at GeV energies, in the TeV band individual sources dominate the energetics.

On the other hand, the detection of VHE neutrinos is an extremely challenging task, in that the weak interaction cross-section properties (e.g. for the deep inelastic scattering $\sigma_{\rm DIS}(1\,{\rm TeV}) \simeq 10^{-35}\,{\rm cm}^2$) combined with the level of fluxes expected from known astronomical sources (given for instance the observation of TeV gamma-ray sources) require the instrumentation of cubic kilometer detectors in order to collect a bunch of neutrinos per year. Since 1960s, large volume of water were suggested by

Reines, Greisen and Markov as a target for detecting astrophysical neutrinos [168]: the ideal solution was identified into a natural and transparent medium, which allows the propagation of the Cherenkov emission induced by the passage of relativistic particles resulting from neutrino interactions. Such a detection technique led to the discovery of cosmic neutrinos from the Sun first, the SN 1987A later and recently from the flaring blazar TXS 0506+056 [10]. At the same time, it also led to the measurement of the atmospheric neutrino flux up to energies of several hundreds of TeV [8] and to the experimental discovery of atmospheric neutrino oscillations [113], providing evidence for a non-zero mass of neutrinos and thus hints for the existence of physics beyond the Standard Model.

1.5.1 VHE Gamma Rays

Gamma rays are high-energy photons which span an energy range of about seven decades in energy and fourteen decades in flux between the low and the high-energy end, corresponding to respectively ~10 MeV and ~100 TeV. Such an extended energy range cannot be explored with a unique detection technique or instrument. In fact, a bright source as the Crab Nebula emits $\sim 10^{-3}$ photons m^{-2} s^{-1} above 100 MeV, while its flux is reduced to $\sim 10^{-7}$ photons m^{-2} s^{-1} at the highest energy. As a consequence, at low energies an effective area of the order of 1 m^2 is adequate, while at high energies (>50 GeV) 10^4 m^2 are needed. It is worth to recall also that the atmosphere prevents gamma rays from reaching the ground. Therefore gamma-ray direct detection requires space-based instruments, whose effective area is however insufficient at the highest energies, where a ground-based technique is needed.

The operation of space-based detectors above 20 MeV is based on the principle of pair-creation in the detector, while ground-based instruments detect the secondary products of the interaction of gamma rays with the atmosphere. In this interaction, a shower of particles is created in the atmosphere: its charged component mostly consists of electrons and positrons (in the case of a gamma-ray or electron-initiated shower) or of electrons and muons (in the case of a proton-initiated shower). A view of the image produced on the camera by a hadronic and an electromagnetic shower is provided in Fig. 1.6a. Two main techniques utilize air showers: (i) IACTs, that detect the optical Cherenkov light induced by the shower of ultra-relativistic particles in the atmosphere, and (ii) water Cherenkov detectors, that reveal the charged particles through the Cherenkov light induced inside water detectors, located on the ground. Since the latter technique requires the particles to reach the ground, these instruments usually operate with a higher energy threshold (>100 GeV). However, they can operate continuously, while IACTs can only operate during dark night time (with a duty cycle of about 10%).

One can derive the basic requirements for IACTs by considering a 1 TeV shower: the light induced by Cherenkov emission arrives in a ring (the so-called *pool*) on the ground with a radius of ~120 m at an altitude of 2000 m. Stereoscopic observations have been demonstrated to be fundamental to improve the angular resolution

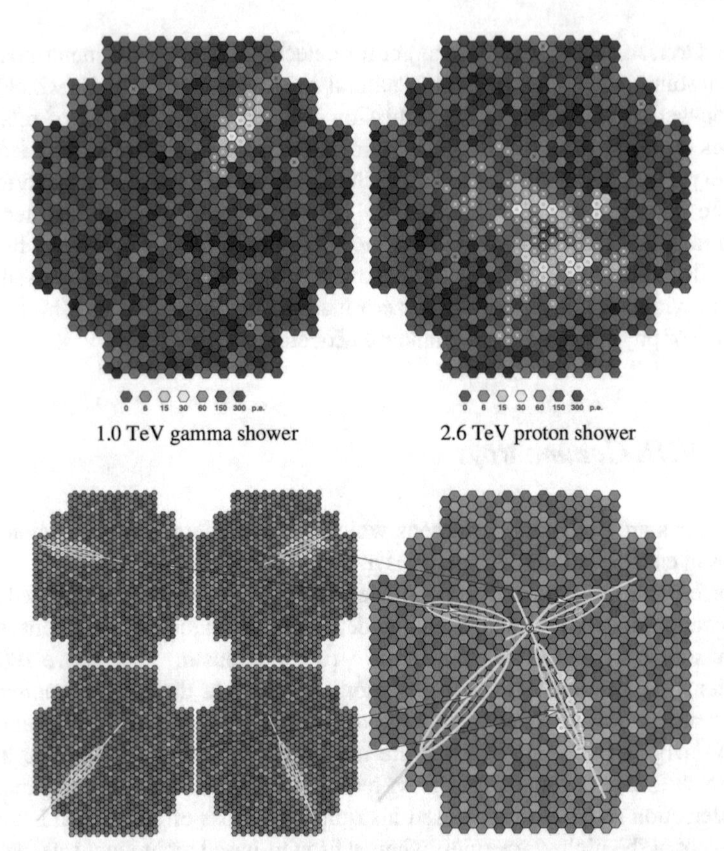

Fig. 1.6 *Top:* Imaging of gamma-initiated (left) and proton-initiated (right) showers. *Bottom:* Reconstruction of the arrival direction of a gamma ray, as performed by a system of IACTs operating in stereoscopic mode. Figure reproduced from Volk & Bernlohr [244], under the CC BY LICENSE

and to enhance the background rejection, given that these allow to significantly suppress hadronic showers in which long-lived muons trigger individual telescopes [33]. Indeed, the angular resolution depends on the number of telescopes N_t used in the reconstruction of the event and up to $N_t \sim 50$ it improves as $N_t^{1/2}$ [244]. An illustration of the event position reconstruction through a stereoscopic vision is given in Fig. 1.6b. The spacing of the telescopes should thus be ~100 m. A closer spacing would provide an improvement of the performances at low energies, but this would reduce the collection area at higher energies (and viceversa). The shower energy is measured as in a calorimeter, by converting the track length of all the particles collected into a total amount of light, which is proportional to the energy of the primary particle. All detection techniques must deal with the suppression of the abundant background of CR protons: this is usually achieved through an active veto for charged particles in the case of space-based detectors, or by considering the difference in shower development between gamma-ray induced electromagnetic showers and proton-induced hadronic showers (mostly shower shape and muon content) in ground-based instruments. Typical suppression factors varies from $\sim 10^{-6}$ for Fermi-LAT, up to $\sim 10^{-5}$ for IACTs and several 10^{-3} for water Cherenkov detectors.

Finally, the various detection techniques are complementary in the energy range of operation: while space-based detectors are ideal for the exploration of the low-energy part of the spectrum starting at ~30 MeV, their detection area is not enough anymore at around ~100 GeV. This is the regime where IACTs onset, as the particle shower produced by a primary photon with energy ≥100 GeV induce enough Cherenkov light in the atmosphere to be revealed by a ~10 m mirror diameter IACT. Note that at ~30 TeV the IACT effective area is not sufficient anymore: in this regime, however, water Cherenkov telescopes can provide the necessary detection areas.

1.5.2 Open Issues and the Future of VHE Gamma-Ray Astronomy

The precious information on the high-energy gamma-ray sky derived in the last decades leaves us some concerns when interpreting the observed radiation, which are still preventing us to derive conclusions on the origin of GCRs. In particular, the main issues are here summarized:

- Why are all SNR spectra softer than expected from the standard DSA theory, and hence what is the spectrum of primary particles?
- What is the exact shape of the cut-off in SNRs?
- What is the efficiency of conversion of SN explosion energy into CRs?
- Can SNRs accelerate protons to PeV energies? If so, for how long?
- How do CRs diffuse away from these sources and eventually reach the Earth?

The current generation of IACTs has demonstrated the great capabilities of such a detection technique to perform imaging and spectrometric studies of VHE gamma-ray sources. The next generation array of IACTs, the Cherenkov Telescope Array (CTA) [22, 24], will be the largest and most sensitive ground-based instrument for VHE gamma-ray astronomy. While a low-energy threshold is required by AGN and GRB studies, in order to access cosmological distances, the coverage of the 100 TeV domain is of primary importance to access the maximum energy region of Galactic emitters, as SNRs and PWNe. To achieve such a broad energy band coverage, CTA will be composed of telescopes of different sizes. Large Size Telescopes (LSTs) allow to access the sub-TeV range: indeed, low-energy photons produce less Cherenkov light, which thus has to be efficiently collected. CTA will adopt several LSTs to access energies as low as 100 GeV: these will be 23 m diameter parabolic reflective surfaces with a 4.5 deg field of view (FoV) and a central camera consisting of 1855 pixels. In order to study transient phenomena, the repointing of LSTs will have to be as fast as 20 s. The energy range from 100 GeV to 10 TeV, the *core* of CTA, will be accessible through Medium Size Telescopes (MSTs), 12 m class telescopes distributed on a regular grid with ~100 m spacing. A large FoV of 7 deg will allow to

(a) CTA Southern array.

(b) CTA Northern array.

Fig. 1.7 Illustrative picture of the future Southern (top) and Northern (bottom) arrays of CTA. Figure from https://www.cta-observatories.org/

rapidly survey the TeV sky, while the large number of MSTs will allow to improve the event reconstruction and to increase effective area. Lastly, to access the multi-TeV domain, where the flux level from non-thermal processes sharply drops, the effective area has to be maximized. A large number of Small Size Telescopes (SSTs), 4 m class telescopes spaced by ~400 m, represents the best solution for the detection of VHE events. In order to scan the whole sky, CTA will consist of two arrays of IACTs, one in the Northern (La Palma, Canary Islands) and one in the Southern (Paranal, Chile) hemisphere, aimed at observing respectively the central part of the Galaxy and extra-galactic sources. An artistic view of the future array is pictured in Fig. 1.7, while the spatial positioning of the telescopes is shown in Fig. 1.8. CTA is expected to achieve improved performances with respect to current generation of IACTs, namely: (i) the angular resolution above 1 TeV should achieve a value as low as 2 arcmin (a factor 3 better than current instruments), in order to resolve the details of complex morphologies and densely populated sky regions; (ii) the effective area will possibly improve by about one order of magnitude, which is crucial to observe the VHE phenomena of the nearby non-thermal Universe; (iii) an energy resolution down to 6% at 1 TeV should be achieved (a factor 2 better than current instruments), together with a broad energy coverage over more than three orders of magnitude; (iv) a temporal resolution within the sub-minute scale should allow to

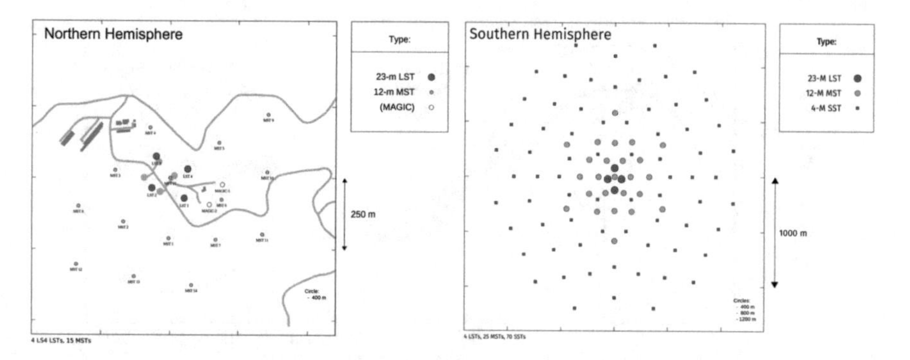

Fig. 1.8 CTA full array footprint. Figure from https://www.cta-observatories.org/

access fast variable phenomena, as source flares (currently resolved on few minutes scale); (v) a sensitivity of the level of a milli-Crab[3] is expected to be achieved in 50 h of observations of a point-like source in the instrument core region (one order of magnitude better than current instruments).

Furthermore, CTA will operate as a proposal-driven observatory, providing a wide and transparent access to its data. The excellent capabilities of this instrument should not only reveal new classes of gamma-ray emitters, but also allow for a deeper understanding of known objects. In particular, CTA is expected to contribute to (i) the understanding of the origin of Galactic and extra-galactic CRs, (ii) the understanding of the nature of particle acceleration, propagation and radiation in extreme astrophysical environments, and (iii) the search for Dark Matter and physics beyond the Standard Model.

1.5.3 VHE Neutrinos

The search for localized sources of neutrino emissions is at the hearth of neutrino astronomy and it is a urgent goal to be pursued in view of the recent discovery of a high-energy 'extraterrestrial' neutrino flux [2]. In fact, the present knowledge of the high-energy diffuse sky indicates that cosmic rays, photons and neutrinos share almost equally the energy budget of the non-thermal Universe, as represented in Fig. 1.9: therefore, each of them constitutes a crucial piece of information in understanding the processes that regulate the Universe. It is worth to keep in mind that each messenger encloses both the information on the source and on the ambient through which it propagates to reach the Earth, where our telescopes are eventually able to detect them. As in the text of the *Syadvada*, we are the blind men, who explore

[3]The Crab Nebula flux at 1 TeV corresponds to 2.8×10^{-11} photons $\text{TeV}^{-1}\,\text{cm}^{-2}\,\text{s}^{-1}$, as measured in [37].

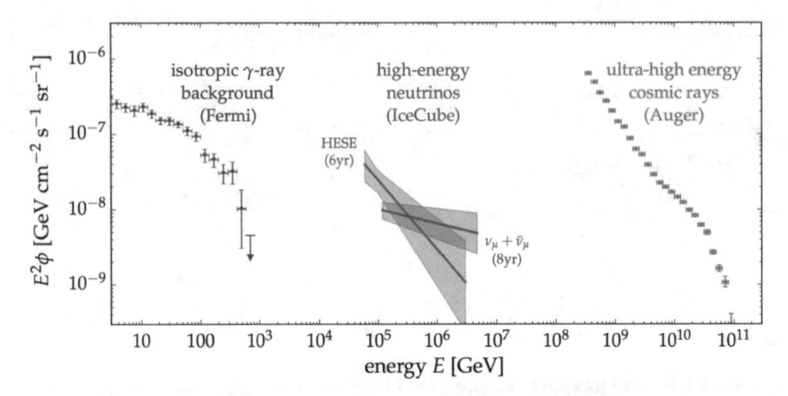

Fig. 1.9 Energy flux into diffuse gamma rays (blue), neutrinos (pink and red) and ultra-high-energy cosmic rays (green). Reprinted from Ahlers & Halzen [44], with permission from Elsevier

the Universe through its different messengers, collecting only partial information and, from this, trying to extract a unified picture of it.

The diffuse neutrino flux observed by IceCube is currently described by a single isotropic component [123]. Such a flux, in principle, should contain contributions from the Galactic sources, the diffuse Galactic Plane emission, nearby or bright extra-galactic sources, plus the diffuse emission from fainter and unresolved extra-galactic sources. In the following, the main features of this astrophysical flux are reviewed: however, in order to understand these results, the different event classes corresponding to different neutrino interaction channels as detected by experiments, and the different event selections, related instead to the analysis technique, have to be discussed.

Neutrino telescopes generally cannot distinguish the weak charge, namely neutrinos from antineutrinos, except in the case of a specific interaction channel, called Glashow resonance, that is only possible for $\bar{\nu}_e$ as $\bar{\nu}_e + e^- \longrightarrow W^-$, where the neutrino threshold energy for the production of the W^- boson in the electron rest frame is 6.3 PeV. However, the event topology allows to separate the neutrino flavors, as events are classified into:

(i) **Track events**: these are charged current (CC) muon neutrino interactions resulting into a muon, that at high energies can travel many kilometers in the water/ice and the bedrock. The long path across the detector allows for directional reconstruction with median angular resolution down to 0.3° at $E_\nu > 10$ TeV, as achieved by the current-generation of water-based telescopes. Such a long path enlarges the detector effective area by more than one order of magnitude at high energies. These features make the track sample the optimal candidate for point-like source searches. However, the measured muon energy provides only a lower limit to the primary neutrino energy, as the energy losses outside of the detector volume are not accessible.

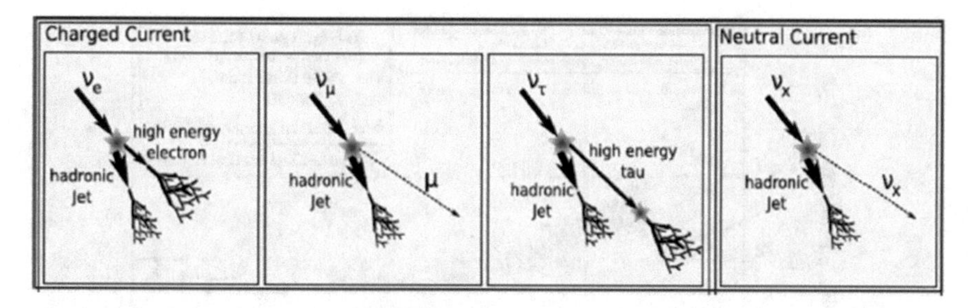

Fig. 1.10 Signatures of neutrino interactions: CC interactions are shown in the left panel, NC in the right panel. Figure reproduced from Tiffenberg [230] with permission by the Local Organizing Committee of ICRC 2009

(ii) **Shower events**: these are neutral current (NC) muon neutrino interactions and CC interactions of both electron and tau neutrinos, which produce a shower of particles of short size (∼10 m in length). Compared to the detector array spacing (∼100 m), these events are thus reconstructed as a light sphere. For this reason, the direction of the primary neutrino can be roughly reconstructed with an angular uncertainty of about 10°. However, the energy deposited into the detector is measured with a 10–15% resolution.

A schematic view of the neutrino interaction channels is given in Fig. 1.10.

Neutrino events are classified for different analyses in the following samples:

(i) **Starting events**: these are collected using the outer layers of the detector as a veto for incoming events, basically in order to remove the downward-going atmospheric background. Such a reduced sample consists predominantly of shower-like events, but it is potentially sensible to all neutrino interactions and flavors. Among the starting events, the first astrophysical neutrino signal was isolated into the High-Energy Starting Event (HESE) sample.
(ii) **Through-going track events**: these events consist in incoming as well as starting tracks. In principle, the upward-going events compose a sample of background-free Earth-filtered events. However, the more abundant are the downward-going events, the higher will be the probability of their mis-identification, polluting the upward-going sample. Different analyses accept different levels of sample purity, going from 99.9% in diffuse searches to around 90% in transient source searches.

In the following, the astrophysical signal observed by IceCube[4] is introduced, and its spectral and angular distribution are discussed.

Spectrum

Evidence for a flux of extraterrestrial neutrinos extending up to several PeV has been observed since 2013: the highest-energy neutrino reported to date has a median

[4]https://icecube.wisc.edu.

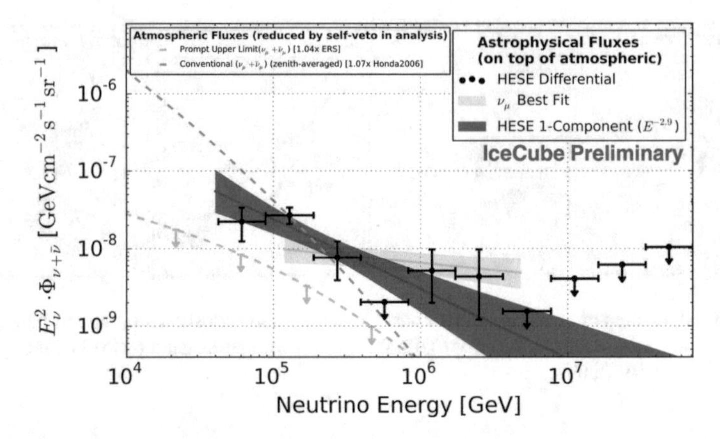

Fig. 1.11 The best-fit per-flavor neutrino (combined with anti-neutrino) flux as a function of energy. IceCube data are black points with 1σ uncertainties. The atmospheric flux has already been subtracted, while the prompt component upper limit is shown separately. The blue band shows the 1σ uncertainty on the HESE flux, while the pink band refers to the through-going muon flux. Figure reproduced from Kopper [153], under the CC BY LICENSE

inferred neutrino energy of 7.8 PeV. The differential energy flux is well described by an isotropic, unbroken power law $\propto E^{-\gamma}$, though some differences in the spectral index γ arise in different analyses. According to the latest results from the IceCube Collaboration [153], the through-going muon sample collected in eight years of data-taking shows a 6.7σ evidence for an astrophysical flux

$$\frac{d\Phi}{dE}(\nu_\mu + \bar{\nu}_\mu) = (1.01 \pm {}^{0.26}_{0.23}) \left(\frac{E}{100\,\text{TeV}} \right)^{-2.19\pm0.10} \times 10^{-18}\,\text{GeV}^{-1}\,\text{cm}^{-2}\,\text{s}^{-1}\,\text{sr}^{-1} \tag{1.68}$$

where the energy range in which 90% of the signal was collected corresponds to [119 TeV–4.8 PeV]. However, considering the HESE sample with six years of data at energies larger than 60 TeV, the astrophysical neutrino flux becomes as steep as

$$\frac{d\Phi}{dE}(\nu + \bar{\nu})_{\text{HESE}} = (2.46 \pm 0.8) \left(\frac{E}{100\,\text{TeV}} \right)^{-2.92} \times 10^{-18}\,\text{GeV}^{-1}\,\text{cm}^{-2}\,\text{s}^{-1}\,\text{sr}^{-1} \tag{1.69}$$

A representation of these two fluxes is given in Fig. 1.11, where also the atmospheric background contribution is reported. The different spectral slopes in the through-going track and HESE samples have received some attention, as they might reveal the presence of two different emission components in the cosmic neutrino flux.

Angular Distribution

Several analyses based on the search for anisotropies in neutrino data have been conducted, but none of them has shown a significant clustering of events in any given direction of the sky or a correlation with any known source. Figure 1.12 shows

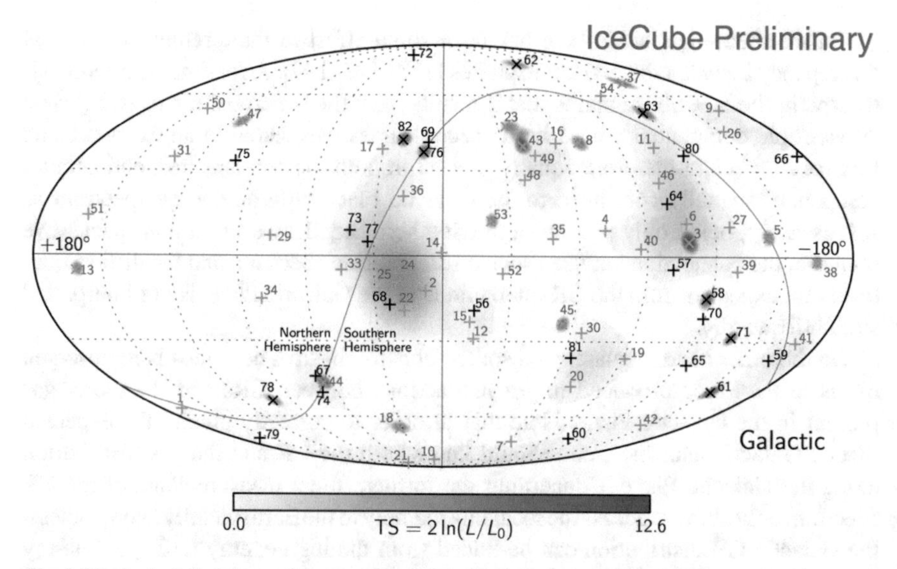

Fig. 1.12 Arrival directions of the neutrino events in Galactic coordinates. Shower-like events are marked with a + and those containing tracks with a x. The color code shows the test statistics (TS) for the point-source clustering test at each location. Figure reproduced from Kopper [153], under the CC BY LICENSE

the spatial distribution of neutrino events in Galactic coordinates: several of them are clustering in the Southern Sky, where most of the Galactic Plane is located, however at the moment they do not constitute a statistically significant excess. Nonetheless, it is timely to investigate the possibility that large-scale anisotropies are present in the signal in order to clarify its origin, in particular whether it is Galactic, extra-galactic or a mixture of both. The most straightforward method to search for large scale anisotropies consists into splitting the data samples into two separate regions of the sky, namely the Northern and the Southern Hemispheres (NH and SH in the following), and compare the corresponding fluxes. In fact, since the Southern Sky contains a larger part of the Galactic Plane, including the Galactic Center, with respect to the Northern Sky, a possible difference might result in the amount of signal coming from our own Galaxy and the amount of extra-galactic emission. When considering the global sample of events, composed of both tracks and showers, the spectral index of the Northern Sky results harder than that of the Southern Sky: with the six years of data, in fact, the excess of events with respect to the atmospheric background results into slopes of $\gamma_{\rm NH} = 2.0^{+0.3}_{-0.4}$ and $\gamma_{\rm SH} = 2.56 \pm 0.12$. However, such a result is not yet considered to be conclusive, because of the different systematics that affect the two event samples. For instance, it should be taken into account that all sensors are directed downward (leading to a non-optimal detection of down-going events), that high-energy neutrinos are absorbed by the Earth (and therefore the statistics for up-going events is reduced) and finally that the rejection of atmospheric neutrino background strongly differs among the two hemispheres [123].

On the theoretical side, there has been some effort in interpreting the spectral discrepancy between the two hemispheres [188, 197–199]. A reasonable hypothesis to explain the IceCube signal is that it results from the superposition of two fluxes: an isotropic component with a hard spectral index associated to an extra-galactic flux ($\propto E^{-2.1}$) and therefore equally present in both the hemispheres, and another component spatially correlated to the Galactic Plane with an energy spectrum as soft as $E^{-2.4}$, being only present in the Southern Sky. However, in this speculative scenario, the connection between such a soft neutrino spectrum and the diffuse neutrino flux expected from the CR interactions in the Galactic Plane is not interpreted straightforwardly.

On the other hand, a guaranteed source of astrophysical neutrinos is represented by those neutrinos produced in *pp* interactions between CRs and the target gas present in the Galactic Plane. Note that in order to correctly predict the expected flux of Galactic neutrinos, one should know both the CR and the gas distribution along the Galactic Plane. Concerning the former, direct measurements of the CR spectrum only allow to access the local flux, namely in the Earth vicinity. Nonetheless, the Galactic CR distribution can be traced from the high-energy (HE) gamma-ray diffuse emission: this information is clearly degenerate with the target gas mass distribution. Pagliaroli et al. [195] considered several scenarios concerning the CR distribution along the Plane reproducing the HE gamma-ray data, and consequently they derived different angular distributions of the neutrino flux. As a result, the Galactic component associated to CR interactions was found to be subdominant with respect to the measured IceCube flux (about $\sim$15%), except in an angular window containing the Galactic Center. At the moment, a two component flux does not appear prominently in the IceCube data: further investigation on this point is thus required and will possibly be achieved with an extended statistics of the data sample. In addition to this, searches for an excess in the Galactic Plane have been performed with data from the ANTARES[5] telescope, whose location in the Mediterranean Sea offers an optimal view of this part of the sky. As reported in [46], no excess of events has been observed: the upper limit set to the neutrino flux is already excluding the diffuse Galactic neutrino emission as the major cause of the spectral discrepancy between the two hemispheres measured by IceCube, in that its contribution would be at most 8.5%. Thus the question on the neutrino flux origin remains open.

1.5.4 *Open Issues and the Future of km^3-Scale Neutrino Detectors*

As was pointed out above, many questions still have to be addressed regarding the observed neutrino flux, namely:

[5]http://antares.in2p3.fr/Overview/performance.html.

- What are the sources of the diffuse neutrino flux observed by IceCube? Why are events not clustering into multiplets?
- Are the CR sources the same as neutrino sources?
- What is precisely the astrophysical neutrino spectrum? Why do we see different spectral indices in different event samples?
- How does the diffuse neutrino flux behave above energies of 10 PeV? Is there any break or cut-off?

The feasibility of neutrino studies with large volume detectors in the deep sea and ice has been demonstrated through the successful deployment and operation of ANTARES and IceCube. The detection of neutrinos is based on the collection of Cherenkov light induced by relativistic particles emerging from a neutrino interaction. The same technology can be used for studying neutrinos from GeV to PeV energies and beyond. The next-generation of deep sea neutrino telescopes, KM3NeT,[6] will consist of multi-km^3 infrastructures instrumented with a three-dimensional grid of photo-sensors. These, together with the readout electronics, are hosted within pressure-resistant glass spheres, the so-called digital optical modules (DOMs). The KM3NeT DOM comprises 31 photo-multiplier tubes (PMTs): with respect to traditional optical modules using single large PMTs, the DOM design has several advantages, as it houses three to four times the photo-cathode area in a single sphere and has an almost uniform angular coverage. In addition, since the photo-cathode is segmented, the identification of more than one photon arriving at the DOM is achieved with higher efficiency and purity. Note also that the directional information improves the rejection of the optical background. The system should provide nanosecond precision on the arrival time of single photons, while the position and orientation of the photo-sensors must be known down to a few centimeters and few degrees, respectively. The DOMs are distributed in space along flexible strings, anchored to the sea floor and held vertical by a submerged buoy. Since the concept of strings is modular by design, the construction and operation of the research infrastructure will allow for a phased and distributed implementation. A collection of 115 strings strings forms a single KM3NeT building block, each string comprising 18 optical modules, as schematically illustrated in Fig. 1.13. The modular design also allows the building blocks to be constructed with different spacings between lines/DOMs, in order to target different neutrino energies. The full KM3NeT telescope will comprise seven building blocks distributed among three sites: Italy, France and Greece [27]. An initial Phase 1 will consist of 24 strings in the Italian site, deployed according to the so-called KM3NeT Astroparticle Research with Cosmics in the Abyss (ARCA) infrastructure, with a large spacing to target astrophysical neutrinos at TeV energies and above, and 7 strings in the French site, according to the specifics of the so-called KM3NeT Oscillation Research with Cosmics in the Abyss (ORCA) infrastructure, to target atmospheric neutrinos in the few-GeV range. Starting from 2015, the first strings have already been deployed in both sites. For the sequent Phase-2.0, three building blocks are planned: two KM3NeT/ARCA blocks and one KM3NeT/ORCA block.

[6] http://www.km3net.org

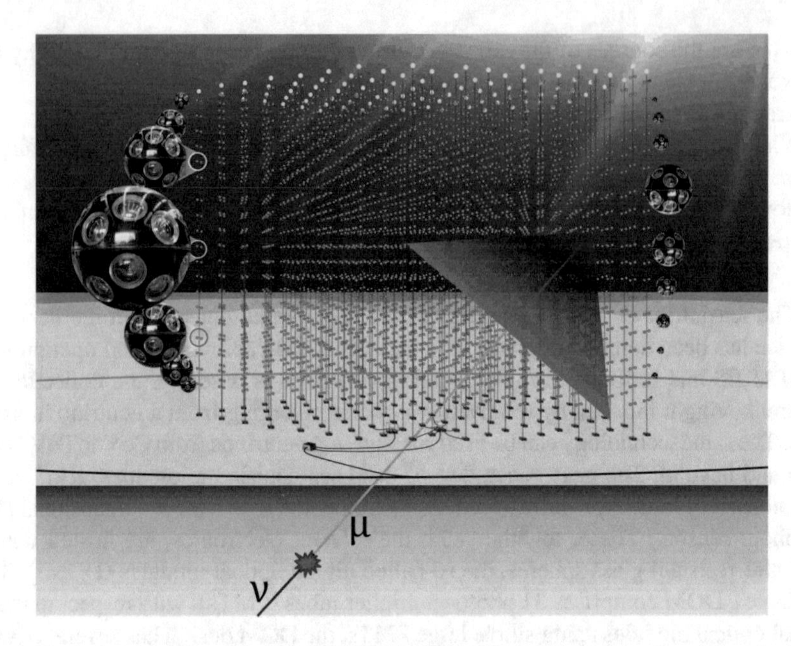

Fig. 1.13 A visual representation of a KM3NeT building block. Figure from http://www.km3net.org

The ARCA building blocks will be configured to fully explore the IceCube signal with different methodology, improved resolution and complementary field of view, including the Galactic Plane, which offers interesting VHE emitters as identified in gamma rays, which are also potential neutrino sources. For KM3NeT/ARCA, each string is about 700 m in height, with DOMs spaced 36 m apart in the vertical direction, while individual strings are spaced on average by 90 m. For KM3NeT/ORCA, on the other hand, a more compact structure is foreseen to access the low-energy range, thus each string is 200 m in height with DOMs spaced 9 m apart in the vertical direction and strings spaced by 20 m in the horizontal direction. A footprint of the spatial distribution of the strings is given in Fig. 1.14. The last Phase-3.0 will include the full array, with six building blocks compliant with the ARCA structure and one block with ORCA.

The main objectives of KM3NeT are: (i) the discovery of high-energy neutrino sources in the Universe, and (ii) the determination of the mass hierarchy of neutrinos. Concerning astronomical studies, the better performances with respect to the current generation of instruments will allow to improve the sensitivity levels of about one order of magnitude. As will be investigated in Chap. 5, this is expected to be already sufficient to claim the detection of individual neutrino sources or to possibly constrain their hadronic content in the case of non-detection. In addition to the increased event statistics connected with a larger instrumentation volume, a necessary requirement for the next-generation instruments is the coverage of a broad energy range. For instance, the instrumentation of radio antennas in ice [51] or on board of balloons,

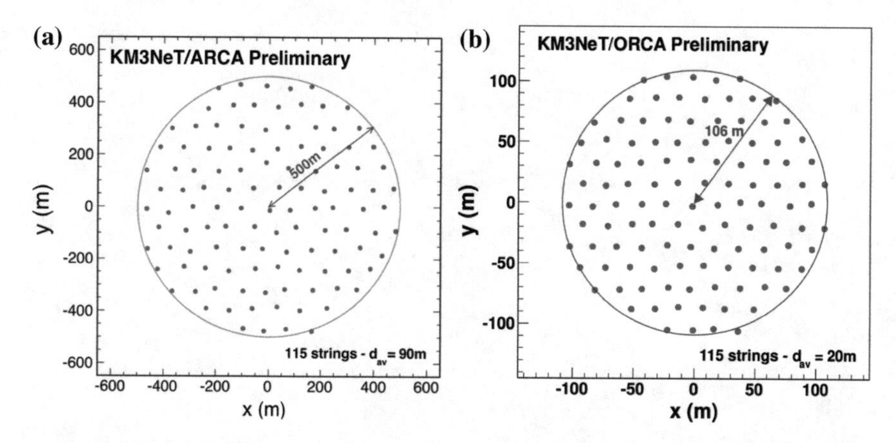

Fig. 1.14 Footprint of the ARCA and ORCA building blocks. Note that the ARCA site will implement two building blocks during its phase 2. Figure reproduced from Adrian-Martinez et al. [27] under the CC BY LICENSE

or alternatively the acoustic detection of neutrinos in water [159], will allow to extend the energy range above 10^{17} eV, thus exploring the UHE domain and possibly unveiling *cosmogenic* neutrinos (also called GZK neutrinos), which are expected to be produced in the interaction among UHECRs and CMB photons.

Chapter 2
Propagation and Radiation of Accelerated Particles in Supernova Remnants with Clumpy Structures

Multi-wavelength observations of SNRs, from the radio band to the VHE gamma-ray domain, can help to solve the leptonic/hadronic degeneracy. This is generally done on a case by case basis. The case of RX J1713.7-3946 is of special interest in this respect. This remnant has been considered for long time the best candidate for an efficient acceleration scenario, mainly due to its high gamma-ray flux. The detection of gamma-ray emission in the energy range [1–300] GeV by the Fermi-LAT satellite [16] has shown an unusually hard spectrum which, at a first glance, seemed to be in a better agreement with a leptonic scenario. Nevertheless, a deeper analysis showed that neither the hadronic nor the leptonic scenarios, taken in their simplest form, could unequivocally explain the observations [119, 182, 254]. To address this issue, it was proposed by Zirakashvili and Aharonian [254] (and later on investigated by [118, 143]) that the very hard energy spectrum at low energies could result also from hadronic emission if the SNR is expanding inside a clumpy medium. The presence of dense structures in the ISM represents a realistic situation, particularly in the Galactic Plane [171], and it affects the acceleration and propagation of non-thermal particles. In such a case, in fact, the magnetic field can result amplified all around clumps because of both the field compression and the MHD instabilities developed in the shock-clump interaction: this makes it difficult for particles at low energies to penetrate inside the clump compared to the most energetic ones. Consequently, the gamma-ray spectrum would be harder than the parent proton spectrum accelerated at the forward shock.

In the following, the realization of such a scenario is investigated. The non homogeneities of the ISM are introduced in Sect. 2.1, which describes the molecular clouds and their relation to CRs. In Sect. 2.2, molecular clumps located in the vicinity of an accelerator are considered, and a focus on their interaction with an SNR shock

---

Part of this chapter has already been published in Celli S., Morlino G., Gabici S. & Aharonian F., 'Supernova remnants in clumpy media: particle propagation and gamma-ray emission', Monthly Notices of the Royal Astronomical Society, Vol. 487, Issue 3, p. 3199–3213 (2019), and it is here reproduced by permission of the Royal Astronomical Society.

S. Celli, *Gamma-ray and Neutrino Signatures of Galactic Cosmic-ray Accelerators*, Springer Theses, https://doi.org/10.1007/978-3-030-33124-5_2

is provided. The numerical techniques adopted for the results shown in Sect. 2.3, namely MHD simulations accounting for the temporal and spatial evolution of the background plasma properties, are discussed in Appendix A. Then, in Sect. 2.4, the center of the discussion is moved towards the temporal and spatial evolution of the density of accelerated particles. The numerical algorithm developed for the solution of the three-dimensional transport equation for CRs propagating into a clumpy medium where a shock is expanding is presented in Appendix B. The diffusion coefficient around the clump is parametrized in order to reproduce the amplified magnetic field resulting from both the regular field compression and MHD instabilities. Once the behavior of the particle density is obtained, the CR spectrum resulting from inside and outside the clumpy regions is derived, showing that the spectrum inside a clump is much harder at early epochs and steepens at later times. The same spectral trend is exhibited in the related gamma-ray emission from proton-proton interactions, as shown in Sect. 2.5. In order to move from the gamma-ray emission of individual clumps to their cumulative contribution, a uniform spatial distribution of clumps is assumed. The case of RX J1713.7-3946 is then discussed in Sect. 2.6: in particular, the possibility to detect clumps embedded inside the remnant through molecular lines is explored, while comments on the fast variability observed in non-thermal X rays are given. Finally, the detectability of clump-shock associations by the major next-generation IACT, CTA, is presented.

2.1 Molecular Clouds and Cosmic Rays

Molecular clouds (MCs) are strongly influenced by the presence of CRs, as low-energy CRs provide their ionization rate (see [194]), which in turn controls both the chemistry of clouds and the coupling of plasma with local magnetic fields, and hence star formation processes. The large majority of CR-H_2 reactions leads to the formation of H_2^+ through the ionization reaction

$$k_{CR} + H_2 \longrightarrow k_{CR} + H_2^+ + e \tag{2.1}$$

where k_{CR} stands for a CR particle of species k (protons, electrons or heavier nuclei). Dissociative and double ionization reactions contribute to the ion fraction within the clouds, reading respectively as

$$k_{CR} + H_2 \longrightarrow k_{CR} + H + H^+ + e \tag{2.2}$$

and

$$k_{CR} + H_2 \longrightarrow k_{CR} + 2H^+ + 2e \tag{2.3}$$

Low-energy protons (in the MeV range) can also access the electron capture reaction

$$p_{CR} + H_2 \longrightarrow H + H_2^+ \tag{2.4}$$

In addition to ionization, CRs are possibly able to heat molecular clouds, since the energy of both primary and secondary electrons produced in the ionization process is mostly converted into heat through inelastic collision with ISM atoms and molecules. Note that, in the case of active clouds, namely those located within shock environments, the timescales relevant to the problem (i.e. the clump evaporation time and the contact discontinuity-embedding time) are much shorter than the clump ionization time by the Galactic CR 'sea', and consequently ionization can be neglected as long as these short timescales are investigated.

Molecular clouds constitute potential sources of both gamma rays and neutrinos, especially if they are located in the vicinity of a powerful accelerator that injects CRs in the ISM. Indeed, high-energy CRs interact with the dense gas and produce neutral pions, which in turn decay into two gamma rays, and charged pions, which generate neutrinos. For this reason, molecular clouds constitute powerful tools to locate the sources of CRs. Formally one should distinguish *passive* clouds, where particle acceleration (either inside it or in their vicinity) is absent and CRs can freely penetrate inside them, from *active* clouds, illuminated by freshly accelerated particles. While the former ones have often been refereed to as *CR barometers*, since they allow to measure the CR intensity in specific regions of the Galaxy [35], the latter ones constitute direct probes of shock environments. The latter will be the object of this chapter. In the context of particle acceleration, a multi-phase ISM was first investigated by Blandford and Cowie [69]: they considered a slope of the particle spectrum inside dense clouds equal to that accelerated at the shock front, without accounting for the magnetohydrodynamics of the shock-cloud interactions. While MHD effects are usually negligible for low ($\sim 10^2$) density contrasts [62], these become relevant in the presence of denser inhomogeneities. The content of this chapter constitutes the first attempt of describing the evolution of the accelerated particle density in the presence of very dense spatial inhomogeneities, also accounting for the hydrodynamical interaction of these structures with a remnant shock.

The importance of the secondary emission from molecular clouds was first realized in connection to the estimate of the cloud masses [67]. This is performed by measuring through spectroscopy the intensity of emission lines of some molecules, as these lines mark the interaction of molecules with shocks [95]. The molecule that is more often identified is the carbon monoxide CO, since this is the second most common molecule in the ISM after the molecular hydrogen. The line intensity is related to the mass of the CO in the cloud, which can be converted to the total mass (largely dominated by the molecular hydrogen) through a conversion factor X_{CO}, which unfortunately is not very well constrained. Despite such an uncertainty, molecular clouds can still be used to measure variations in the CR intensity in the Galaxy: indeed, spectral measurements are not affected by the uncertainty in the mass determination.

As the amplitude and duration in time of the CR overdensity around a given source depend on how rapidly CRs diffuse in the turbulent Galactic magnetic field [117], a detailed study of the effects of CR propagation within molecular clouds is fundamental to both interpret correctly the observed VHE source spectra and infer the CR diffusion properties, as will be presented in Sect. 2.2. In at least two cases, molecular cloud observations have been used to constrain the properties of

CR diffusion in the Galaxy. For the SNRs W 28 [15, 41] and W 44 [233], some gamma-ray emission has been detected outside of the SNR shell, in coincidence with the position of dense gas clouds. In another case, the SNR IC 443, the centroids of the GeV and TeV emissions have been observed as not coincident, but significantly displaced [13]. To interpret these observations, CR escape from the SNR shells and energy-dependent propagation of CRs have been often invoked. Note that, due to the steepness of the Galactic CR spectrum, an excess in the CR intensity with respect to the CR background would appear more easily at higher (~TeV) than at lower (~GeV) energies. Additionally, in the GeV energy domain this kind of studies is complicated by the very intense diffuse emission from the Galactic Disc, that constitutes an important background in the search for gamma-ray sources.

2.2 Shock Propagation Through a Clumpy Medium

Observations of the ISM have revealed a strong non homogeneity, particularly inside the Galactic Plane. On the scales of the order of few parsecs, dense molecular clouds constitute structures, mainly composed of H_2 molecules, while their ionized component is composed by C ions. Typical temperatures and masses are respectively of the order of $T_{MC} \simeq 10^2$ K and $M_{MC} \gtrsim 10^3\ M_\odot$. On smaller scales (of the order of a fraction of a parsec), colder and denser molecular clumps are present: these are characterized by typical temperatures of $T_c \simeq 10$ K, masses of the order of $M_c \simeq 0.1 - 1\ M_\odot$ and therefore number densities of the order of $n_c \gtrsim 10^3\ \mathrm{cm}^{-3}$. In the following, the lower bound is considered as a reference value for the density of target gas inside individual clumps. These clumps are mostly composed by neutral H_2 molecules while the most relevant ions are HCO^+. Typical values for the ionized density are at least equal to $n_i \leq 10^{-4} n_c$, while the average mass of ions is $m_i = 29 m_p$ and that of neutrals is $m_n = 2 m_p$. Therefore the ion-to-neutral mass density ϵ in clumps amounts to [120]

$$\epsilon = \frac{m_i n_i}{m_n n_n} \leq 1.5 \times 10^{-3} \ll 1 \tag{2.5}$$

Shocks propagating through inhomogeneities of the ISM are able to generate MHD instabilities, which modify the thermal properties of the plasma and might be able to disrupt the clumps because of thermal conduction [89, 192]. Therefore the dynamical interaction between the shock emitted at the supernova explosion and the medium surrounding the star is an essential ingredient for the understanding of star formation processes [102, 131, 135]. In particular, the environment where type II SNe explode is likely populated by molecular clumps: indeed, given the fast evolution of massive stars, these are expected to explode in an environment rich of molecular clouds, the same that generated the star. Moreover, given the massive progenitor, strong winds in the giant phase of the stellar evolution accelerate the fragmentation of clouds into clumps, while creating a large cavity of hot and rarified gas around them. If the mass of the cool clouds does not exceed the Jeans mass,

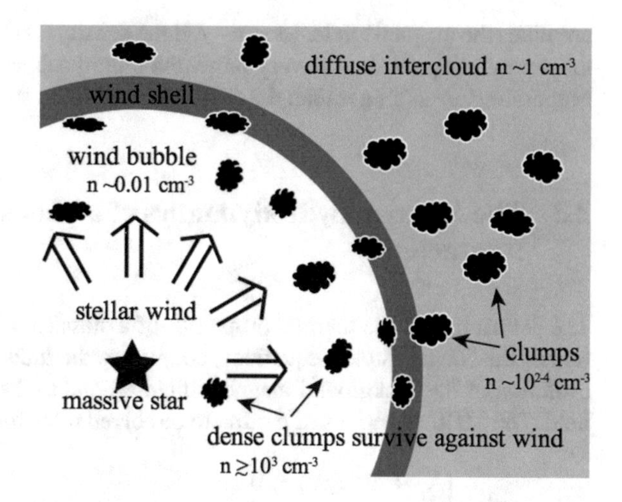

Fig. 2.1 Schematic view of a wind bubble expanding in a cloudy ISM. This constitutes the initial configuration where the SNR shock will later expand. The diffuse intercloud gas is swept by the stellar wind, while dense cloud cores and clumps can survive in the wind. Density in the wind bubble is much smaller than the inter-cloud gas density. Figure from Inoue et al. [143] ©AAS. Reproduced with permission

these can remain in pressure equilibrium with the surrounding hot phase [185]. A schematic visualization of this environment is given in Fig. 2.1. The physical size of these ISM inhomogeneities results from MHD simulations which include interstellar cooling, heating and thermal conduction. These have been conducted by Inoue et al. [143], who found that the characteristic length scale of clumps amounts to $R_c \simeq 0.1$ pc, corresponding to the smaller scale where thermal instability is effective. Such a scale will be fixed in the following. This scenario could be similar to the one in which the remnant of RX J1713.7-3946 is evolving [219]. Previous works in this direction [151] have shown that hydromagnetic instabilities arise when the upstream medium is not homogeneous [110, 127, 209]. In particular, in the presence of density inhomogeneities, vorticity may develop after the passage of the shock. In such a situation, Rayleigh–Taylor and Kelvin–Helmholtz instabilities arise, due to respectively the density contrast and the velocity shear within the clumps and the surrounding medium, generating a strong enhancement of the local magnetic field. The generated turbulence then cascades to smaller spatial scales through a Kolmogorov-like process [143] on timescales $\tau_{\text{cascade}} = R_c/v_A \simeq 50$ yr (where the Alfvén speed of MHD waves v_A is computed in a low density environment as that of a rarefied cavity). Interestingly, in the presence of a cosmic-ray precursor in front of the shock, vorticity and magnetic amplification might also be realized in the upstream [99].

In the next section, using MHD simulations, a simplified scenario is studied where a single clump much denser than the circumstellar plasma is engulfed by a shock. In such a configuration, two different kinds of processes produce magnetic field amplification. Around the clump, the regular field results to be compressed reaching a value up to ∼10 times the original value, in a layer that is found to be about half of the clump size, i.e. 0.05 pc, similar to the results obtained by Inoue et al. [143]. This region will be referred to as the *clump magnetic skin*. In addition, in the region behind the clump, turbulence develops and the associate vorticity further amplifies the magnetic field. Moreover, if the density contrast between the clump and the surrounding medium is very large, the clump can survive for long time before evaporating, even longer than the SNR age, as will be shown in the next section.

2.3 The Magnetohydrodynamics of a Shock-Clump Interaction

The description of the thermal properties of a classical fluid follows from the solution of the Navier–Stokes equations, coupled to the induction equation for the time evolution of the background magnetic field $\mathbf{B_0}$ and to the Gauss's law for the same field. The MHD equations of motion to be solved read, for ideal non-resistive fluids, as

$$\begin{cases} \frac{\partial \rho}{\partial t} + \nabla \cdot (\rho \mathbf{v}) = 0 \\ \frac{\partial \mathbf{v}}{\partial t} + \mathbf{v} \cdot \nabla \mathbf{v} = -\frac{1}{\rho}\left(\nabla P + \frac{1}{4\pi}(\nabla \times \mathbf{B_0}) \times \mathbf{B_0}\right) \\ \frac{\partial}{\partial t}\left(\mathcal{E} + \frac{B_0^2}{8\pi}\right) = -\nabla \cdot \left[(\mathcal{E} + P)\,\mathbf{v} + \frac{1}{4\pi}\mathbf{B_0} \times (\mathbf{v} \times \mathbf{B_0})\right] \\ \frac{\partial \mathbf{B_0}}{\partial t} = \nabla \times (\mathbf{v} \times \mathbf{B_0}) \\ \nabla \cdot \mathbf{B_0} = 0 \end{cases} \tag{2.6}$$

where ρ is the fluid density, $\mathbf{v}$ its velocity, P its pressure, $\mathcal{E} = \rho v^2/2 + P/(\gamma - 1)$ its kinetic plus internal energy and γ is its adiabatic index, namely the ratio between the specific heat at constant pressure and that at constant volume ($\gamma = 5/3$ for a monoatomic non-relativistic gas with no internal degrees of freedom). A review of the hydrodynamical equations governing the fluid evolution is provided for the interested reader in Appendix A.

In order to introduce a shock discontinuity, as well as the presence of a clump, a numerical approach has been adopted, through the PLUTO code (see [174]). The shock-cloud interaction is implemented among one of the possible configurations provided by this code. A three-dimensional simulation, in cartesian coordinates, is adopted in the following. The finite difference scheme for the solution of Eq. (2.6) is based on an unsplit 3rd order Runge–Kutta algorithm with an adaptive time step subject to the Courant condition $C = 0.3$. The interested reader is referred to Appendix A for a detailed discussion on the different configurations explored within the numerical simulation of the shock-clump interaction with the PLUTO code.

In order to investigate a situation as much similar as that of high-mass star SN explosion, like RX J1713.7-3946, the upstream region is simulated as a low density medium with $n_{up} = 10^{-2}\,\text{cm}^{-3}$, which gets compressed by the shock in the downstream region up to $n_{down} = 4 \times 10^{-2}\,\text{cm}^{-3}$. A strong shock is moving in the direction of the z-axis, with a sonic Mach number $M = v_s/c_s \simeq 37$, as expected for this remnant, given its measured shock speed of $v_s = 4.4 \times 10^8\,\text{cm}\,\text{s}^{-1}$ and an upstream temperature of $T = 10^6\,\text{K}$, which is typical for bubbles inflated by stellar winds. A representation of the shock-clump interaction is provided in Fig. 2.2: at the

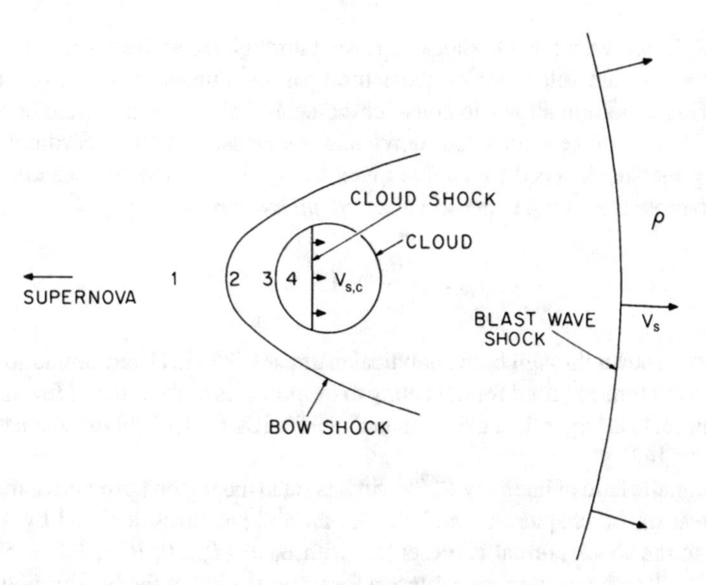

Fig. 2.2 Schematic representation of the shock waves that occur when a blast wave produced by a SN overtakes a clump: the gas in region 1 is the hot gas behind the blast wave, region 2 is just behind the bow shock, point 3 is a stagnation point at the surface of the clump and region 4 contains the clumpy gas which has passed through the shock. The blast wave moves at velocity v_s into the ambient medium of density ρ. Figure from McKee & Cowie [170] ©AAS. Reproduced with permission

interaction point between the forward shock and the clump, two more shocks are generated, namely the shock transmitted inside the clump and the shock reflected back inwards. A flat density profile clump is set in the upstream as initial condition, with a density as high as $n_c = 10^3\,\mathrm{cm}^{-3}$. Therefore, a density contrast

$$\chi = \frac{n_c}{n_{\mathrm{up}}} = 10^5 \tag{2.7}$$

is assumed. Consequently, if the shock speed is $\mathbf{v} = v_s\widehat{z}$, it propagates inside the clump with a velocity $\mathbf{v_{s,c}} = v_{s,c}\widehat{z}$ equal to [143, 151]

$$v_{s,c} = \frac{v_s}{\sqrt{\chi}} = 1.4 \times 10^6\,\mathrm{cm\,s}^{-1} \tag{2.8}$$

Boundary conditions are set as outflow in all directions, except in the downstream boundary of the z-direction, where an injection flow is set. A uniform grid with size $2\,\mathrm{pc} \times 2\,\mathrm{pc} \times 2\,\mathrm{pc}$ and spatial resolution of $\Delta x = \Delta y = \Delta z = 0.01$ pc is set. A spherical clump of radius $R_c = 0.1$ pc is located in $x_0 = y_0 = z_0 = 1$ pc. All the evolution is followed in the clump reference frame, as further discussed in Appendix A.

In the MHD simulation the clump is assumed to be fully ionized: this is not the real condition, since molecular clumps are mainly composed by neutrals, as discussed in

Sect. 2.2. However, while the shock is passing through the ionized part of the clump, the heated ions are able to ionize the neutral part on a timescale of the order of few years. This condition allows to consider the neutral clump as if it were completely ionized. In this process ions cool down and the pressure drops accordingly, hence reducing the shock speed to a value given by $v_{s,c}$ [151]. The time needed for the shock to cross the clump is the so-called *clump crossing time*

$$\tau_{cc} = \frac{2R_c}{v_{s,c}} \simeq 1.4 \times 10^4 \text{ yr} \tag{2.9}$$

It has been shown through both analytical estimates [85, 151] and simulations [191, 192] that the time required for the clump to evaporate is of the order of few times τ_{cc}. This timescale is larger than the estimated age for RX J1713.7-3946, which amounts to $T_{\rm SNR} \simeq 1625$ yr.

A magnetic field of intensity $B_0^{\rm up} = 5\mu$G is set in the region upstream of the shock. In the rest of the chapter, a simulation of an oblique shock inclined by 45° with respect to the shock normal is presented, with $\mathbf{B_0} = (B_{0x}, 0, B_{0z})$, $B_0^{\rm up} = 5\mu$G and $B_{0x} = B_{0z}$. The interesting time interval for the evolution of the background plasma is about 300 years from the first shock-clump interaction, as will be described in Sect. 2.5. Within this time interval, results from the MHD simulations are shown in Figs. 2.3, 2.4, 2.5 and 2.6a corresponding to plasma mass density, vorticity ω, magnetic energy density and plasma velocity, respectively. The plasma vorticity is here defined as $\omega = \nabla \times \mathbf{v}$. These results can be summarized as follows:

(i) The clump maintains its density contrast, although the density distribution tends to smoothens, as seen in Fig. 2.3;
(ii) During the whole simulated time, the shock has not yet crossed the clump, as represented by the plasma velocity field lines in Fig. 2.6a;
(iii) Comparing Figs. 2.4 and 2.5, one can see that the magnetic field in the clump skin results amplified from compression and stretching due to the strong vorticity developed in the plasma. The magnetic amplification is most effective where the vorticity is the largest, meaning that the shear amplification is more important than pure compression as already pointed out by, e.g., [145, 164]. The resulting magnetic field energy density in the clump skin is ~100 times larger than in the regular downstream region: the amplification factor obtained directly follows from the simulation set up, namely from the shock Mach number, which is set in order to reproduce the conditions expected to operate in RX J1713.7-3946. The fact that the magnetic field around the clump is mainly directed along the tangential direction implies that it is difficult for accelerated particles to diffuse orthogonally to the clump surface, along the radial direction, as will also be discussed in Sect. 2.4.2.
(iv) In the region immediately behind the clump, a long tail develops where eddies are generated, that will eventually become turbulent, as observed in Fig. 2.6a. As a consequence, the vorticity amplifies the magnetic field in a turbulent dynamo-like process: also in this region, the magnetic field is amplified up to a factor

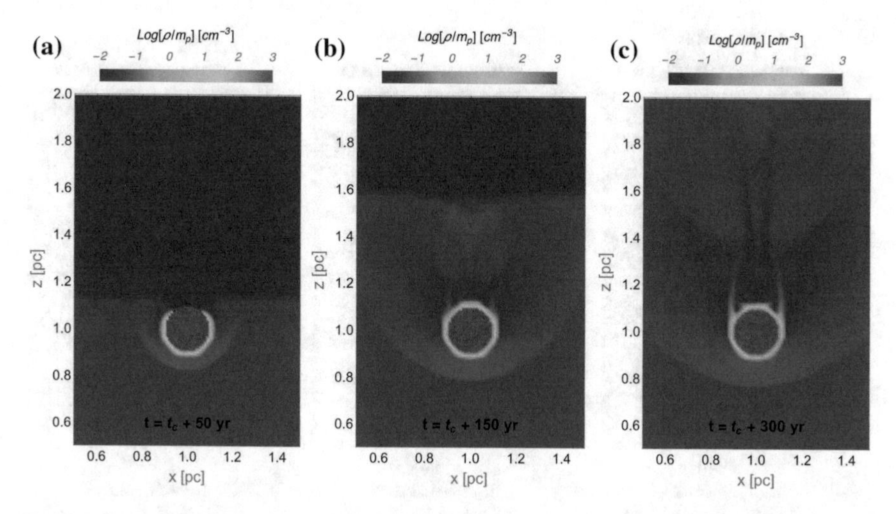

Fig. 2.3 Mass density plot of the 3D MHD simulation in the oblique shock configuration with density contrast $\chi = 10^5$. The plots show a 2D section along $y = 1$ pc, passing through the centre of the clump. From left to right, the mass density is shown for 50, 150 and 300 yr after the first shock-clump contact, occurring at $t = t_c$. Figure from Celli et al. [79], reproduced by permission of the Royal Astronomical Society

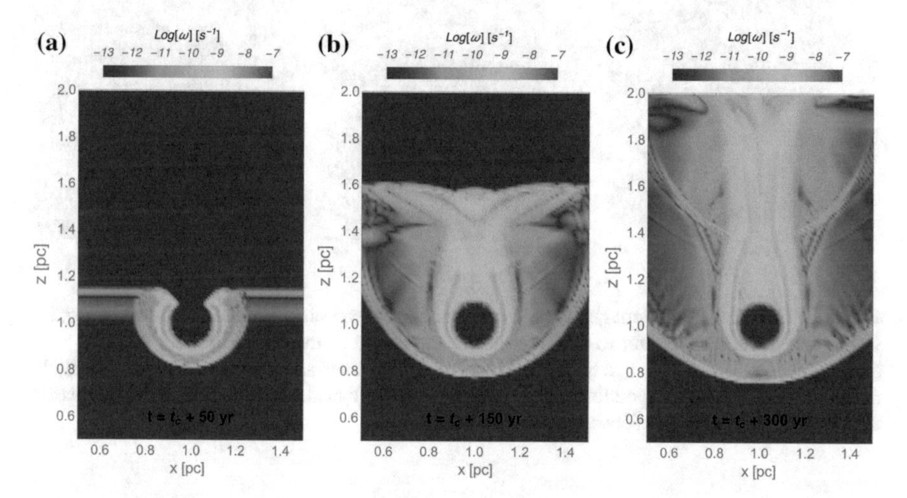

Fig. 2.4 Vorticity $\omega = \nabla \times \mathbf{v}$ of the plasma for the same simulation shown in Fig. 2.3. Figure from Celli et al. [79], reproduced by permission of the Royal Astronomical Society

$\sim$10. However, this region is not particularly relevant for the CR propagation inside the clump, but it can be important when considering the synchrotron X-ray emission from electrons (see Sect. 2.6.2).

In order to understand whether the magnetic amplification is peculiar of the chosen configuration, MHD simulations were also performed changing the initial magnetic

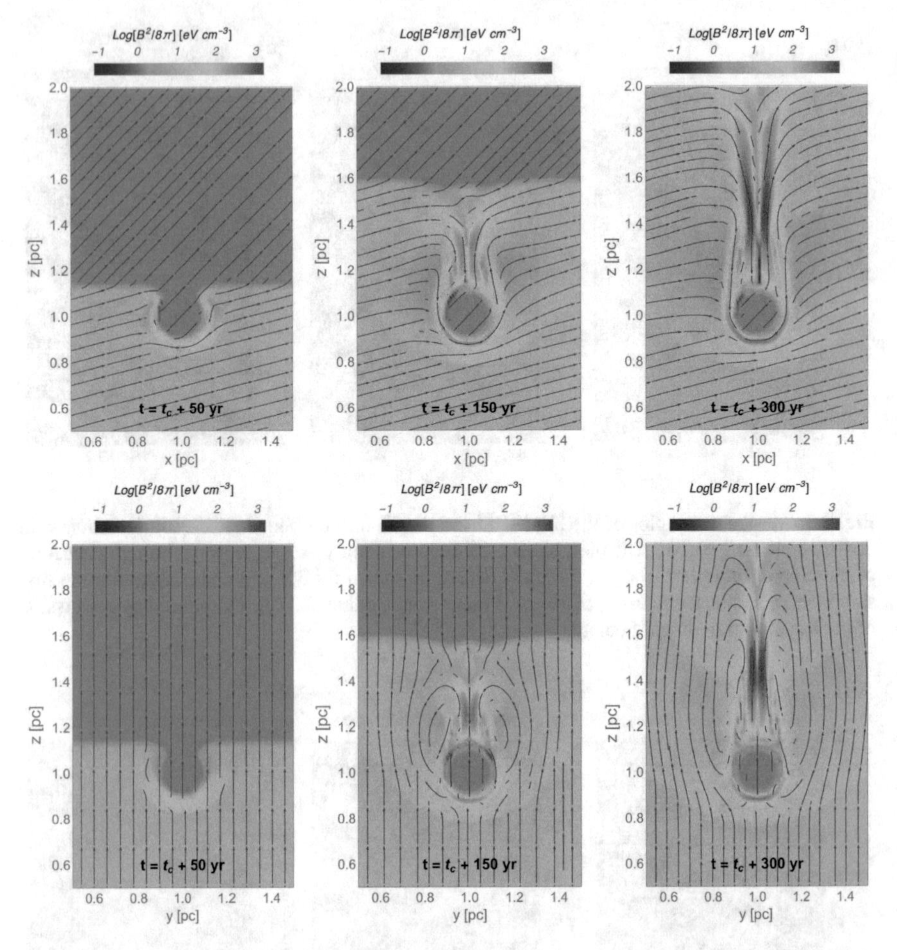

Fig. 2.5 Energy density of magnetic field (color scale) for the same simulation shown in Fig. 2.3. Upper and lower panels refer to a 2D cut across the centre of the clump in the $x - z$ and $y - z$ planes, respectively at $y = 1$ pc and $x = 1$ pc. The stream lines show the direction of the regular magnetic field in the corresponding planes. Figure from Celli et al. [79], reproduced by permission of the Royal Astronomical Society

field orientation and the density contrast, as described extensively in Appendix A. Concerning the orientation, the only case where the amplification results negligible is when the magnetic field is purely parallel to the shock normal: in this situation, the magnetic field around the clump is not compressed and also the shear is less effective, while the amplification in the tail is observed as due to plasma vorticity (for more details, see Appendix A). To explore, instead, the effect of the density contrast simulations with $\chi = 10^2$, 10^3 and 10^4 were investigated. In these cases, for a fixed shock speed, the clump is expected to evaporate on shorter timescales than the one given by Eq. (2.9). An effective amplification of the magnetic field around the

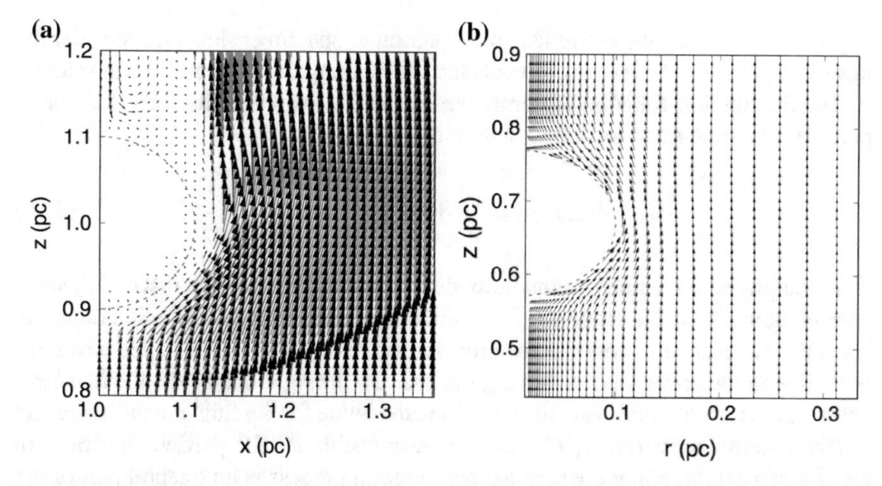

Fig. 2.6 *Left*: velocity field resulting from the MHD simulation in the oblique shock configuration with density contrast $\chi = 10^5$, at a time $t = t_c + 300$ yr and in a 2D section along $y = 1$ pc passing through the center of the clump. The bow-shock produced by the clump in the shocked ISM is also visible which, however, is always very mild (see Eq. (2.16)). *Right*: analytical velocity field adopted in the numerical solution of the proton transport equation. The field is fully tangential to the clump surface and directed along the z-direction in the far field limit. The red dashed line defines the clump size, without considering its magnetic skin. Figure from Celli et al. [79], reproduced by permission of the Royal Astronomical Society

clump is obtained if $\chi \gtrsim 10^3$. For less massive clumps, where the evaporation time is comparable to the remnant age, two differences arise: (i) once heated, these clumps would contribute to the thermal emission of the remnant through some X-ray emission, and (ii) the resulting gamma-ray spectrum would not manifest a pronounced hardening. Hence, detailed spectroscopic and morphological observations are crucial for providing a lower limit on the density contrast of the CSM.

2.4 Particle Transport in Presence of Clumps

The propagation of accelerated particles into the magnetized ISM is described by the *transport equation*, through the temporal and spatial evolution of the CR density function in the phase space $f(\mathbf{x}, \mathbf{p}, t)$, as derived in Ginzburg and Syrovatsky [128] and Skilling [217]

$$\frac{\partial f}{\partial t} + \mathbf{v} \cdot \nabla f = \nabla \cdot [D \nabla f] + \frac{1}{3} p \frac{\partial f}{\partial p} \nabla \cdot \mathbf{v} + Q_{\mathrm{CR}} \tag{2.10}$$

where D is the diffusion coefficient, p is the particle momentum and $Q_{\mathrm{CR}} = f_{\mathrm{inj}} v_s \delta(\mathbf{x} - \mathbf{x_s})$ is the injection flux, acting at the shock position $\mathbf{x} = \mathbf{x_s}$. Without any loss of accuracy, we can reduce the equation to one dimension, and consider the upstream plasma to be moving towards the shock along the z direction with velocity $\mathbf{v}$. The distribution function is normalized in such a way that the total number of particles at a time t and location z with momentum p is

$$N(z, t, p) = \int 4\pi p^2 dp f(z, t, p) \tag{2.11}$$

Here the physical meaning of the individual terms of Eq. (2.10) is discussed, since each of them affects the space and time variation of the particle distribution function. The left hand side term contains the time derivative of f and the so-called *advective* term, due to the presence of a moving plasma. On the right hand side, the first term describes the *spatial diffusion*, then a term responsible for the fluid *compression* and finally a *source* term $Q(z, t, p)$, which is responsible for the particle injection. In a self-consistent treatment, where the acceleration process is an integral part of the processes that lead to the formation of the shock, injection would result from the microphysics of the particle motions at the shock so that one would not need to include a dedicated term to describe it. However, in the approach adopted here, the injection term only determines an arbitrary normalization of the particle spectrum. Note that in Eq. (2.10) the diffusion term in momentum space has been neglected, as the second order Fermi acceleration is not expected to be relevant in the context of this work. A further comment on this is given in Sect. 2.4.4.

Given the system symmetry, Eq. (2.10) will be solved in cylindrical coordinates, through a finite difference method. A grid of 2 pc × 2 pc is set, with a spherical clump located at $(r_0, z_0) = (0, 0.67)$ pc. A logarithmic step is used in the radial dimension, while a uniform spacing is fixed along the shock direction. The spatial resolution of the grid is set in such a way that, for each simulated momentum, the proton energy spectrum reaches a convergence level better than 5%. An operator splitting scheme is set, based on an the Alternated Direction Implicit (ADI) method, flux conservative and upwind, second order in both time and space, subject to a Courant condition $C = 0.8$. The interested reader is referred to Appendix B for a detailed discussion on the numerical algorithm developed by the author for the solution of the time-dependent transport equation. The initial condition of the simulation includes the presence of a shock precursor in the upstream region (everywhere but inside the clump, which starts empty of CRs): it represents the equilibrium solution of the diffusive-advective transport equation, such that

$$f(r, z, p, t = 0) = f_0(p) \exp\left[-\frac{(z - z_s)v_s}{D(p)}\right] \tag{2.12}$$

Here $f_0(p)$ represents the particle spectrum at the shock location z_s, as regulated by the acceleration process. Following the test-particle approach of DSA, a p^{-4} power-law spectrum is set. An exponential cut-off in momentum p_{cut} is introduced in order to take into account the maximum attainable energy, resulting in

$$f_0(p) \propto p^{-4} \exp\left[-\frac{p}{p_{\rm cut}}\right] \tag{2.13}$$

Boundary conditions are such that a null diffusive flux is set on every boundary, except in the upstream of the z-direction, where the precursor shape is set.

When the forward shock hits the clump, the transmitted shock is assumed to not accelerate particles because it is very slow and it is propagating in a highly neutral medium. Moreover, the reflected shock which propagates back into the remnant is neglected, since it would not contribute to the acceleration of particles because of its low Mach number $\lesssim 2$. This can be derived by using the Rankine–Hugoniot conditions for strong shocks, as described in Appendix A, where the downstream temperature reads as (k being the Boltzmann constant)

$$T_2 = \frac{3}{16}\frac{m_p}{k}v_s^2 \tag{2.14}$$

and consequently the sound speed in the downstream $c_{s,2}$ is equal to

$$c_s = \sqrt{\gamma\frac{P}{\rho}} = \sqrt{\gamma\frac{kT}{m_p}} \rightarrow c_{s,2} = \sqrt{\frac{5}{3}\frac{3}{16}}v_s = \sqrt{\frac{5}{16}}v_s \tag{2.15}$$

In this way, the sonic Mach number of the reflected shock can be computed as

$$M_2 = \frac{v_2}{c_{s,2}} = \frac{\frac{3}{4}v_s}{\sqrt{\frac{5}{16}}v_s} = \frac{3}{\sqrt{5}} \sim 1.3 \tag{2.16}$$

Note that the Mach number downstream is independent of both the shock velocity and the Mach number upstream. Thus, though the reflected shock is still supersonic ($M_2 > 1$), it is not strong ($M_2 \simeq 1$). This means that the reflected shock is not relevant for particle acceleration, since it leads to a compression ratio of

$$r_2 = \frac{(\gamma+1)M_2^2}{(\gamma-1)M_2^2+2} = 1.5 \tag{2.17}$$

and consequently to a very steep power-law index of the accelerated CR spectrum equal to

$$\alpha = \frac{r_2+2}{r_2-1} = 7 \tag{2.18}$$

Given the result shown in Sect. 2.3, an analytical velocity field is set, irrotational and divergence-less through all the space (except at the shock and clump surfaces). This is obtained by solving the Laplace equation for the velocity potential in cylindrical coordinates, with the boundary conditions that in the far field limit the velocity field is directed along the shock direction and equal to $\mathbf{v} = v_{\rm down}\widehat{z} = \frac{3}{4}v_s\widehat{z}$, while at the clump surface the field is fully tangential. The resulting solution reads as

$$\begin{cases} v_r(r,z) = -\frac{3}{2}\frac{R_c^3 rz}{(r^2+z^2)^{5/2}} v_{\rm down} \\ v_z(r,z) = \left[1+\frac{1}{2}R_c^3\frac{(r^2-2z^2)}{(r^2+z^2)^{5/2}}\right] v_{\rm down} \end{cases} \tag{2.19}$$

With such a choice of the velocity field, the adiabatic compression term in Eq. (2.10) vanishes. Moreover, a null velocity field inside the clump is set, since $v_{s,c} \ll v_s$, as well as in the upstream region. A schematic view of the velocity vector field adopted is given in Fig. 2.6b. Comparing it with the results from MHD simulation obtained in Sect. 2.3 and showed in Fig. 2.6a, one can see that the two velocity profiles are quite similar except for the turbulent region behind the clump.

Furthermore, a stationary space-dependent and isotropic Bohm diffusion is assumed through all the space, so that

$$D_{\rm Bohm}(\mathbf{x}, p) = \frac{1}{3} r_L(\mathbf{x}, p) v(p) = \frac{1}{3}\frac{pc}{ZeB_0(\mathbf{x})} v(p) \tag{2.20}$$

where $r_L(\mathbf{x}, p)$ is the Larmor radius of a particle with charge Ze in a background magnetic field $B_0(\mathbf{x})$. As a reference value, the diffusion coefficient for protons propagating in the standard CSM amounts to $D_{\rm Bohm} = 3.3 \times 10^{22} (p/10\,{\rm GeV}\,c^{-1}) (B/10\,\mu G)^{-1}\,{\rm cm^2\,s^{-1}}$. In the following only relativistic protons are considered, with momenta $p \in [1\,{\rm GeV/c} - 1\,{\rm PeV/c}]$, since this is the energy interval relevant for the production of HE and VHE gamma rays. A space-dependent magnetic field $B_0(\mathbf{x})$ is set, defining four regions in the space:

(i) The unshocked CSM, in the far upstream;
(ii) The shocked CSM, where $B_0 \equiv B_{\rm CSM} = 10\,\mu$G;
(iii) The clump interior, with a size of $R_c = 0.1$ pc, where diffusion is expected to be significantly faster and consequently not efficient due to the ion-neutral friction, such that $B_0 \equiv B_c = 1\,\mu$G.
(iv) The clump skin, with a size of $R_s = 0.5 R_c = 0.05$ pc around the clump itself, where the amplification of the magnetic field is realized such that $B_0 \equiv B_s = 100\,\mu$G;

It is worth to recall here the relevant timescales for particle propagation in such environment. The penetration time inside the clump, i.e. the time taken by particles to diffuse in the amplified field B_s of the magnetic skin, amounts to

$$\tau_{\rm diff,s}(p) = \frac{R_s^2}{4D_{\rm Bohm}(B_s, p)} \simeq 535 \left(\frac{p}{1\,{\rm TeV/c}}\right)^{-1} {\rm yr} \tag{2.21}$$

On the other hand, the advection time is energy independent and, considering the characteristic size of the clump, it can be evaluated as

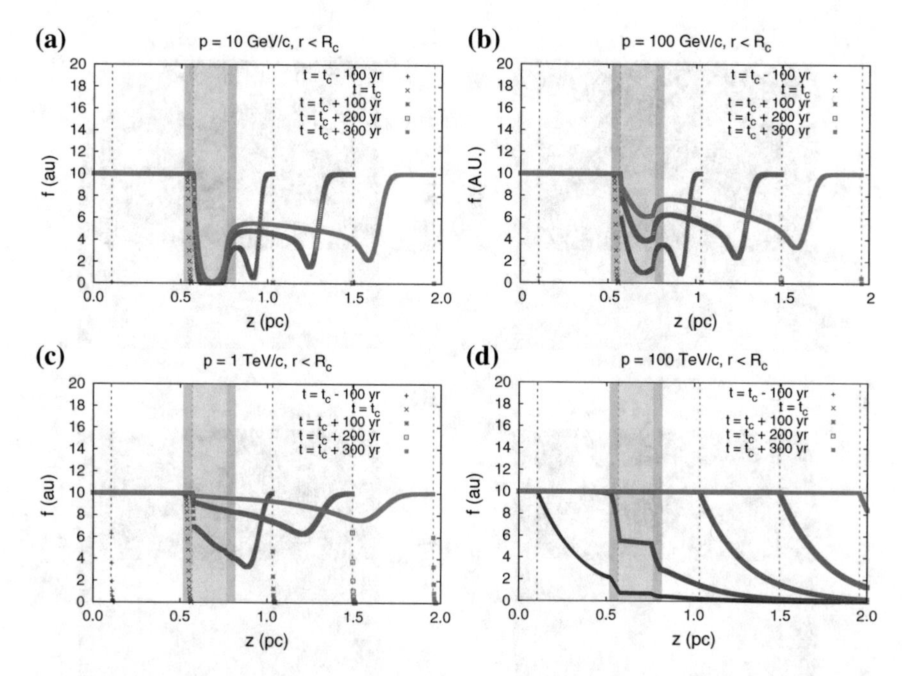

Fig. 2.7 Density profiles of the accelerated particles in the shock region along z (at fixed radial position inside the clump) for CR protons of momentum **a** 10 GeV/c, **b** 100 GeV/c, **c** 1 TeV/c and **d** 100 TeV/c. Vertical dashed lines represent the shock position at the time indicated in the legend. The light pink band defines the clump interior, while the dark pink band defines its magnetic skin. Figure from Celli et al. [79], reproduced by permission of the Royal Astronomical Society

$$\tau_{\rm adv} = \frac{2R_c}{v_s} \simeq 43\,\text{yr} \tag{2.22}$$

Note that the advection and diffusion timescales, as estimated in Eqs. (2.21) and (2.22), are comparable for particles with momentum $p \simeq 10\,\text{TeV/c}$: hence, the propagation of particles with lower energies is dominated by advection, while for particles at larger energies it is dominated by diffusion.

As a result of the numerical simulation, the density profile of accelerated particles diffusing in the region of interaction between the shock and the clump is shown in Fig. 2.7a–d for particles of different energy. The distribution function is flat in the downstream region, while a precursor starts at the shock position, as defined in Eq. (2.12). As expected, low-energy particles penetrate inside the clump at much later times with respect to high-energy ones. The time evolution of the CR distribution function for different CR momenta is also shown: Fig. 2.8a, c, e refer to 10 GeV particles, Fig. 2.8b, d, f concern 100 GeV particles, Fig. 2.9a, c, e are for 1 TeV particles, while Fig. 2.9b, d, f refer to 10 TeV particles. It is visible that low-energy particles are not able to fill uniformly the clump interior on the temporal scale relevant for the gamma-ray emission (around few hundred years, as explained in Sect. 2.5).

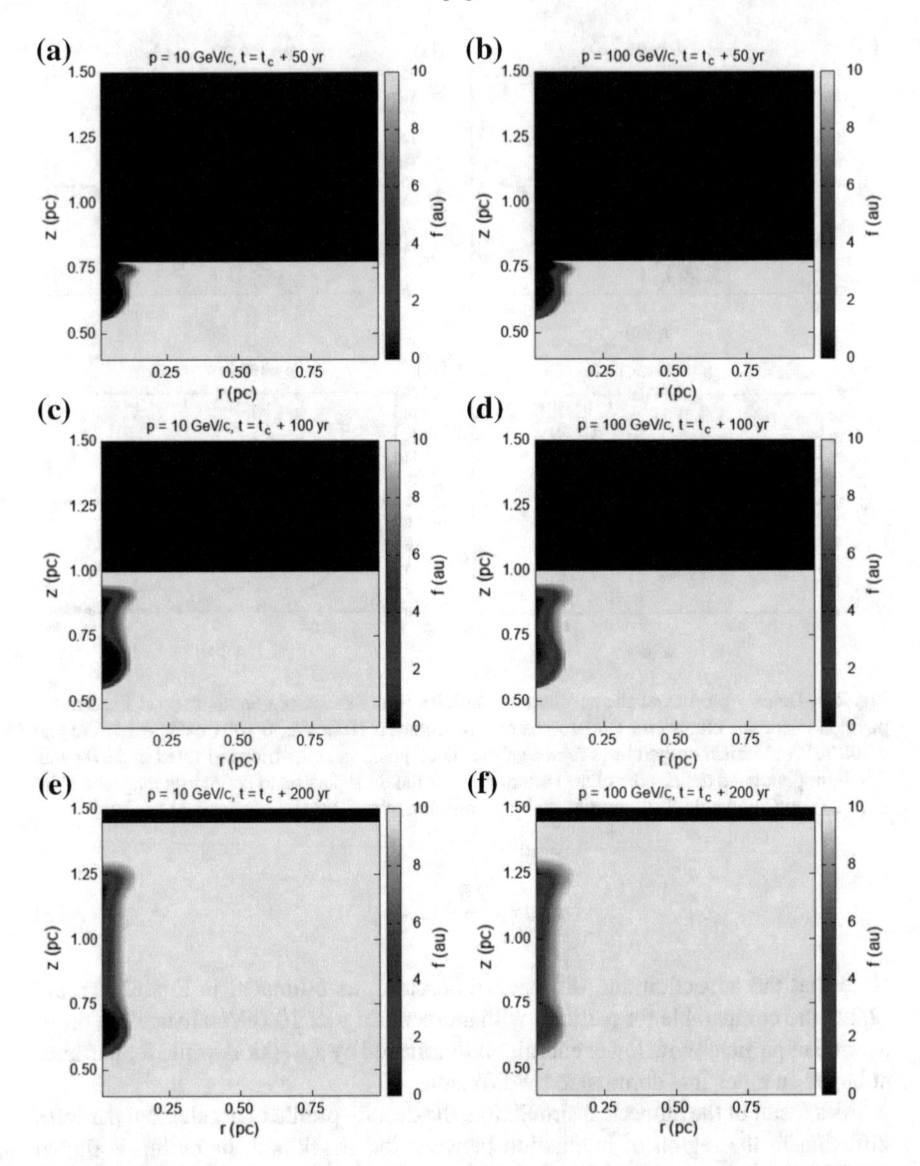

Fig. 2.8 Distribution function of CR protons of momentum **a**, **c**, **e** 10 GeV/c and **b**, **d**, **f** 100 GeV/c at different times with respect to t_c: panels **a** and **b** refer to $t = t_c + 50\,\text{yr}$, **c** and **d** to $t = t_c + 100\,\text{yr}$, while **e** and **f** to $t = t_c + 200\,\text{yr}$. Figure from Celli et al. [79], reproduced by permission of the Royal Astronomical Society

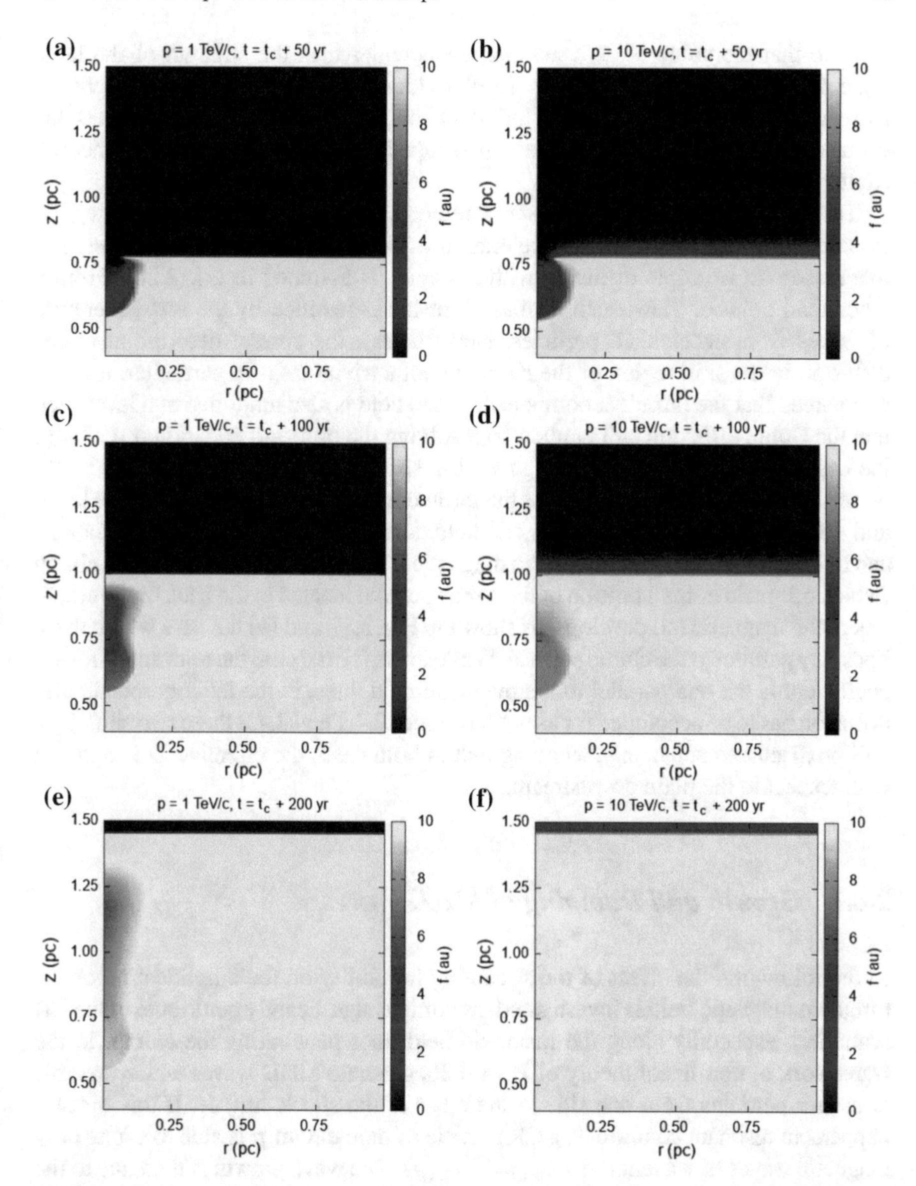

Fig. 2.9 Distribution function of CR protons of momentum **a**, **c**, **e** 1 TeV/c and **b**, **d**, **f** 10 TeV/c at different times with respect to t_c: panels **a** and **b** refer to $t = t_c + 50\,\text{yr}$, **c** and **d** to $t = t_c + 100\,\text{yr}$, while **e** and **f** to $t = t_c + 200\,\text{yr}$. The precursor presence in front of the shock is well visible at 10 TeV. Figure from Celli et al. [79], reproduced by permission of the Royal Astronomical Society

Note that the value of B_c used above is smaller than the strength of the large scale magnetic field expected in molecular clouds. Such a smaller value is chosen as representative of the effective turbulent magnetic field, which determines the diffusion coefficient in Eq. (2.20), and is damped by the presence of the ion-neutral friction (see Sect. 2.4.1).

The diffusion coefficient in the shock region should be close to the Bohm regime in order to obtain an effective acceleration of protons to multi-TeV energies. For this reason an isotropic diffusion with $\delta B \simeq B_0$ is assumed in Eq. (2.20). From a theoretical point of view, such a small diffusion is justified by the self-generation of waves from accelerated particles. Nevertheless, the correct description of the diffusion in the skin region of the clump is not a trivial task. In particular, it is not guaranteed that the turbulent component of the field is also amplified at a level such that the Bohm diffusion still applies. In resolving the transport equation, Eq. (2.10), the case of Bohm diffusion through all the space is assumed, namely $\delta B_s \simeq B_s$. Nevertheless, if the amplification of the turbulence does not occur at the same level, and $\delta B_s < B_s$, the large scale magnetic field dominates and two opposite situations can be envisioned: (*a*) the case where magnetic field lines penetrate inside the clump (which occurs in a small portion of the clump surface located in the back of the cloud, where the magnetic tail develops, as shown in Fig. 2.5), and (*b*) the case where these lines stay parallel to the clump surface. While in the former case the relevant diffusion coefficient is the one parallel to the magnetic field lines, in the latter perpendicular diffusion has to be accounted for as well. In Sects. 2.4.1 and 2.4.2 these two situations will be discussed separately, showing that in both cases the effective D is reduced with respect to the plain downstream.

2.4.1 Growth and Damping of MHD Waves

In the following, the effect of the streaming instability on the amplification of the turbulent magnetic field is investigated, as it might significantly contribute to the CR scattering, especially along the magnetic field lines penetrating the clump. In the framework of non-linear theory of DSA, CRs generate MHD waves which are able to scatter particles from one side to the other of the shock surface. If this process happens in resonant conditions, a CR particle of momentum p is able to excite only magnetic waves of wavenumber $k_{\rm res} = 1/r_L(p)$. The wave growth is then due to the streaming of CRs. Thus the CR density, obtained as a solution of Eq. (2.10), affects the amount of turbulence that is generated, which in turn modifies the diffusion properties of the system as [217]

$$D(\mathbf{x}, p, t) = \frac{1}{3} r_L(p) v(p) \frac{1}{\mathcal{F}(k_{\rm res}, \mathbf{x}, t)} \tag{2.23}$$

where $\mathcal{F}(k, \mathbf{x}, t)$ is the turbulent magnetic energy density per unit logarithmic bandwidth of waves with wavenumber k, normalized to the background magnetic energy density as

$$\left(\frac{\delta B(\mathbf{x},t)}{B_0}\right)^2 = \int \mathcal{F}(k,\mathbf{x},t)d\ln k \tag{2.24}$$

Given the strong non linearity of the problem, it is computationally prohibitive to solve in a self-consistent way the system composed by the transport equation and by the time evolution of the wave power density, which satisfies the following equation

$$\frac{\partial \mathcal{F}}{\partial t} + \mathbf{v_A} \cdot \nabla \mathcal{F} = (\Gamma_{\mathrm{CR}} - \Gamma_{\mathrm{D}})\mathcal{F} \tag{2.25}$$

in the limit where the background fluid motion can be neglected [172]. Here, Γ_{CR} is the growth rate of MHD waves, Γ_{D} is the damping rate and v_A is the Alfvén speed. Instead, Eq. (2.10) will be solved in a stationary given magnetic field and, once the f is known, the contribution of the growth and damping of resonant waves due to CR streaming will be evaluated a posteriori.

The growth rate of the streaming instability strongly depends on the CR density gradient. It is therefore expected to be more pronounced in the clump skin, where magnetic field amplification makes diffusion very efficient, thus increasing the CR confinement time in this region. This rate can be expressed as [217]

$$\Gamma_{\mathrm{CR}}(k) = \frac{16}{3}\pi^2 \frac{v_A}{B_0^2 \mathcal{F}(k)} \left[p^4 v(p) \nabla f\right]_{p=p_{\mathrm{res}}} \tag{2.26}$$

where p_{res} is the resonant momentum.

The amplified magnetic field can in turn be damped by non-linear damping (NLD) due to wave-wave interactions, or by ion-neutral damping (IND) due to momentum exchange between ions and neutrals as a consequence of the charge exchange process. Since the clump under exam is not isolated, the typical timescale of the system should also be accounted for. In fact, if the age of the clump is shorter than the time for damping to be effective, then waves can grow freely for a timescale equal to the clump age.

The dominant mechanism of wave damping in the clump magnetic skin is the NLD, since the plasma is assumed to be completely ionized. The damping rate can thus be expressed as [204]

$$\Gamma_{\mathrm{D}}(k) = \Gamma_{\mathrm{NLD}}(k) = (2c_k)^{-3/2} k v_A \sqrt{\mathcal{F}(k)} \tag{2.27}$$

where $c_k = 3.6$ is the so-called Kolmogorov constant. In stationary conditions (when the system age is not a limiting factor), the wave growth rate (due to streaming instability) equals the damping rate, namely $\Gamma_{\mathrm{CR}} = \Gamma_{\mathrm{D}}$. By equating Eqs. (2.26) to (2.27), the power in the resonant turbulent momentum results in

$$\mathcal{F} = \left[\frac{16}{3} \frac{\pi^2}{B_0^2} \left(p^4 v(p) \frac{\partial f}{\partial r} \right)_{p=p_{\mathrm{res}}} r_L \right]^{2/3} 2c_k \tag{2.28}$$

In Eq. (2.28), the CR density gradient is computed within the clump skin, along the radial dimension. In order to verify whether the stationarity assumption is correct or not, one has to insert Eq. (2.28) into Eq. (2.27): by setting v_A in $B_0 = 10\,\mu$G and a typical ion density for a clump of $n_i = 10^{-4} n_c = 10^{-1}$ ions cm^{-3}, one obtains that stationary is not valid for CR momenta larger than $p \geq 1$ TeV/c, where the clump age constraints the damping mechanism. Therefore, in this case, the power in turbulence is computed by equating the growth rate of the MHD waves, as reported in Eq. (2.26), to the inverse of the clump age. This gives

$$\mathcal{F} = \frac{16}{3} \pi^2 \frac{v_A}{B_0^2} \left[p^4 v(p) \frac{\partial f}{\partial r} \right]_{p=p_{\mathrm{res}}} \tau_{\mathrm{age}} \tag{2.29}$$

The result of the computation is shown in Fig. 2.10a. The turbulence generated in the clump magnetic skin is such that CRs with momentum between 100 GeV/c and 1 TeV/c are closer to the Bohm diffusive regime.

On the contrary, in the clump interior, where neutral particles are abundant, the most efficient damping mechanism is IND (see [187, 255]). Waves dissipate energy because of the viscosity produced in the charge exchange between ions and neutrals, such that previously neutrals start to oscillate with the waves. The frequency of ion-neutral collision is [100, 157]

$$\nu_{\mathrm{c}} = n_{\mathrm{n}} \langle \sigma v \rangle = 8.4 \times 10^{-9} \left(\frac{n_{\mathrm{n}}}{1\,\mathrm{cm}^{-3}} \right) \left(\frac{T_{\mathrm{c}}}{10^4\,\mathrm{K}} \right)^{0.4} \tag{2.30}$$

where an average over thermal velocities is considered. The rate of IND depends on the wave frequency regime, namely whether ions and neutrals are strongly coupled or not. Defining the wave pulsation $\omega_k = k v_A$ in a collision-free medium, then the study of the dispersion relation defines different regimes for ion-neutral coupling depending on the value of the ion-to-neutral density ratio defined in Eq. (2.5). These regimes are as follows:

(i) If $\epsilon < 1/8$, there's a range of ω_k for which waves can not propagate, that is a range of k for which ω_k is a purely imaginary number. This range is for

$$4\epsilon < \frac{\omega_k^2}{\nu_c^2} < \frac{1}{4} \tag{2.31}$$

which, within our assumptions ($\epsilon = 1.5 \times 10^{-3}$), equals to CR momenta in 15 GeV/c $< p <$ 95 GeV/c;

(ii) In the intervals $\epsilon \ll 1$ and $(\omega_k/\nu_c)^2 \ll 4\epsilon$, then

$$\Gamma_{\mathrm{IND}}(k) = -\frac{\omega_k^2}{2\nu_c} = -\frac{k^2 v_A^2}{2\nu_c} \tag{2.32}$$

(iii) If $\epsilon \ll 1$ and $(\omega_k/\nu_c)^2 \ll 1/4$, then

$$\Gamma_{\mathrm{IND}} = -\frac{\nu_c}{2} \tag{2.33}$$

(iv) If $\epsilon \gg 1$, then

$$\Gamma_{\mathrm{IND}}(k) = -\frac{\nu_c}{2}\left[\frac{(\omega_k^2/\nu_c^2)}{(\omega_k^2/\nu_c^2)+\epsilon^2}\right] \tag{2.34}$$

Again, the damping time Γ_{D}^{-1} should be compared with the clump age τ_{age}. For $p = 10\,\mathrm{GeV/c}$ the IND time is shorter than the clump age: by setting the equilibrium condition $\Gamma_{\mathrm{D}} = \Gamma_{\mathrm{IN}}$ through Eq. (2.33), one obtains

$$\mathcal{F} = \frac{16}{3}\pi^2 \frac{v_A}{B_0^2}\frac{2}{\nu_c}\left[p^4 v(p)\frac{\partial f}{\partial r}\right]_{p=p_{\mathrm{res}}} \tag{2.35}$$

For $p = 100\,\mathrm{GeV/c}$, the dominant damping mechanism is still IND. The equilibrium condition $\Gamma_{\mathrm{D}} = \Gamma_{\mathrm{IN}}$ is now set through Eq. (2.32) to get

$$\mathcal{F} = \frac{16}{3}\pi^2 \frac{1}{B_0^2}\left[p^4 v(p)\frac{\partial f}{\partial r}\right]_{p=p_{\mathrm{res}}} \frac{2\nu_c}{k_{\mathrm{res}}^2 v_A} \tag{2.36}$$

On the other hand, for $p \geq 1\,\mathrm{TeV/c}$, the clump age is the limiting factor since IN damping requires a longer time. As shown in Fig. 2.10b, IND is very effective in damping waves resonant with CR particles of momentum lower than 10 GeV/c. Nonetheless, a strong suppression of the diffusion coefficient is reached between 100 GeV/c and 1 TeV/c.

2.4.2 Perpendicular Diffusion

As shown in Fig. 2.5, the large scale magnetic field is compressed and stretched around a large fraction of the clump surface. In this region the Alfvénic turbulence produced by CR-driven instabilities is not enough to reach the Bohm limit, since $\delta B_s \simeq (0.1 - 0.3)\,B_s$ is obtained at most, as shown in Fig. 2.10a. As a consequence, if additional pre-existing turbulence is not amplified at the same level of the regular field, the penetration inside the clump requires a perpendicular diffusion. As streaming instability does not affect perpendicular diffusion, in this section particle diffusion is discussed neglecting its role. According to quasi-linear theory, the diffusion perpendicular to the large scale magnetic field, $D_\perp$, is related to parallel diffusion, $D_\parallel$, through [77]

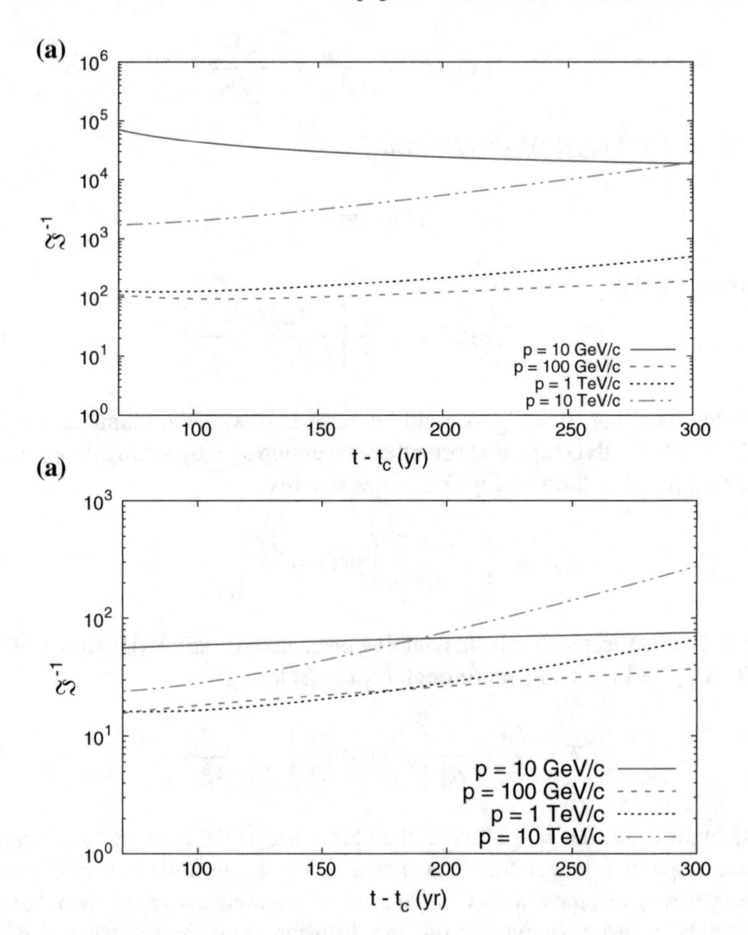

Fig. 2.10 Ratio between self-generated diffusion coefficient $D_{\rm self}$ and Bohm diffusion coefficient $D_{\rm Bohm}$, as a function of the clump age for different particle momenta. From Eq. (2.23), it follows that $\mathcal{F}^{-1} = D_{\rm self}/D_{\rm Bohm}$: $D_{\rm self}$ is obtained by imposing $1/\Gamma_{\rm CR} = \min(1/\Gamma_D, \tau_{\rm age})$. *Top*: results in the clump skin, where $\Gamma_D = \Gamma_{\rm NLD}$ is considered. *Bottom*: results for the clump interior, where $\Gamma_D = \Gamma_{\rm IND}$ is assumed. Figure from Celli et al. [79], reproduced by permission of the Royal Astronomical Society

$$D_\perp = D_\parallel \frac{1}{1 + (\lambda_\parallel / r_L)^2} \tag{2.37}$$

where $\lambda_\parallel$ is the particle mean free path along the background field B_0. Since $\lambda_\parallel = r_L(\delta B/B_0)^{-2}$, the perpendicular diffusion coefficient results in

$$D_\perp = D_\parallel \frac{1}{1 + (\delta B/B_0)^{-4}} \tag{2.38}$$

Hence, the radial diffusion into the clump is strongly suppressed with respect to the azimuthal diffusion along the field lines provided that a tiny amplification of the magnetic field is realized. In the following, it is assumed that in the region downstream of the shock $(\delta B(k)/B)_{\rm down} \simeq 1$ at all scale k resonant with accelerated particles. The results concerning MHD simulations presented in Sect. 2.3 show that the magnetic field in the clump skin is amplified, reaching a value $\sim$10 times larger than that in the unperturbed downstream, so that $B_s = 10B_{\rm down}$ (Fig. 2.5). If the turbulence in the skin is amplified as well, then the Bohm diffusion limit might be reached: this is the case for isotropic diffusion, where the distinction among parallel and transverse diffusion is lost ($D_\perp = D_\parallel$) and a strong suppression of the diffusion coefficient is realized. However, if the turbulence in the clump skin remains at the same level as the unperturbed downstream, then $(\delta B/B)_s = 0.1(\delta B/B)_{\rm down} \simeq 0.1$: this implies that in the skin parallel diffusion holds with $D_{\parallel,s} = 10D_{\parallel,{\rm down}}$, while for the perpendicular diffusion, using Eq. (2.38), one gets $D_{\perp,s} = 10^{-3}D_{\parallel,{\rm down}}$. Therefore, in this regime, particle penetration inside the clump is even more suppressed than in the case of isotropic diffusion.

2.4.3 Proton Spectrum

Once the proton distribution function is known from the solution of Eq. (2.10), it is possible to obtain the proton energy spectrum inside the clump $J_c(p,t)$ at different times with respect to the first shock-clump contact, that in the following is indicated as $t = t_c$. The average spectrum inside the clump reads as

$$J_c(p,t) = \frac{1}{V_c}\frac{d^3N_c(t)}{dp^3} \tag{2.39}$$

where $V_c = 4\pi R_c^3/3$ is the clump volume and $d^3N_c(t)/dp^3$ is the number of protons inside the clump at a time t per unit volume in momentum space. The spectrum can be computed summing upon all the discretized bins which define the clump volume. In this way, one obtains

$$J_c(p,t) = \frac{2\pi}{V_c}\sum_{i\in clump} f_i(r_i, z_i, p, t)r_i\Delta r_i\Delta z_i \tag{2.40}$$

Results are shown in Fig. 2.11, where a proton cut-off momentum of $p_{\rm cut} = 70\,{\rm TeV/c}$ was set in order to reproduce the very-high-energy gamma-ray data of RX J1713.7-3946. The spectrum of particles from younger clumps is much harder than the one accelerated at the shock as defined in Eq. (2.13). This is explained by the prevention of penetration of low-energy CRs into the clump due to the amplified magnetic field at the skin and because of the linear dependency of the diffusion coefficient with the particle momentum. In this way, the entrance of CRs into the clump is delayed.

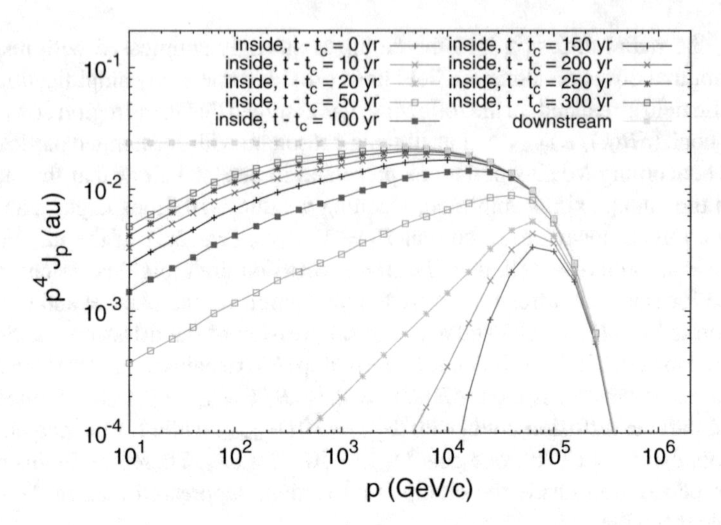

Fig. 2.11 Proton spectrum inside clumps of different ages. The downstream particle energy density, which is constant in time, is also reported. Figure from Celli et al. [79], reproduced by permission of the Royal Astronomical Society

The spectral index of protons below 100 GeV/c is as hard as $\alpha = -3.50$ when the clump age is 50 yr, moving to $\alpha = -3.54$ when the clump age is 150 yr, and finally $\alpha = -3.57$ when the clump is 300 yr old. On the other hand, CRs with $p \gtrsim 100$ TeV/c are quite unaffected by the presence of the clump.

2.4.4 A Possible Role of the Second Order Fermi Acceleration

As previously anticipated, the second order Fermi acceleration term has been neglected in the solution of the transport equation. However, it might become relevant in the presence of amplified magnetic fields. Here it is shown that its effect on the final gamma-ray spectrum of RX J1713.7-3946 is negligible.

The idea that stochastic re-acceleration might play a significant role once the turbulence has developed was put forward in Drury [96] as a possible explanation of the deviations observed in the radio spectral index of middle-aged SNRs with respect to what expected if only first order acceleration were acting. More recently, it was pointed out [200] that only fast-mode waves can provide momentum diffusion fast enough to significantly modify the spectra of particles in the shocked downstream. In the present work, the question arises whether the possible increase of Alfvénic turbulence in the clump skin can significantly scatter particles in momentum space, leading to a modification of the proton spectrum. As was previously shown in Sect. 2.3, the regular magnetic field is amplified in the clump skin; if the turbulence is also amplified above some level, particles residing in this region for a long time might start diffusing in momentum space too. It is possible to estimate this effect by using the acceleration timescale for second order acceleration, $\tau_{acc} = p^2/D_{pp}$, where D_{pp} is the diffusion coefficient in momentum space, that is connected to the diffusion coefficient in the physical space D_{xx} by the relation $D_{pp}D_{xx} \propto p^2 v_A^2/9$ (see [96]). By comparing τ_{acc} with the minimum between the diffusion time in the clump skin and the age of the clump, it turns out that the latter timescale is the most relevant one. Hence, from the condition $\tau_{acc} < t_{age}$, it follows that second order acceleration can be relevant when

$$pc\beta \sim 25 \left(\frac{B_0}{100\,\mu\text{G}}\right)^3 \left(\frac{\delta B/B_0}{0.1}\right)^2 \left(\frac{n}{0.04\,\text{cm}^{-3}}\right)^{-1} \left(\frac{t_{age}}{100\,\text{yr}}\right) \text{GeV}\,, \tag{2.41}$$

where $B_0 \simeq 100\,\mu$G was set, as estimated in Sect. 2.3 within the clump skin. As a consequence, if $\delta B/B_0 \lesssim 0.1$, only particles with energies below ~25 GeV are affected, implying that the gamma-ray spectrum would be modified only below few GeV, a region which is not strongly constrained by Fermi-LAT data in the case of RX J1713.7-3946 (see Sect. 2.6). It is worth to note that such a level of turbulence would be compatible with the scenario where CRs perpendicularly penetrate the clump (see Sect. 2.4.2). Moreover, if the clump distribution is spatially uniform, as assumed in the subsequent section, younger clumps are the most numerous and contribute the most to the final gamma-ray spectrum, further reducing the relevance of second order acceleration.

On the contrary if the turbulence is strongly amplified, such that $\delta B/B_0 \gg 0.1$, than second order acceleration may be important. Interestingly, the effect on the particle spectrum in the case of Bohm diffusion would be a hardening [173], hence one may even speculate that second order acceleration may be responsible for the production of hard CR spectrum as inferred from gamma-ray observations (see Fig. 2.11). Nevertheless, it is worth recalling that second order acceleration strongly depends on the angular distribution of magnetic turbulence hence, before drawing any conclusions, a detailed study of the turbulence in the clump skin would be needed. A possible further indication on the effectiveness of second order acceleration could come from radio observations of single clumpy structures, in order to probe electron energies down to ~1 GeV.

2.5 Gamma Rays from a Uniform Clump Distribution Inside the Shell

If the number of clumps is large enough, they could be the main source of gamma-ray emission, due to hadronic inelastic collisions of CRs with the ambient matter. In such a case the gamma-ray spectrum would reflect the spectrum of particles inside the clumps rather than the one produced by the shock acceleration. In this section, the total gamma-ray spectrum due to hadronic interactions is computed, assuming that clumps are uniformly distributed over the CSM where the shock expands, with number density $n_0 = 0.2$ clumps pc^{-3}. For such a density, the effect of an SNR shock impacting on a clump distribution can be described, with good accuracy, as the result of individual shock-clump interactions. Note that the average distance among clumps is much larger than the clump size and the simulation box.

The emissivity rate of gamma rays from a single clump, given the differential flux of protons inside the clump $\phi_c(T_p, t)$ and the density of target material, is

$$\epsilon_c(E_\gamma, t) = 4\pi n_c \int dT_p \frac{d\sigma_{pp}}{dE_\gamma}(T_p, E_\gamma)\phi_c(T_p, t) \tag{2.42}$$

where T_p is the particle kinetic energy, $d\sigma_{pp}/dE_\gamma$ is the differential cross section of the interaction while $\phi_c(T_p, t)$ is obtained from the spectrum in Eq. (2.40). For the pp cross-section, the analytical parametrization provided by the LibPPgam library (see [146]) will be adopted and, specifically, the parametrization resulting from the fit to Sibyll 2.1.

In order to evaluate the cumulative distribution resulting from a fixed distribution of clumps, all clumps satisfying the following two conditions will be included: (I) they should survive the shock passage (not evaporated); (II) they should be located between the position of the contact discontinuity (CD) $R_{\rm cd}$ and the shock position R_s. Indeed it is assumed that, once a clump passes through the CD, either it is destroyed by MHD instabilities or it soon gets emptied of CRs. Therefore one should consider the minimum time between the evaporation time $\tau_{\rm ev}$, and the time elapsed between the moment that the clump crosses the forward shock and the moment that it crosses the contact discontinuity $\tau_{\rm cd}$. As estimated in Sect. 2.2, the evaporation time is of the order of few times the cloud crossing time (see Eq. (2.9)). In the following, the conservative value of $\tau_{\rm ev} = \tau_{\rm cc}$ is considered. For the parameters chosen, this time is always larger than the SNR age. The CD radial position, instead, can be estimated by imposing that all the compressed matter is contained in a shell of size $\Delta R = R_{\rm SNR} - R_{\rm cd}$, so that

$$\frac{4}{3}\pi R_{\rm SNR}^3 n_{\rm up} = 4\pi R_{\rm SNR}^2 \Delta R n_{\rm down} \tag{2.43}$$

which for strong shock amounts to $\Delta R = R_{\rm SNR}/12$. Therefore, the time that a clump takes to be completely engulfed in the CD is

$$\tau_{\rm cd} = \frac{(2R_c + \Delta R)}{\frac{3}{4}v_s} \tag{2.44}$$

The oldest clumps in the remnant shell will therefore have an age

$$T_{\rm c,max} = \min(\tau_{\rm ev}, \tau_{\rm cd}) \tag{2.45}$$

As already mentioned, a uniform spatial distribution of clumps is considered inside the remnant of age $T_{\rm SNR}$. Therefore, the total number of clumps at a distance among

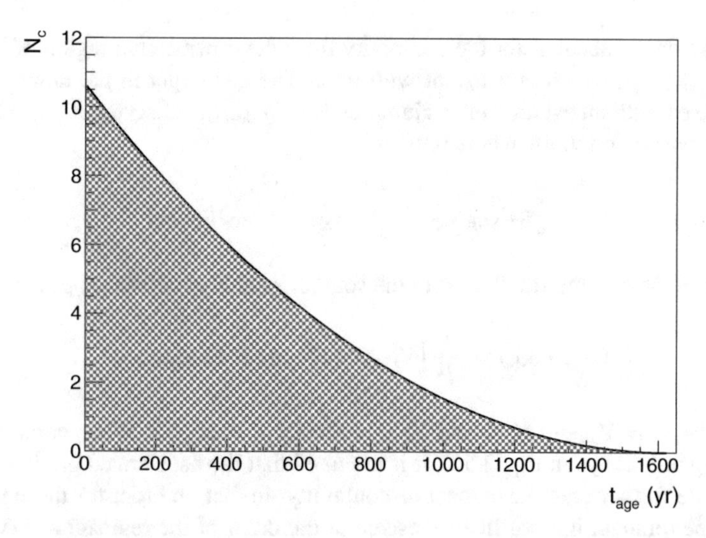

Fig. 2.12 Number of clumps N_c with age $t_{\rm age}$ embedded within the source, as resulting from the assumption of uniform distribution in Eq. (2.46). Clumps located between the contact discontinuity and the shock position have an age $t_{\rm age} < 300\,{\rm yr}$

r and $r + dr$ from the source is equal to

$$\frac{dn(r)}{dr} = 4\pi n_0 r^2 \implies \frac{dn(t)}{dt} = 4\pi n_0 r^2(t) v_s(t) \tag{2.46}$$

Furthermore, a constant shock speed is assumed as for a shock evolving in the ejecta-dominated phase. We are interested in the number of clumps with a given age $t_{\rm age}(r) = T_{\rm SNR} - t_{\rm c}(r)$. The number of clumps with an age between $t_{\rm age} - \Delta t$ and $t_{\rm age}$, namely $N_{\rm c}(t_{\rm age})$, is equal to the number of clumps that the shock has encountered between $T_{\rm SNR} - t_{\rm age}$ and $T_{\rm SNR} - t_{\rm age} + \Delta t$, that is

$$N_{\rm c}(t_{\rm age}) = 4\pi n_0 \int_{T_{\rm SNR}-t_{\rm age}}^{T_{\rm SNR}-t_{\rm age}+\Delta t} r(t')^2 v_s(t') dt' \tag{2.47}$$

Figure 2.12 shows the distribution of clumps with fixed age as described by Eq. (2.47). The total number of clumps with $t_{\rm age} \le T_{\rm c,max}$ is equal to $N_c \simeq 440$, which corresponds to a total mass in clumps inside the remnant shell equal to $M_c \simeq 45 M_\odot$.

Consequently the total gamma-ray emissivity due to these clumps is

$$\epsilon_c(E_\gamma, T_{\rm SNR}) = \sum_{t_{\rm age}=0}^{T_{\rm c,max}} N(t_{\rm age}) \epsilon_\gamma(E_\gamma, t_{\rm age}) \tag{2.48}$$

One also has to account for the emissivity from the downstream region of the remnant $\epsilon_{\rm down}(E_\gamma)$, which is constant with time. The gas target in the downstream is considered with an average inter-clump density $\langle n_{\rm down}\rangle$, satisfying the condition of mass conservation in the whole remnant

$$\frac{4}{3}\pi R_{\rm SNR}^3 n_{\rm up} = \frac{4}{3}\pi (R_{\rm SNR}^3 - R_{\rm cd}^3)\langle n_{\rm down}\rangle \tag{2.49}$$

Therefore the gamma-ray flux from the source, located at a distance d, is computed as

$$\phi_\gamma(E_\gamma, T_{\rm SNR}) = \frac{1}{d^2}\left[V_{\rm c}\epsilon_c(E_\gamma, T_{\rm SNR}) + V_{\rm down}\epsilon_{\rm down}(E_\gamma)\right] \tag{2.50}$$

where $V_{\rm down} = V_{\rm shell} - N_{\rm c}V_{\rm c}$ and $V_{\rm shell} = 4\pi(R_{\rm SNR}^3 - R_{\rm cd}^3)/3$. When computing the flux of gamma rays in Eq. (2.50), it is assumed that the accelerated particles do not propagate further than the contact discontinuity: in fact, in order for these particles to fill the remnant interior in a timescale of the order of the remnant age (which is equivalent to requiring that $L_{\rm diff} = R_{\rm SNR}$), the level of diffusion in the downstream of the shock should be comparable to the average Galactic one, which is likely not the case in such a turbulent region. Hence, the model applies to the situation when the target clumps are not probed by low-energy CRs on a timescale equal to the age of the system.

2.6 Application to RX J1713.7-3946

The Galactic supernova remnant RX J1713.7-3946 (also called G 347.3-0.5) represents one of the brightest TeV emitters in the sky. The origin of its gamma-ray flux in the GeV-TeV domain (see [12, 16]) has been object of a long debate, since both hadronic and leptonic scenarios are able to reproduce, under certain circumstances, the observed spectral hardening. The presence of accelerated leptons is guaranteed by the detected X-ray shell (see [219, 225]), which shows a remarkable correlation with the TeV gamma-ray data, indicating a strong link between the physical processes responsible for these emission components. Meanwhile, a clear signature of accelerated hadrons, that would come from neutrinos, is still missing. RX J1713.7-3946 has been used as a standard candidate for the search of a neutrino signal from Galactic sources [148, 181, 238]. One of the arguments against the hadronic scenario has been claimed because of the absence of thermal X-ray lines (see [103, 149]). In the scenario where the remnant is expanding into a clumpy medium, the non observation of a thermal X-ray emission is naturally explained by the low density plasma between clumps. On the other hand, since clumps remain mostly unshocked and therefore cold, they would not be able to emit thermal X rays.

Clearly, the distribution of gas in RX J1713.7-3946 is crucial to establish the origin of the observed gamma rays. The target material required by pp interactions may

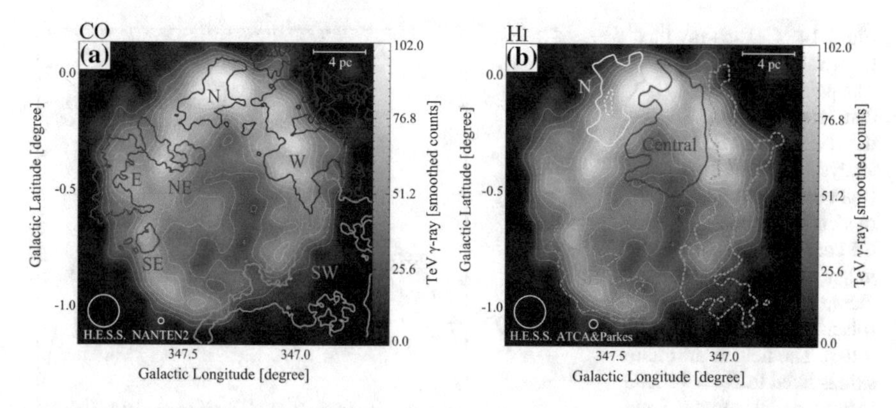

Fig. 2.13 The image and white contours show the TeV gamma-ray distribution. Colored contours schematically draw the locations of the identified **a** ^{12}CO(J=2−1) clouds and **b** HI clouds. Figure from Fukui et al. [115] ©AAS. Reproduced with permission

be present in any chemical form, including both the molecular and the atomic gas. High-resolution mm-wave observations of the interstellar CO molecules with NANTEN [114] revealed the presence of molecular clouds in spatial correlations with TeV gamma rays, in the northwestern rim of the shell. The densest cores of such clouds have been detected in highly excited states of the molecular gas, manifesting signs of active star formation, including bipolar outflow and possibly embedded infrared sources [207]. Other density tracers, such as Cesium, have also confirmed the presence of very dense gas in the region ($n > 10^4\,\mathrm{cm}^{-3}$), as indicated in a recent MOPRA survey [169]. Furthermore, a combined analysis of CO and HI [115] has shown a counterpart of the southeastern rim of the gamma-ray data in atomic hydrogen, as visible in Fig. 2.13. The multi-wavelength observations point towards the clear presence of a non-homogeneous environment, where the young SNR is expanding: in fact, spatial correlations with non-thermal X rays [208] indicate the presence of enhanced X-ray emission in the northwestern part of the remnant, where dense cloud cores are located. Such emission is likely connected to the turbulent fluid motion due to the shock-cloud interaction, which amplifies the local magnetic field.

The estimated distance of the remnant is about $d \simeq 1\,\mathrm{kpc}$ [114, 177], while the radial size of the detected gamma-ray shell today extends up to $R_s \simeq 0.6\,\mathrm{deg}$ [12]. The remnant is supposed to be associated to the Chinese detected type II SN explosion in 393 AD [245]; this would assign to the remnant an age of $T_{\mathrm{SNR}} \simeq 1625\,\mathrm{yr}$. The age, distance and detected size yield an average shock speed of about $\langle v_s \rangle \simeq 6.3 \times 10^8\,\mathrm{cm\,s}^{-1}$. Measurements of proper motion of X-ray structures indicate that the shock speed today should be $v_s \leq 4.5 \times 10^8\,\mathrm{cm/s}$ [232], meaning that the shock has slightly slown down during its expansion. This is expected in SNR evolution [230] during both the ejecta-dominated and the Sedov–Taylor phases, as will be introduced in Sect. 3.1. RX J1713.7-3946 is nowadays moving towards the Sedov phase, therefore one can safely assume a constant shock speed through the time evolution up to now, with a value of $v_s = 4.4 \times 10^8\,\mathrm{cm/s}$ [118]. At this speed, the

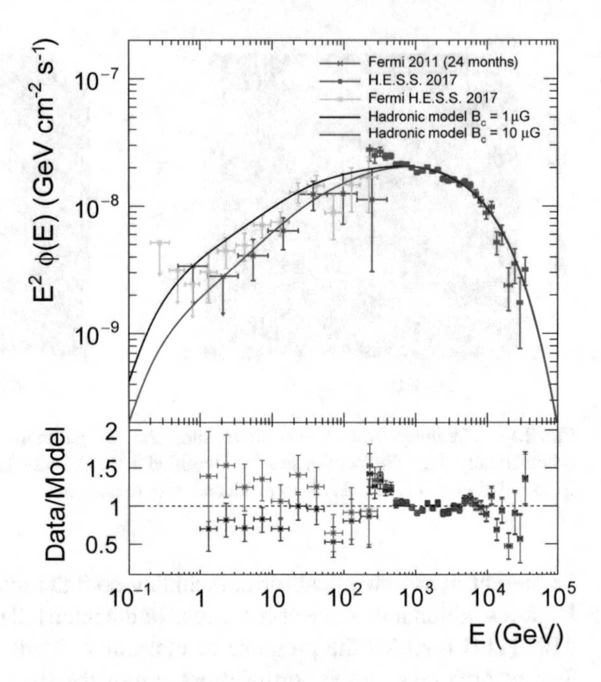

Fig. 2.14 Gamma-ray flux from SNR RX J1713.7-3946. The points are Fermi-LAT data (magenta), H.E.S.S. data (violet) and H.E.S.S. analysis of Fermi-LAT data (light blue). The hadronic models (solid lines) refer to the configuration with a magnetic field inside the clump equal to $B_c = 1\,\mu$G (black) and to $B_c = 10\,\mu$G (blue). The field in the clump skin is fixed to $B_s = 100\,\mu$G in both models. Figure from Celli et al. [79], reproduced by permission of the Royal Astronomical Society

time that the CD takes to completely engulf a clump is, following Eq. (2.44), $\tau_{\rm cd} \simeq$ 300 yr. On the other hand, the evaporation time would be much longer, indicating that the relevant clumps contributing to the gamma-ray emission are younger than $T_{\rm c,max} = 300$ yr.

With the parameters representing RX J1713.7-3946, as defined above, the gamma-ray flux of the remnant shell is computed through Eq. (2.50). The normalization factor k results from a χ^2 minimization procedure among the Fermi-LAT and H.E.S.S. data and the model. Two different configurations are investigated. The first model explores a configuration with the magnetic field inside the clump reduced by a factor of 10 with respect to the CSM value, in order to account for the effect of IND, therefore it is set to $B_c = 1\,\mu$G. The second model, instead, explores a situation where no IND is acting, therefore the magnetic field inside the clump is set to be $B_c = 10\,\mu$G, as in the CSM. Results are shown in Fig. 2.14. The GeV data from two years of data-taking of the Fermi-LAT satellite [16] are reported, together with the H.E.S.S. TeV data [12] and with the H.E.S.S. Collaboration analysis of five years of Fermi-LAT data, as reported in Abdalla et al. [12]. The two models predict slightly different trend in the GeV emission of the remnant. A more pronounced hardening in the case of $B_c = 10\,\mu$G better reproduces the GeV data, while a flatter trend is visible in case diffusion would act less efficiently inside the clump. In this respect, electrons are more suitable to derive constraints on the magnetic field properties of the remnant. A more quantitative study on secondary electrons from pp interactions will be discussed elsewhere.

The normalization constant k, obtained by fitting the gamma-ray data, defines the amount of ram pressure $P_{\rm ram} = \rho_{\rm up} v_s^2$ that is instantaneously converted into CR pressure. The latter is defined, for relativistic particles, as

$$P_{\rm CR} = \frac{k}{3} \int_{m_p c}^{\infty} 4\pi p^2 dp f_0(p) pc \,. \tag{2.51}$$

The efficiency of the pressure conversion mechanism from bulk motion to accelerated particles equals to $\eta = P_{\rm CR}/P_{\rm ram} \simeq 2\%$. Such a value is somewhat smaller than the efficiency estimated by other works in the context of hadronic scenarios, where usually $\eta \simeq 10$–20% [see, e.g. 118, 182]. Compared to Morlino et al. [182] the main differences are due to the highest total target mass used here ($\sim$45 $M_\odot$ vs. $\sim$15 $M_\odot$) which is close to the total mass in clumps estimated by HD simulations [143]. A weaker effect is also due to the fact that adiabatic losses are here neglected, which lead to a smaller acceleration efficiency by less than a factor of two. Also, comparing this result with Gabici and Aharonian [118], few more differences arise. Here, a constant shock speed and a $\propto p^{-4}$ acceleration spectrum are considered, while Gabici and Aharonian [118] used a time dependent shock velocity and a steeper acceleration spectrum $\propto p^{-4.2}$. The latter assumption implies a number density of accelerated protons at 100 TeV smaller by a factor of $\sim$10 (for the given acceleration efficiency). For this reason, Gabici and Aharonian [118] adopted a larger target mass in clumps, $\sim$500 $M_\odot$.

2.6.1 Observing Clumps Through Molecular Lines

An interesting possibility to detect active clumps is provided by radio observations. Secondary electrons emit synchrotron radiation in the radio domain. On top of this continuum, the molecular gas emits lines. For instance, rotational CO lines are often observed in these systems [114]. In the case of RX J1713.7-3946, a few arcsecond angular resolution is needed to probe the spatial scales of clumps: such a small scale can currently be achieved only through the superior angular resolution of the Atacama Large Millimeter Array (ALMA). A precise pointing is however required, since the instrument field of view of $\lesssim 35''$ would not entirely cover a region as extended as the remnant RX J1713.7-3946.

In the following, the radio flux for the $J = 1 \rightarrow 0$ rotational line of the CO molecule is evaluated. This transition is located at $\nu = 115\,{\rm GHz}$ (band 3 of ALMA receivers) and radiates photons with a rate equal to $A_{10} = 6.78 \times 10^{-8}$ Hz. Assuming a CO abundance of $n_{CO}/n_c = 7 \times 10^{-5}$ and a clump density of $n_c = 10^3\,{\rm cm}^{-3}$, the expected flux from an individual clump amounts to

$$F = \frac{h A_{10}}{4\pi d^2} N_{CO} = 3.15 \times 10^{-3}\,{\rm Jy} \tag{2.52}$$

where h is the Planck constant and N_{CO} is the number of CO molecules contained in each clump ($d = 1\,\text{kpc}$ is assumed). The flux level obtained in Eq. (2.52) is well within the performances of the nominal ten 7-m diameter antennas configuration of the ALMA observatory.

Finally, a well known tracer of shock interaction with molecular clouds is constituted by the SiO molecule. Si ions are generally contained in the dust grains, which are destroyed by the passage of a shock, and can form SiO in gas phase [132, 133]. This tracer has been successfully traced in the supernova remnant W 51C [101]. In the case of RX J1713.7-3946, existing observations toward the north-west rim, in the so-called Core C, do not show evidence for significant amount of SiO emission [169]. This result can be interpreted in different ways: either the Core C is located outside of the remnant or, if it is inside, the shock is not yet penetrated into the densest cores where the dust grains are typically located. A further possibility is that the shock propagating inside the core is not strong enough to efficiently sputter Si ions from the grains. Indeed Gusdorf et al. [133] showed that a shock velocity $\gtrsim 25\,\text{km}\,\text{s}^{-1}$ is necessary for an efficient sputtering and the shock speed inside the clump could be lower than such a threshold if $\chi \gtrsim 10^5$ (see Eq. (2.8)). As a consequence, a positive SiO detection could shed light on the value of the density contrast χ.

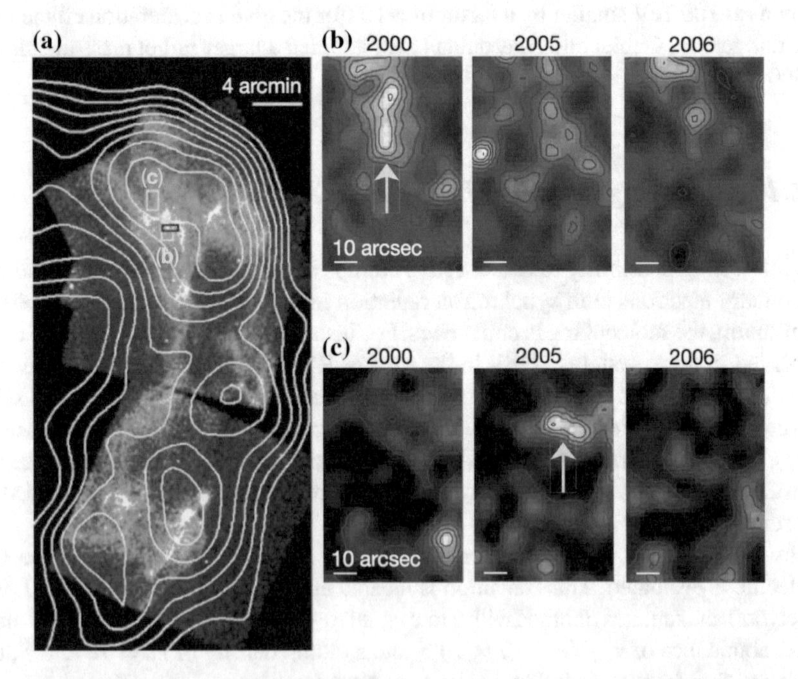

Fig. 2.15 Chandra X-ray images of the western shell of SNR RX J1713.7-3946. Panel **a**: TeV gamma-ray contours from H.E.S.S. overlaid. Panel **b**: a sequence of X-ray observations (30-ks exposure times) in July 2000, July 2005 and May 2006 for a small box (labeled **b**) in panel **a**. Panel **c**: same as panel **b**, for the region identified as **c** in panel **a**. Figure from Uchiyama et al. [232], reprinted by permission from Springer Nature

2.6.2 Clumps and the X-Ray Variability

The origin of bright hot spots in the non-thermal X-ray image of RX J1713.7-3946 [232], decaying on the timescale of one year, remains an unexplained puzzle. Several time-variable compact features have been identified in Chandra data, mostly in the northwest part of the shell, as reported in Fig. 2.15. The size of the fast variable X-ray hot spots is about $\theta \simeq 20$ arcsec, which corresponds to a linear size of $L_x = d\theta \simeq 3 \times 10^{17}$ cm at a distance $d \simeq 1$ kpc. Similar time-dependent features have also been identified in the young SNR CasA [231]. It has been already suggested that the X-ray variability could be connected to the clump scenario [143]. The striking similarity between the physical size of the observed hot spots and the MHD instabilities which are formed in shock flows around clumpy structures suggests a possible intrinsic physical link.

In the context of shock propagation into a non-uniform ambient medium, as discussed above, a natural interpretation addresses the X-ray variability to the electron synchrotron cooling in the amplified large scale magnetic field. It was shown in Fig. 2.5, that the spatial scale where amplification takes place is of the order of the clump size (though it specifically depends on the clump age). The timescale over which electrons lose energy is $t_{\rm sync} \simeq 12.5(B/{\rm mG})^{-2}({\rm E/TeV})^{-1}$ yr. The typical energy of synchrotron photons is $E_{\rm sync} = 0.04(B/{\rm mG})({\rm E/TeV})^2$ keV, hence the energy loss time-scale at the observed frequency is

$$t_{\rm sync} \simeq 2.4 \left(\frac{B}{\rm mG}\right)^{-3/2} \left(\frac{E_{\rm syn}}{\rm keV}\right)^{-1/2} {\rm yr}\,. \tag{2.53}$$

For a density contrast of $\chi = 10^5$ (see Eq. (2.7)), MHD simulations show that amplification of magnetic field can bring the background field up to 10 times above the downstream value. Hence a mG magnetic field needed to explain the observed timescale would require a magnetic field in the downstream of $\sim$100 μG, which is a reasonable value if the CR induced magnetic field amplification is effective at the shock.

2.6.3 Resolving the Gamma-Ray Emission

Detailed morphological and spectroscopic studies of SNRs are among the highest priority scientific goals of the forthcoming Cherenkov Telescope Array [21, 24]. Despite its great potential, CTA will be not able to resolve the gamma-ray emission from individual clumps. For an SNR at a distance of 1 kpc, the angular extension of a clump with a typical size of $\sim$0.1 pc does not exceed 20 arcsec, which is one order of magnitude smaller than the angular resolution of CTA. Nevertheless, the component related to the superposition of gamma-ray emission from several clumps aligned along the line of sight in principle could be detected. Hence, one should estimate the number of clumps $dN_{\rm p}/d\rho$ which overlap along the line of sight l, when

Table 2.1 CTA map of the SNR: first column gives the region number, second column is center of the observation, third column is the lower value in ρ, fourth column is the upper value in ρ, last column is the number of clumps included in each observation

Region	ρ	from	to	N_c
0	0	$-\sigma_{\rm CTA}$	$\sigma_{\rm CTA}$	1.0
1	$5\sigma_{\rm CTA}$	$4\sigma_{\rm CTA}$	$6\sigma_{\rm CTA}$	1.1
2	$10\sigma_{\rm CTA}$	$9\sigma_{\rm CTA}$	$11\sigma_{\rm CTA}$	1.4
3	R_{cd}	$R_{cd} - \sigma_{\rm CTA}$	$R_{cd} + \sigma_{\rm CTA}$	2.6
4	R_s	$R_s - \sigma_{\rm CTA}$	$R_s + \sigma_{\rm CTA}$	0.6

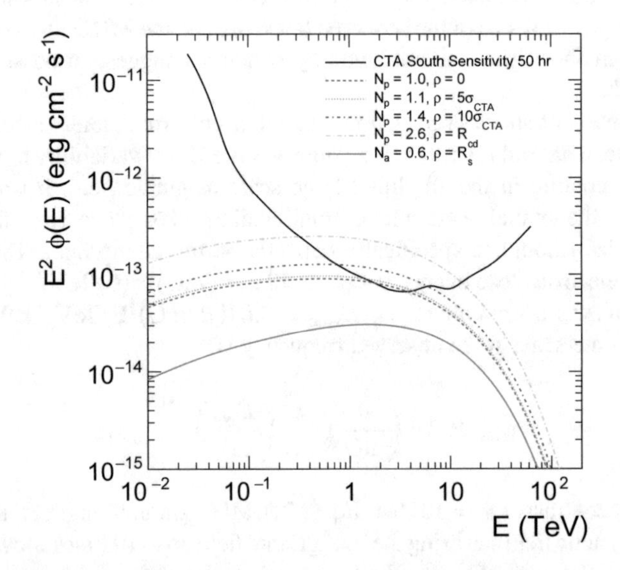

Fig. 2.16 CTA Southern array sensitivity curve for point-like sources located in the center of the field of view (zenith $\theta = 20$ deg, pointing average) for an observation time equal to 50 h (black solid line). Also shown are the gamma-ray fluxes due to the overlapping clumps (indicated in the legend) in different circular sky regions of radius $\sigma_{\rm CTA}$, located at a distance ρ from the SNR center. Figure from Celli et al. [79], reproduced by permission of the Royal Astronomical Society

observing the annular region extending from a distance $\rho = \sqrt{r^2 - l^2}$ up to $\rho + d\rho$, with respect to the center of the SNR. Assuming a uniform distribution of clumps, as described in Sect. 2.5, and integrating it along the line of sight, one obtains

$$\frac{dN_{\rm p}}{d\rho}(\rho) = 2n_0\left(\sqrt{R_{\rm s}^2 - \rho^2} - \max\left[0, \sqrt{R_{\rm cd}^2 - \rho^2}\right]\right) \tag{2.54}$$

where the emission is assumed to come from the shocked ISM located between the contact discontinuity and the forward shock [180]. Considering a uniform map of the whole remnant by CTA, the gamma-ray flux is computed from several circular regions centered at a given ρ and with radius equal to $\sigma_{\mathrm{CTA}} = 0.037\,\mathrm{deg}$[1] (corresponding to the instrument point spread function at $E_\gamma = 10\,\mathrm{TeV}$) [22]. The number of overlapping clumps at the different radial distances ρ from the SNR center is reported in Table 2.1, where ρ is considered to be spanning from 0 to R_s. Moreover, one should take into account the different ages of clumps, since they produce gamma rays with different spectral shapes, as shown in Sect. 2.4.3. These fluxes are represented in Fig. 2.16, where also shown is the sensitivity curve of the CTA Southern array, for a 50 h observation of a point-like source located at the center of the instrument field of view. The predicted flux clearly shows that CTA will be able to resolve the gamma-ray emission from clumps contained in a circle of radius equal to its high-energy point spread function (PSF) over about one decade in energy. However, the gamma-ray fluxes expected at different pointing positions strongly reflect the number of overlapping clumps, which represent the main contributors to the emission. Such a number is maximum in correspondence of $\rho = R_{\mathrm{cd}}$, where $N_{\mathrm{p}} = 2.6$. Given the limited number of overlapping clumps in each pointing region, large fluctuations are expected, according to the Poissonian statistics. Therefore, the detection of such fluctuations constitutes a characteristic signature of the presence of clumps. The amount of the fluctuations depends on the clump density n_0 in the CSM. In fact, once the mass of the target gas is fixed, more massive clumps with $n_c = 10^4\,\mathrm{cm}^{-3}$ would require a lower clump density and therefore would produce much stronger fluctuations on the scale of σ_{CTA}. These kinds of morphological studies are hence crucial to derive constraints on the number density of clumps in the remnant region. Large fluctuations on the scale of σ_{CTA} are not expected if the SNR is expanding into a uniform medium or into a medium where the density contrast is such that the clump evaporates soon after the shock crossing, namely if $\tau_{\mathrm{ev}} \ll T_{\mathrm{SNR}}$. The latter condition can be rearranged using Eqs. (2.8) and (2.9) to give an upper limit for the density contrast which reads $\chi \ll (T_{\mathrm{SNR}} v_s / 2R_c)^2 \simeq 10^3 (R_c/0.1\,\mathrm{pc})^{-2}$. Such a small density contrast also implies a contained amplified magnetic field and, as a consequence, a flattening of the gamma-ray spectrum.

[1] http://www.cta-observatory.org/science/cta-performance/.

Chapter 3
Particle Escape from Supernova Remnants

Understanding the escape mechanism of accelerated particles is a key ingredient to establishing a connection between SNRs and the origin of CRs. In fact, it is often believed that the spectrum of particles released into the Galaxy by a single SNR corresponds to the instantaneous spectrum accelerated at the shock. Such conclusion depends on several subtleties of the acceleration process, i.e. (i) the amount of time that particles spend inside the SNR, during which they would suffer severe adiabatic losses, (ii) the rate at which particles are released from the SNR, which depends on the intensity level of magnetic turbulence, (iii) how does the acceleration efficiency vary in time during the remnant evolution. In the approximate scenario where particles are confined inside the remnant until it dissolves into the ISM, these would lose a substantial fraction of their energy because of the adiabatic expansion of the shocked plasma; hence the requirement to reach PeV energies in the released spectrum becomes more severe than it already is, in that the maximum energy achievable during the acceleration should be $\gg$PeV. This scenario does not appear to be a realistic one. On the other hand, it appears unavoidable that, as the shock slows down due to the accumulated mass, particles start to diffuse away from the shock and the probability that they might return to the shock from upstream is reduced. Thus, CRs are free to escape since the shock is no longer able to confine them inside the expanding shell.

Unfortunately, a comprehensive theory of the escape process is still missing, due to the high non linearity of the process. In fact, in the current formulation of DSA, the turbulence needed to confine CRs around the shock discontinuity is thought to be provided by the very same particles: these are supposed to induce plasma instabilities (see reviews by [71, 98]), the most relevant being the resonant [218] and the non-resonant [59] streaming instability. In particular, the latter is thought to dominate during the initial stage of the blast wave expansion, when its speed is $\gg 1000$ km/s,

Part of this chapter has already been published in Celli S., Morlino G., Gabici S. & Aharonian F., 'Exploring particle escape in middle-aged supernova remnants through gamma rays', Monthly Notices of the Royal Astronomical Society, vol. 490, Issue 3, p. 4317-4333 (2019), and it is here reproduced by permission of the Royal Astronomical Society.

S. Celli, *Gamma-ray and Neutrino Signatures of Galactic Cosmic-ray Accelerators*, Springer Theses, https://doi.org/10.1007/978-3-030-33124-5_3

while the former should dominate at later times. The escape of particles onsets when the particle-confining turbulence starts to fade out. Now, because the turbulence intensity level depends on the CR energy density which is, in turn, influenced by the shock speed, one would expect the escape process to be a function of the shock speed. It is also worth mentioning that the non-resonant instability requires a net current to be triggered, which would be provided by the same escaping particles. Moreover, as the return current, that is established in the plasma to restore the charge neutrality once particle escape has started, depends on the density of the CSM and on the shock speed, the conditions to achieve PeV energies are most likely satisfied only during the first few decades following the explosion of stars occurring in dense CSM winds, as those typical in type II SN explosions [61, 75]. On the contrary, type Ia SNe as well as core-collapse SNe expanding into wind-excavated bubbles should be able to produce only $\sim$100 TeV protons, as they occur in a less dense ISM.

Qualitatively, one can envision the escape mechanism as a two step process. In the free expansion phase, when the shock speed is approximately constant, the maximum momentum is expected to increase with time: during this stage, only a fraction of particles at the maximum energy is able to escape at each time, providing the turbulence able to scatter back to the shock particles with slightly smaller energy.[1] When the SNR enters the Sedov-Taylor (ST) phase and the shock speed starts to slow down significantly, both the shock acceleration efficiency and the intensity level of turbulence decrease, and the particle confinement is reduced: hence, the particles at the maximum energy start escaping. In this phase, only the resonant instability is important and the maximum energy decreases with time. In fact, even though particles continue to amplify the turbulence they scatter off, this eventually becomes not enough for self-confinement. This description is further complicated when accounting for the processes which damp the amplified magnetic waves, like the MHD cascade or the ion-neutral friction (valid when the plasma is not fully ionized), which would cause an even faster escape rate of particles.

Beyond the theoretical description of the escape process, one could legitimately wonder whether it is possible to test such a scenario by means of observations, particularly in the HE and VHE gamma-ray domain. The answer to this question mainly depends on the transport regime of particles outside of the remnant. If the diffusion coefficient immediately outside of the shock region is at the level of the average Galactic one, particles would escape immediately far away from the SNR. In this case, assuming a Galactic diffusion coefficient of $D_{\mathrm{Gal}}(p) = 10^{28}(p/10\ \mathrm{GeV/c})^{1/3}\ \mathrm{cm^2\ s^{-1}}$, the time taken by the run-away particles to reach a distance 100 pc away from the shock position would be

[1]Actually, the description of the free-expansion stage is more complicated: on one hand, the shock speed is not constant but it slowly decreases with time [85], suggesting that the maximum momentum should also decrease with time; on the other hand, non-linear magnetic field amplification should increase the maximum energy up to $\sim$1 PeV during a fraction of the free-expansion phase. Nevertheless, the behavior of particle escape during this phase is not particularly relevant because it has a marginal impact on the final gamma-ray spectrum.

$$\tau_{\rm diff} = \frac{L^2}{6D_{\rm Gal}(p)} \simeq 4.8 \times 10^3 \left(\frac{L}{100\ \rm pc}\right)^2 \left(\frac{p}{10\ \rm TeV/c}\right)^{-1/3}\ \rm yr \quad (3.1)$$

namely a fraction of a middle-aged remnant lifetime. Nevertheless, this simplified picture has been challenged by some recent works [93, 106, 167]: in fact, escaping particles might continue to excite the magnetic instability up to a distance from the source where their density drops below the one of the Galactic CR 'sea'. The results obtained by these authors show that, in the absence of ion-neutral damping, a suppression of the diffusion coefficient by about two orders of magnitude is achieved in a region extending $\sim$20 pc from the shock. Hence, one would naively expect a transition region between the SNR shock and the region far away, where the diffusion coefficient gradually increases from the shock level towards the Galactic value. On the other hand, if a relevant fraction of neutral hydrogen is present around the source, the suppression of the diffusion coefficient is expected to be less prominent [186].

If the diffusion coefficient is suppressed, escaping particles diffuse around the remnant for a time longer than Eq. (3.1). Two relevant observational consequences can be envisioned. A first effect is that particles located outside of the SNR and interacting with target gas would produce a diffuse gamma-ray halo [93]. At the moment, such halos have not been detected yet around SNRs, though they have been observed around young pulsars, as Geminga [18]. Intriguingly, future observations of TeV halos might potentially lead to the discovery of new sources: to this extent, next-generation telescopes (as CTA) might eventually be able to detect single halos. Even if the observation of individual halos will remain a challenging target for pointing telescopes, the overlapping from many of these sources along the single line of sight across the Galactic Plane could provide a non negligible contribution to the diffuse Galactic gamma-ray emission, as described by [93]. The second relevant consequence is that, during the diffusion process, escaping particles can occasionally re-enter the SNR interior even if their energy is large enough that they do not experience the shock discontinuity anymore (hence they would not be affected by the acceleration process). Nevertheless, once inside, they can contribute to the gamma-ray emission through hadronic interactions. In this scenario, a spectral break is expected in the gamma-ray spectrum above the energy corresponding to the maximum energy achievable by particles accelerated at the remnant age. The latter effect will be the subject of this chapter. In particular, the particle escape will be modeled by assuming that the maximum momentum is an arbitrary function of the shock speed and the consequence on the resulting gamma-ray emission from the remnant interior will be analysed.

The following discussion will be limited to middle-aged SNRs, mainly because of two reasons: (i) the amount of escaping particles should be large enough to produce more evident observational effects, and (ii) the remnant evolution can be well approximated by the ST model, which allows to provide a simple analytical model for the description of particle propagation. Therefore, the treatment presented does not apply to young SNRs that are still evolving in the ejecta-dominated phase. However, escape of very energetic particles could be relevant even in this kind of systems. For instance, the bright SNR RX J1713.7-3946 (presented in Chap. 2) is believed to be nowadays undergoing the transition towards the ST phase. Thus, PeV particles could have already escaped the shock and possibly be interacting with the surrounding molecular clouds [76]. As that, the recent identification of a VHE gamma-ray shell extending beyond the X-ray shell [12] might be a signature of particles escaping the shock. Note that the presence of a large target density would enhance the gamma-ray emission and consequently the corresponding VHE halo might be potentially observable even as an individual source by the next-generation instruments. However, a different treatment than that presented in the following is required to describe the particle escape from young SNRs, since the time-dependency of the maximum momentum in such a case is very uncertain.

The chapter is structured as follows: Sect. 3.1 introduces SNRs and their temporal evolution, in order to characterize the relevant evolutionary phase for escape to be efficient within this class of accelerators. The Sedov-Taylor phase is identified as the most relevant in terms of particle escape. Then, in Sect. 3.2 several viable assumptions on the maximum momentum evolution are presented. In particular, in Sect. 3.2.1 a time-limited escape is first introduced, namely a situation where the escape is limited to particles beyond a maximum momentum that depends on time, $p_{\max}(t)$. Alternatively, a spatially-limited escape might be assumed, as described in Sect. 3.2.2 following the results from [73]. The spatial boundary where the particle distribution function vanishes, namely where particles can escape the system, is basically a surface located at some position upstream of the shock $z_0 = \chi R_s$ (with $\chi \sim 0.1$), regardless of the particle energy [71]. Such a boundary is justified by the fact that the CRs themselves are supposed to amplify the magnetic field at the shock and produce the turbulence they need to be scattered off: hence one can generally envision a situation where far enough from the shock the self-generated turbulence is no more sufficient to confine high-energy particles. Lastly, Sect. 3.2.3 presents a scenario where the temporal dependence of the maximum momentum is connected to the excitation of the non-resonant instability by CRs, as discussed by [75]. As spatially-limited escape models provide a steady maximum momentum, while gamma-ray spectral observations of middle-aged SNRs show a maximum momentum generally lower than that ascribable to young SNRs, one would be naturally lead to consider a scenario with a time-limited maximum momentum as a more appropriate description of reality. Hence, Sect. 3.3 introduces a phenomenological description of such an escape scenario, through the solution of the temporal and spatial dependent CR transport equation applied to extended sources. The content of this section represents an original derivation by the author, aimed at characterizing the VHE emission spectra of middle-aged SNRs.

Few words of caution are mandatory: in order to describe self-consistently a situation where the escape is limited to particles beyond a maximum momentum that depends on time $p_{\max}(t)$, the knowledge of the diffusion coefficient is required, which in turn is determined by the level of magnetic turbulence generated by the accelerated particles themselves. In fact, the value of $p_{\max}$ depends on some crucial but barely known aspects of the problem, namely the nature of the CR-driven instability and the level of wave damping. In order to characterize these aspects, a non-linear approach should be adopted, which is however beyond the intent of this work. A linear solution assuming a steady diffusion coefficient is instead presented. This method allows to:

(i) estimate the contribution of the run-away particle flux from SNRs to the GCR flux, as evaluated and discussed in Sect. 3.4, and (ii) interpret the GeV-TeV data collected from several middle-aged SNRs (specifically IC 443, W 51C and W 28N), where escape is believed to be relevant, as explained in Sect. 3.5. In particular, the latter modeling is achieved provided that a diffusion coefficient smaller by a factor $\sim$10–100 than the average Galactic one is effective close to these sources. Hence, the direct comparison between the model and gamma-ray observations provides constraints to the temporal evolution of the maximum momentum in specific sources. Thanks to the advent of CTA, a statistical approach based on a population study of middle-aged SNRs might be attempted in the future, in order to limit the parameter space of the model and infer the properties of particle escape in this class of accelerators.

3.1 The Temporal Evolution of SNRs

A supernova remnant results from the interaction of ambient gas with stellar material ejected by a supernova. As it is often difficult to establish the SN origin of an SNR, as either connected to thermonuclear or core-collapse SNe, SNRs have a different classification which is mostly based on their morphology. This comprises three categories: *shell type SNRs*, *plerions* and *composite SNRs*. In the first case, the SNR is characterized by a limb bright shell, associated to the shocked plasma that has been heated behind the blast wave (see Appendix A). On the other hand, plerions have a bright center but do not show a shell: in such systems (as the Crab Nebula for instance), the emission is powered by the pulsar wind and not by the supernova explosion. In the composite case, a pulsar nebula is surrounded by an SNR shell.

The evolution of the interaction among the ISM gas and the ejecta can be characterized in terms of several distinct stages [248]: (i) the ejecta-dominated phase (ED or free-expansion), in which the mass of the SN ejecta prevails on the swept-up mass; (ii) the Sedov-Taylor phase (ST or adiabatic), where the swept-up mass becomes larger than the ejecta mass while radiative losses are still not significant; (iii) the pressure-driven snowplow phase (PDS or radiative), in which radiative cooling becomes energetically relevant; and (iv) the merging phase, where the temperature behind the shock and the shock speed become comparable to the ISM values. The particle acceleration is believed to be more efficient during the initial stages of the remnant evolution, when the expansion proceeds at high speed. The ED stage starts with the explosion of the stellar progenitor and the subsequent expansion of the ejecta: as the expansion velocity of the ejecta is much larger than the sound speed in the ambient gas, these are anticipated by a shock wave, also called the *blast-wave shock*. The main result due to the presence of such a shock on the ambient medium is that it will result accelerated, compressed and heated. In turn, the shocked ambient medium will push back on the ejecta, producing the effect of decelerating, compressing and heating them. At the same time, a second shock wave, called the *reverse shock*, is generated since the shocked ambient medium stars to expand back into the ejecta much faster than the sound speed in the ejecta. A *contact discontinuity* will separate the cold, high density, shocked ejecta from the hot, low density, shocked ambient medium. The unshocked ejecta freely expand until they are hit by the reverse shock, which accelerates the ejecta inward. As a consequence, the post-shock ejecta will retain a net outward velocity, lower than the pre-shock value. To summarize, assuming a spherically symmetrical system, for increasing values of the radial coordinate one will encounter, in this order, the unshocked ejecta, the reverse shock, the shocked ejecta, the contact discontinuity, the shocked ambient gas, the forward shock and the unshocked ambient gas.

It is clear that the reverse shock is the agent that communicates the existence of the ambient medium to the ejecta, and it is responsible for the initial deceleration of the ejecta. In the early-time limit of the ED stage, before the reverse shock has attained a significant velocity, the ejecta behave approximately as a spherical piston freely expanding into the ambient gas. At the beginning of the ST stage, after the majority of the ejecta energy has been transferred to the ambient gas, the flow is an adiabatic blast wave. In both limits, the system evolves self-similarly [237]: these cases are called ' intermediate asymptotic' [230]. The passage towards the ST phase is achieved when the condition that the mass in the ejecta $M_{\rm ej}$ equals the mass swept by the remnant is satisfied. From this condition, it is possible to determine the radius $R_{\rm Sed}$ at which the ST stage starts: indeed, for an explosion in a constant density ISM $\rho_{\rm ISM}$, this condition reads as

$$\frac{4}{3}\pi\rho_{\rm ISM}R_{\rm Sed}^3 = M_{\rm ej} \implies R_{\rm Sed} = \left(\frac{3M_{\rm ej}}{4\pi\rho_{\rm ISM}}\right)^{1/3} \tag{3.2}$$

As a consequence, the Sedov time of a remnant expanding into a homogeneous medium reads as

$$t_{\rm Sed} = \frac{R_{\rm Sed}}{V_{\rm ej}} = 0.495\, E_{\rm SN}^{-1/2} M_{\rm ej}^{5/6} \rho_{\rm ISM}^{-1/3} \tag{3.3}$$

where $V_{\rm ej}$ represents the velocity of the ejecta (a quarter of the shock speed), $E_{\rm SN}$ is the kinetic energy released by the SN and the numerical multiplicative factor 0.495 was obtained in [230] as a continuity solution among the two asymptotic regimes of the free-expansion and deep Sedov-Taylor phase of the remnant hydrodynamical evolution. For characteristic values of $E_{\rm SN} = 10^{51}$ erg, $\rho_{\rm ISM} = 1\,(m_{\rm p}/{\rm g})\,({\rm g/cm^3})$ and $M_{\rm ej} = 1\,M_\odot$, one obtains $R_{\rm Sed} \simeq 2$ pc and $t_{\rm Sed} \simeq 200$ yr. Note that the peculiar conditions realized around the supernova explosion might significantly alter the remnant dynamical evolution. For instance, in the case of a fast wind excavating a low density environment (such as would be the case for Wolf–Rayet pre-supernova star), the SN explosion might take place in a bubble of hot and dilute gas, delaying the onset of the adiabatic phase. For this reason, the values reported above should be considered merely as order of magnitude estimates.

It is possible to derive the maximum energy that a particle can reach at the shock by equating the acceleration time $t_{\rm acc}$ to the age of the SNR, $t_{\rm age}$. If particles are accelerated at the shock via diffusive shock acceleration [97], then the acceleration time reads as $t_{\rm acc} \sim D/v_{\rm s}^2$ (see Eq. (1.37)), namely it will depend on the diffusion coefficient and on the shock speed. Concerning the former, one can assume that Bohm diffusion applies at the shock, hence $D \propto E/B$ (where E is the particle energy and B the magnetic field strength, here considered to be constant in time). Concerning the latter, if both the ejecta and the ambient medium are uniform in density, then the forward shock speed is found to slowly decrease with time during the ED stage [84]. However, it can be described without significant loss of accuracy as a constant speed shock, according to

$$\begin{cases} R_s(t) \simeq v_{Sed} t \\ v_s(t) \simeq \text{const} = v_{Sed} \end{cases} \tag{3.4}$$

where v_{Sed} refers to the value that the shock speed assumes in correspondence of the Sedov time. Consequently, one derives a particle maximum energy of $E_{max} \propto t_{age}$. This simple reasoning shows that, during the ED phase, the maximum energy increases linearly with time. Note that the highest energy CRs diffuse ahead of the shock one diffusion length $\lambda \sim D(E_{max})/v_s$ before returning to it when advected downstream. This length is generally some fraction $\chi \simeq 0.1$ of the SNR radius $R_{SNR} = v_s t_{age}$. Thus, by adopting a Hillas-like criterion for CR confinement in the SNR, namely by setting the diffusion length comparable to the SNR size, one would conclude that all the accelerated particles are effectively confined and the escape process is ineffective. Actually, CR escape may occur also during the free-expansion phase [205] due to the slow decrease of the shock velocity with time. However, the number of escaping particles is negligible during the ED phase, as the number of particles involved in the shock acceleration is relatively low at this stage.

On the other hand, during the ST stage, the forward shock evolves into a uniform medium following

$$\begin{cases} R_s(t) = \left(\frac{\xi_0}{\rho_{ISM}} E_{SN}\right)^{1/5} t^{2/5} \\ v_s(t) = \frac{2}{5}\left(\frac{\xi_0}{\rho_{ISM}} E_{SN}\right)^{1/5} t^{-3/5} \end{cases} \tag{3.5}$$

where the dimensionless constant ξ_0 depends on the plasma adiabatic index: $\xi_0 = 2.026$ for a non-relativistic monoatomic gas with $\gamma = 5/3$ [214, 227]. The ST solution can be generalized for a gaseous medium with a power-law density profile $\rho(r) \propto r^{-s}$: $R_s \propto t^{\beta}$ and $v_s \propto \beta R_s/t$, with $\beta = 2/(5-s)$. A relevant case is that with $s = 2$, corresponding to an SNR shock moving into the progenitor stellar wind, where the blast wave would evolve with $\beta = 2/3$. The ST solution does not take into account the structure of the SN ejecta itself, thus it represents a good approximation to the stage when the swept-up mass has exceed largely the ejecta mass. An analytical model that describes the structure and evolution of SNRs accounting for the velocity profile of the ejecta is provided by [84], where the early evolution of the SNR with freely expanding ejecta is described. The transition among the two regimes, from the ED to the ST stage, can be smoothly parametrized by means of an analytical formulation derived by [230] (see Fig. 3.1). Once the SNR has entered the Sedov phase, the particle diffusion length will increase in time as $\lambda \sim D(E_{max})/v_s \sim t^{3/5}$, i.e. faster than the evolution of the SNR radius (scaling like $R_s \sim t^{2/5}$). Therefore, particles

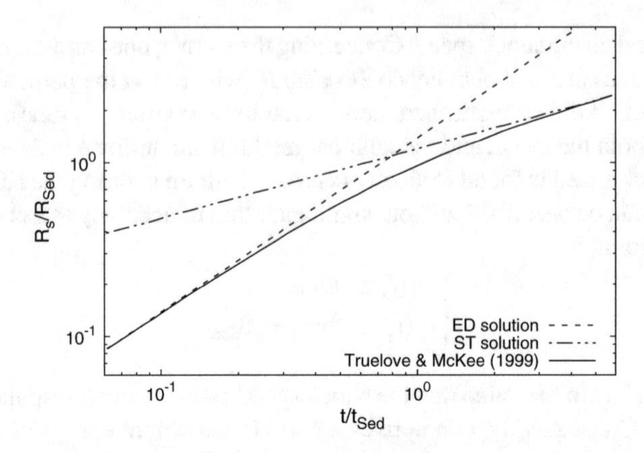

Fig. 3.1 Position of the blast-wave shock versus time for uniform ejecta expanding into a uniform ambient medium. The ejecta evolve non self-similarly from the self-similar limits of free expansion ($R_s \propto t$) in the early ED stage to the power-law decelerating expansion ($R_s \propto t^{2/5}$) in the ST stage. R_{Sed} and t_{Sed} refer to the Sedov radius and time, respectively. Figure from Truelove & McKee [230] ©AAS. Reproduced with permission

with energy E_{max}, which were formerly confined during the ejecta-dominated phase, will start to violate the Hillas criterion, thus escaping the SNR. Therefore, the ability of the remnant to confine particles is reduced by the decrease in time of the shock velocity [116].

After the ST stage, the blast-wave shock eventually decelerates to the point that it becomes radiative: this occurs once the shock has slowed down to $v_s \simeq 200\,\mathrm{km/s}$, for downstream temperature of $\sim 5 \times 10^6$ K [239]. A thin shell of radiatively cooled shocked ambient medium forms behind the blast-wave shock. The pressure of the hotter shocked ambient gas will drive this shell deeper into the interior of the remnant. At this point, the remnant enters in the PDS phase: consequently, the shock radius will evolve with time as $R_s(t) \propto t^{1/4}$. The energy losses become dynamically important, and momentum conservation (rather than energy conservation) governs the hydrodynamical evolution of the shock. However, it is worth mentioning that efficient CR acceleration might alter such an evolution, as it affects the equation of state and the energy loss process (in that particle escape is realized).

3.2 Time Evolution of the Maximum Energy

As mentioned above, the maximum momentum p_{max} of accelerated particles is determined by the confinement condition, namely that the diffusion length λ of the particles does not exceed the characteristic size of the system R_s:

$$\lambda = \frac{D(p_{max})}{v_s} \leq R_s \tag{3.6}$$

If diffusion operates in the Bohm regime, and particles are subject to a magnetic field B_s at the shock, then $D(p_{\max}) \propto p_{\max}/B_s$. During the ST phase one would expect $p_{\max} \propto t^{-1/5} B_s$. Actually, the drop of $p_{\max}$ with time is expected to be even faster, in that also the magnetic field is expected to decrease in intensity with time. Thus a parametrization in the form of $p_{\max} \propto t^{-\delta}$ provides a general description of the phenomenon, as will be discussed in Sect. 3.2.1. However, few other possibilities can be exploited in the parametrization of $p_{\max}$. For instance, particle escape can be assumed in correspondence of a predefined position in the shock upstream: despite introducing an abrupt cut in the particle density function, which likely does not provide an accurate description of reality, such assumption naturally guarantees the existence of a maximum momentum in the particle spectrum. This approach will be discussed in Sect. 3.2.2. Alternatively, a third model of the temporal evolution of $p_{\max}$ is investigated in Sect. 3.2.3, where the current of escaping particles is considered as responsible for the excitation of the non-resonant instability, which in turn leads to the formation of resonant modes. As these modes are able to confine particles close to the shock, the maximum energy results affected by the development of these instabilities. As already mentioned, the non-resonant modes should dominate the amplification of MHD waves driving the initial evolution of the SNR ($t \leq 100$ yr). Hence, one would not expect that this mechanism provides the correct description of the escape process during the ST phase.

3.2.1 *Time-Limited Maximum Energy*

Within a time-limited scenario for the maximum energy, one should consider all the different evolutionary stages of a remnant lifetime. As introduced above, during the ED phase, the maximum momentum at the shock location $p_{\max,0}(t)$ is expected to grow linearly with time as

$$p_{\max,0}(t) = p_{\mathrm{M}} \left(\frac{t}{t_{\mathrm{Sed}}} \right) \qquad t < t_{\mathrm{Sed}} \tag{3.7}$$

where p_{M} represents the maximum value of the momentum at the beginning of the ST phase. Later, during the ST phase, the maximum momentum is expected to decrease with time. In this section, this decrease is assumed to follow a power-law dependency with slope $\delta > 0$, as

$$p_{\max,0}(t) = p_{\mathrm{M}} \left(\frac{t}{t_{\mathrm{Sed}}} \right)^{-\delta} \qquad t \geq t_{\mathrm{Sed}} \tag{3.8}$$

With such a definition, the escape time of particles with momentum p in the ST phase is

$$t_{\mathrm{esc}}(p) = t_{\mathrm{Sed}} \left(\frac{p}{p_{\mathrm{M}}} \right)^{-1/\delta} \tag{3.9}$$

satisfying the condition that high-energy particles escape at earlier times. From Eq. (3.9), the escape radius cam be introduced as

$$R_{\rm esc}(p) = R_{\rm s}(t_{\rm esc}(p)) \tag{3.10}$$

A simple theoretical argument to estimate the value of δ in the test-particle DSA scenario consists into equating the remnant age to the acceleration time $t_{\rm acc} \sim D/v_{\rm s}^2$. Then, by setting the diffusion coefficient as $D = D_{\rm B}\mathcal{F}^{-1}$, with $D_{\rm B} = r_{\rm L}(p)v(p)/3$ corresponding to the Bohm diffusion and $\mathcal{F}$ for the turbulent magnetic energy density, one obtains

$$p_{\rm max,0}(t) \propto \mathcal{F}(t)v_{\rm s}^2(t)t \tag{3.11}$$

If there is no amplification of the magnetic field and diffusion depends only on the pre-existing magnetic turbulence (which does not depend on time), then the only time dependence resides in the shock speed: during the ST phase, one should expect $\delta = 1/5$ as a minimum value. On the other hand, if the turbulence is amplified by the resonant streaming instability, then $\mathcal{F}(t) \propto P_{\rm CR} \propto v_{\rm s}^2(t)$, leading to $\delta = 7/5$. Finally, if the turbulence relevant for particle scattering is amplified through non-resonant instability, then $\mathcal{F}(t) \propto v_{\rm s}(t)P_{\rm CR} \propto v_s^3(t)$, leading to $\delta = 2$. Moreover, if some damping mechanism, like MHD cascade or ion-neutral friction, is effective in damping waves, an even larger value of δ is foreseen. Constraints on the parameter δ can be derived through gamma-ray spectral observations of middle-aged SNRs, as described in Sect. 3.5. In this respect, future observations of CTA will be very valuable, as they will allow to perform population study of SNRs, providing hints on the escape process dynamics within this class of objects.

3.2.2 *Space-Limited Maximum Energy: The Escape Boundary*

As anticipated, the consideration of a spatial escape boundary is justified by the fact that CRs themselves are supposed to amplify the magnetic field at the shock and generate the turbulence they need to scatter off via resonant [158, 218] or non-resonant [59] streaming instability. In fact, depending on the intensity of the CR flux at a given position in the space, the level of the magnetic turbulence is expected to decrease with the distance upstream of the shock. This suggests the existence of a spatial region beyond which CRs cannot be effectively scattered and become free to escape: in the following, the limit of this region will be referred to as a spatial boundary.

By introducing a spatial boundary for particle escape, it is possible to determine the shape of the cut-off in the spectrum of escaping particles, as this is no longer a delta-function in energy, like in the case of a momentum-limited escape boundary. Such a feature is crucial in order to investigate the radiative signatures related

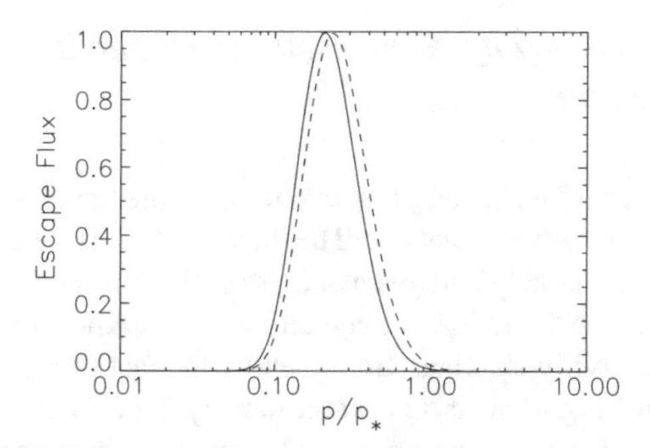

Fig. 3.2 Escape flux, in the test-particle regime, as a function of momentum. The curves refer to two different values of the shock compression ratio: $r = 4$ (solid line) and $r = 7$ (dashed line). The x-axis is in units of the reference momentum p^* (see text), while units along the y-axis are arbitrary. Figure reproduced from Caprioli et al. [73] by permission of Oxford University Press on behalf of the Royal Astronomical Society

to CR acceleration at shocks. Moreover, the spatial boundary allows to check—a posteriori—the consistency among the assumed position of the escape boundary and the value of the amplified magnetic field (and consequently of the diffusion coefficient). The spectrum of the escaped particles can be derived as a solution of the one-dimensional stationary transport equation: by imposing a free escape boundary upstream of the shock, where the distribution function vanishes, [73] derived that it exists a characteristic momentum p^*, which depends on the location of the boundary, such that

- for $p < p^*$, the distribution function at the shock front is a power law in momentum with slope $\alpha \equiv 3r/(r-1)$. This is the standard result of the test-particle approximation (see Eq. (1.34)). For a compression ratio $r = 4$, then $f_0(p) \propto p^{-4}$.
- for $p > p^*$, the distribution function at the shock front contains an exponentially-suppressed term. In particular, the harder is the acceleration spectrum and the higher is the maximum momentum that particles are able to reach.

As a consequence, the flux of escaping particles is found to be narrowly peaked around the momentum p^*, as represented in Fig. 3.2. This implies that only very energetic particles are allowed to leave the system, since these cannot diffuse back to the shock and thus get lost in the ISM. It is clear that, given the stationarity hypothesis assumed to solve the transport equation, the space-limited escape approach predicts a maximum momentum that is constant with time and a spectral slope which is also constant. However, middle-aged SNRs (10^4 to 2×10^4 yr old) show spectra which are systematically steeper than younger remnants (10^2–10^3 yr old). In order to account for this, it is necessary to adopt a time-dependent approach in deriving the solution of the CR transport equation.

3.2.3 Escape Limited by Non-resonant Magnetic Amplification

An alternative scenario for the temporal evolution of the maximum momentum was presented in [75]. The authors considered the possibility that the escaping CRs excite magnetic instabilities, namely both resonant Alfvénic waves and non-resonant modes [59], so as to achieve both efficient magnetic field amplification and particle scattering. The two instability channels are driven by the fact that CRs stream at super-alfvénic speed, carrying an electric current density $\mathbf{J}_{\rm CR}$. By doing so, they induce a reaction in the background plasma, that is a return current composed of moving electrons aimed at restoring a null net current. Hence, in both the instability channels, the $\mathbf{J}_{\rm CR} \times \mathbf{B}$ force drives the motions of the background fluid. The main difference between the resonant and non-resonant linear instabilities is that while the collective effect of CRs generates non-resonant modes (namely their strong drift), individual CRs are rather responsible for resonant modes. In order to better explain this point, let us consider the vectors $\mathbf{J}_{\rm CR}$ and $\mathbf{B}$ as composed by unperturbed zero-th order components, $\mathbf{J}_{\rm CR,0}$ and $\mathbf{B}_0$, and first order perturbed components, $\mathbf{J}_{\rm CR,1}$ and $\mathbf{B}_1$. In this scheme, the $\mathbf{J}_{\rm CR} \times \mathbf{B}$ force will have two first order components, namely $\mathbf{J}_{\rm CR,0} \times \mathbf{B}_1$ and $\mathbf{J}_{\rm CR,1} \times \mathbf{B}_0$. The resonant instability is driven by $\mathbf{J}_{\rm CR,1} \times \mathbf{B}_0$, where $\mathbf{J}_{\rm CR,1}$ is particularly enhanced if the CRs react resonantly to a magnetic field with wavelength equal to their Larmor radius (as previously discussed in Chap. 1). On the other hand, the non-resonant instability is driven by the other first order contribution to the force, that is $\mathbf{J}_{\rm CR,0} \times \mathbf{B}_1$. In this case, only the zero-th order current $\mathbf{J}_{\rm CR,0}$ is relevant and there is no requirement for a resonance with the CR Larmor radius [213]. Since the return current induced in the background plasma by the CR streaming is carried by electrons, it will develop instability modes on small scales. Note that the CRs will have a Larmor radius much larger than the wavelength of the spiral perturbations in a zero-th order uniform magnetic field, and consequently their streaming will be basically undeflected by the perturbed field. However, as the $\mathbf{J}_{\rm CR} \times \mathbf{B}$ force acts towards the centre of the spiral, the corresponding reactive force will act on the background plasma by expanding the spiral. As a result, the perturbed magnetic field will be stretched and strengthen, and consequently the $\mathbf{J}_{\rm CR} \times \mathbf{B}$ force will increase in a positive feedback loop that drives the instability. For high-speed shocks, the growth of these modes is so fast that the non-resonant instability occurs as the most rapidly growing instability driven by CR streaming.

By considering the non-resonant instability developed by the CR streaming from a remnant expanding into a homogeneous medium and by assuming that a constant fraction of energy is instantaneously transferred to the escaping particle flux, [75] derived the following implicit equation in the maximum energy $E_{\rm max}(t)$

$$\frac{E_{\rm max}(t)}{E_{\rm min}} \ln\left(\frac{E_{\rm max}(t)}{E_{\rm min}}\right) = \frac{e}{10cE_{\rm min}} \xi_{\rm CR} v_{\rm s}^2(t) R_{\rm s}(t) \sqrt{4\pi\rho_{\rm ISM}} \qquad (3.12)$$

holding for an E^{-2} acceleration spectrum, $E_{\rm min}$ being the minimum energy provided by in the acceleration mechanism (here after assumed equal to the proton rest mass energy). Its derivation is achieved by combining Eqs. (2) and (9) of [75] (by setting $m = 0$, corresponding to expansion into a homogeneous medium). In this case, the escape time of particles with energy E during the ST phase is dictated by the following equation

$$t_{\rm esc}(E) = \left[\frac{4e}{125c} \xi_{\rm CR} \sqrt{\pi \rho_{\rm ISM}} \left(\frac{\xi_0 E_{\rm SN}}{\rho_{\rm ISM}} \right)^{3/5} \frac{1}{E \ln(E/E_{\rm min})} \right]^{5/4} \tag{3.13}$$

However, for a spectrum like $E^{-(2+\beta)}$ (with $\beta \neq 0$), the equation regulating the temporal evolution of the maximum energy $E_{\rm max}(t)$ reads as

$$\frac{1+\beta}{\beta} \left(\frac{E_{\rm max}(t)}{E_{\rm min}} \right)^{1+\beta} \left[1 - \left(\frac{E_{\rm min}}{E_{\rm max}(t)} \right)^{\beta} \right] = \frac{e}{10c E_{\rm min}} \xi_{\rm CR} v_{\rm s}^2(t) R_{\rm s}(t) \sqrt{4\pi \rho_{\rm ISM}} \tag{3.14}$$

while the escape time is

$$t_{\rm esc}(E) = \left[\frac{4e}{125c} \frac{\xi_{\rm CR}}{E_{\rm min}} \sqrt{\pi \rho_{\rm ISM}} \left(\frac{\xi_0 E_{\rm SN}}{\rho_{\rm ISM}} \right)^{3/5} \frac{\beta}{1+\beta} \left(\frac{E_{\rm min}}{E} \right)^{1+\beta} \left(\frac{1}{1-(E_{\rm min}/E)^{\beta}} \right) \right]^{5/4} \tag{3.15}$$

This approach defines a maximum momentum which varies with time according to the remnant evolutionary stage: in the following, the shock radius and position will be considered in their evolution through the ST stage. The results are presented in the next section, where these equations are solved by combining a Newton-Raphson algorithm with a bisection method for the identification of the $E_{\rm max}(t)$ function roots. Note that these equations show explicitly the fact that the maximum energy depends on the acceleration efficiency, since the higher is the efficiency and the larger is the current of escaping particles.

3.3 A Simplified Model for Particle Propagation

This section is based on the assumption that a boundary in momentum regulates the efficiency of particle escape, namely at any given time the shock is able to confine only particles up to a certain maximum momentum $p_{\rm max}(t)$: efficient particle escape is achieved for particles with larger momenta, which are therefore no more subject to both the acceleration at the shock and the advection within the background plasma, since they can freely diffuse in both the shock upstream and downstream. This approach was also considered by [165, 205], who adopted a step-function in momentum in order to compute analytically the spectrum of run-away particles and quantify their contribution to the Galactic CR flux. Consequently, only narrow momentum distributions of particles at the escape time are considered. A more realistic situation should however account for broader particle spectra, requiring a numerical treatment to the escape problem. This will not be the object of this work, where rather an approach consistent with past computations will be adopted in order to explore the behavior of particle escape in the close vicinity of the SNR and investigate how the theoretical assumptions can be tested with gamma-ray observations.

Since the escape phenomenon is not instantaneous, the particles released by the shock may well be contained (but not confined) within the shock radius for some time after the escape time, depending on the diffusion coefficient operating there. In other words, particle escape starts when the turbulence at the scale of the resonance is so low that the diffusion length $\lambda \sim D(p)/v_s$ is increased to the level that it results larger than the characteristic system length scale. However, escaped particles are not obliged to reside outside of the remnant, since nothing is preventing them from diffusing inside of it, even though they do not 'belong' to the shock anymore. In fact, they can still scatter off some turbulence. In order to study this scenario, the CR transport equation is solved assuming spherical symmetry for the system evolution. An analogous scenario has been previously investigated by [121, 189, 190, 205, 212]: in all of these works, however, the main concern was the evaluation of the contribution of the run-away particle flux from SNRs to the GCR flux or alternatively to distant molecular clouds. For this purpose, the escape mechanism was considered instantaneous, without accounting for the fact that escaped particles might well be residing within the remnant for few escape times before they get released in the ISM. In the following, a description of the escape process including the contribution from these particles too is provided with the aim of: (i) describing the observed radiation within SNRs as the sum between the escaped and the confined flux, and (ii) evaluating the SNR contribution to the GCR flux. To this extent, the escape time $t_{\rm esc}(p)$ for particles with momentum p, as introduced in Eq. (3.9), represents the characteristic timescale regulating the process. Particle propagation is then divided into two distinct temporal phases:

- An initial stage, for $t < t_{\rm esc}(p)$, where all CRs with momentum $< p$ are confined within the shock: they contribute to what in the following is indicated as the *confined distribution function*, $f_{\rm conf}(t, r, p)$;
- A subsequent stage, for $t \geq t_{\rm esc}(p)$, where all CRs with momentum $\geq p$ are able to escape the shock: they contribute to what in the following is indicated as the *non-confined distribution function*, $f_{\rm esc}(t, r, p)$.

The distribution function of confined particles is obtained as a solution of a simplified transport equation, including advection and adiabatic losses [205]

$$\frac{\partial f}{\partial t} + \mathbf{v} \cdot \nabla f = \frac{p}{3}\frac{\partial f}{\partial p}\nabla \cdot \mathbf{v} \tag{3.16}$$

Here, the diffusion term is neglected because particles are assumed to be strongly tighten to the plasma. On the contrary, the distribution function of non-confined particles is solution of a simplified transport equation, including only diffusion

$$\frac{\partial f}{\partial t} = \nabla \cdot [D \nabla f] \tag{3.17}$$

where D is the diffusion coefficient upstream of the shock. In fact, since particles detached from the plasma, they are not advected anymore with it. Both Eqs. (3.16) and (3.17) do not account for energy losses, and the respective solutions will apply to protons. The two particle populations, namely the confined and the non-confined one, are matched through the solution derived at the escape time and the particle acceleration spectrum assumed at the shock. Then, the evolution of $t_{esc}(p)$ regulates the flux of non-confined particles.

3.3.1 The Distribution Function of Confined Particles

The goal here is to derive, assuming spherical symmetry, the distribution function of confined particles $f_{conf}(t, r, p)$, with fixed momentum $p \leq p_{max}(t)$ at time t and position $r \leq R_s(t)$ inside the shock location. Such a function solves the following simplified transport equation, in which diffusion is neglected:

$$\frac{\partial f}{\partial t} + v \frac{\partial f}{\partial r} = \frac{p}{3} \frac{1}{r^2} \frac{\partial}{\partial r} \left(r^2 v \right) \frac{\partial f}{\partial p} \tag{3.18}$$

where the plasma velocity profile is

$$\begin{cases} v = v(r, t) & r \leq R_s \\ v = 0 & r > R_s \end{cases} \tag{3.19}$$

For $r \leq R_s$ the transport equations reads

$$\frac{\partial f}{\partial t} + v \frac{\partial f}{\partial r} = \frac{p}{3} \frac{1}{r^2} \left(2 r v + r^2 \frac{\partial v}{\partial r} \right) \frac{\partial f}{\partial p} \tag{3.20}$$

whose boundary conditions are the Rankine–Hugoniot conditions, which affect the DSA spectrum at the shock. A linear approximation of the ST solution for the downstream velocity field of the plasma is adequately represented by

$$v(r, t) = \frac{3}{4} v_s(t) \frac{r}{R_s(t)} \tag{3.21}$$

as proposed by [193] and also adopted in [205]. It is worth to mention here that the velocity profile given in Eq. (3.21) does correctly reproduce the expected trend of gas pressure as a function of radius only if the pressure is dominated by relativistic

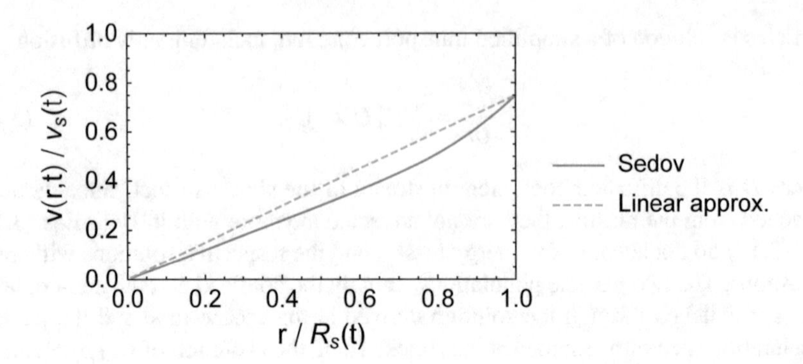

Fig. 3.3 Velocity profile of the downstream plasma for an SNR evolving in the ST phase and expanding into a uniform medium. The exact solution, given by the Sedov profile [214], is reported as a solid blue line, while the approximate linear expression reported in Eq. (3.21) is given by the orange dashed line

particles: in fact, in the case of an adiabatic expansion, the internal pressure amounts to a constant fraction of the energy density [147], namely $P(r) \propto r^{-3}$. For completeness, Fig. 3.3 shows the deviation of the velocity profile assumed in Eq. (3.21) from the true ST solution.

Substituting Eq. (3.21) into Eq. (3.20), one gets

$$\frac{\partial f}{\partial t} + \frac{3}{4} v_s(t) \frac{r}{R_s(t)} \frac{\partial f}{\partial r} = \frac{3}{4} p \frac{v_s(t)}{R_s(t)} \frac{\partial f}{\partial p} \tag{3.22}$$

which can be rewritten as

$$\frac{R_s(t)}{v_s(t)} \frac{\partial f}{\partial t} + \frac{3}{4} r \frac{\partial f}{\partial r} - \frac{3}{4} p \frac{\partial f}{\partial p} = 0 \tag{3.23}$$

Introducing the auxiliary variables

$$\begin{cases} \eta \equiv \ln r \\ \chi \equiv \ln p \\ d\tau \equiv dt\, v_s(t)/R_s(t) \implies \tau = \ln R_s(t) \end{cases} \tag{3.24}$$

Equation (3.22) can be rewritten as

$$\frac{\partial f}{\partial \tau} + \frac{3}{4} \frac{\partial f}{\partial \eta} - \frac{3}{4} \frac{\partial f}{\partial \chi} = 0 \tag{3.25}$$

This equation can be solved with the method of characteristics, which introduces a variable z such that the distribution function is constant along it, namely $df/dz = 0$. Then, the characteristic equation will allow to obtain $t = t(z)$, $r = r(z)$ and $p =$

$p(z)$, while the boundary condition will completely define the solution. The set of ordinary differential equations which applies in the spatial region downstream of the shock reads as

$$\frac{d\tau}{dz} = 1, \qquad \frac{d\eta}{dz} = \frac{3}{4}, \qquad \frac{d\chi}{dz} = -\frac{3}{4} \tag{3.26}$$

so that

$$\frac{df}{dz} = \frac{\partial f}{\partial \tau}\frac{d\tau}{dz} + \frac{\partial f}{\partial \eta}\frac{d\eta}{dz} + \frac{\partial f}{\partial \chi}\frac{d\chi}{dz} = 0 \tag{3.27}$$

By solving each of them, one obtains

$$\begin{cases} \tau(z) = z + A_\tau \\ \eta(z) = \frac{3}{4}z + A_\eta \\ \chi(z) = -\frac{3}{4}z + A_\chi \end{cases} \tag{3.28}$$

where A_τ, A_η, A_χ are integration constants. Summing up the last two identities, one gets

$$\eta + \chi = \ln(rp) = \text{const} \tag{3.29}$$

which implies that $d(rp)/dz = 0$. Since $f(z)$ is also constant along z, then f will be constant along the product rp. Thus

$$f(rp) = f(r_0 p_0) \tag{3.30}$$

where r_0 and p_0 represent respectively the position and momentum of the plasma element, at the time t_0 when it has been crossed by the shock. In order to get unambiguously the solution, one needs to impose as a boundary condition that the spectrum at the shock is the one resulting from the test-particle DSA approach

$$f(t_0, R_s(t_0), p_0) = f_0(t_0, p_0) = A(t_0)p_0^{-4}\theta[p_{\max,0}(t_0) - p_0] \equiv A(t_0)p_0^{-4}\theta[p_{\max,0}(t_0) - p_0] \tag{3.31}$$

where the normalization constant $A(t_0)$ defines the efficiency of conversion of the shock ram pressure into CR pressure (further details on this assumption are given in the next section), while the maximum momentum at the shock evolves with time according to Eq. (3.8). Here, the acceleration spectrum is assumed to be non vanishing only for $p_0 < p_{\max}(t_0)$, in order to derive the exact spectral features of the escape mechanism—without any interference from pre-existing cut-offs due to the acceleration scenario. From Eq. (3.30), it follows that

$$rp = r_0 p_0 \implies rp = R_s(t_0)p_0 \tag{3.32}$$

Hence, one needs to find the position $r_0 = R_s(t_0)$ where the plasma element was located at time t_0, the same element being in r at time t. This position is derived by considering the plasma speed as in Eq. (3.21) since

$$\frac{dr}{dt} = v(r,t) = \frac{3}{4} v_{\rm s}(t) \frac{r}{R_{\rm s}(t)} \tag{3.33}$$

Using the shock motion during the ST phase, one obtains

$$\frac{dr}{dt} = \frac{3}{10}\frac{r}{t} \implies \int_{R_{\rm s}(t_0)}^{r} \frac{dr'}{r'} = \frac{3}{10}\int_{t_0}^{t}\frac{dt'}{t'} \implies r = R_{\rm s}(t_0)\left(\frac{t}{t_0}\right)^{3/10} \tag{3.34}$$

Moreover, during the ST phase

$$\frac{R_{\rm s}(t)}{R_{\rm s}(t_0)} = \left(\frac{t}{t_0}\right)^{2/5} \tag{3.35}$$

so that Eq. (3.34) becomes

$$r = R_{\rm s}(t_0)\left(\frac{R_{\rm s}(t)}{R_{\rm s}(t_0)}\right)^{3/4} \tag{3.36}$$

Introducing the plasma position at time t into Eq. (3.32), one obtains

$$R_{\rm s}(t_0)\left(\frac{R_{\rm s}(t)}{R_{\rm s}(t_0)}\right)^{3/4} p = R_{\rm s}(t_0)\, p_0 \tag{3.37}$$

from which the adiabatic compression in momentum is derived as

$$p_0 = p\left(\frac{R_{\rm s}(t)}{R_{\rm s}(t_0)}\right)^{3/4} \tag{3.38}$$

showing that the momentum of a plasma element decreases with time because of the shell adiabatic expansion (namely for $t > t_0$ then $p < p_0$).

In this way, the distribution function of confined particles, i.e. particles that at a time t have a momentum $p < p_{\rm max,0}(t)$, is

$$f_{\rm conf}(t,r,p) = f_0\left(t_0(t,r),\, p\left(\frac{R_{\rm s}(t)}{R_{\rm s}(t_0)}\right)^{3/4}\right) \tag{3.39}$$

where the time $t_0(t,r)$ can be obtained through Eqs. (3.34) and (3.5) as

$$t_0(t,r) = t\left(\frac{R_{\rm s}(t_0)}{r}\right)^{10/3} = t\, R_{\rm s}(t_0)^{10/3} r^{-10/3} = t\left(\frac{\xi_0 E_{\rm SN}}{\rho_{\rm ISM}}\right)^{2/3} t_0^{4/3} r^{-10/3} \tag{3.40}$$

and hence

$$t_0(t,r) = \frac{r^{10}}{t^3}\left(\frac{\rho_{\rm ISM}}{\xi_0 E_{\rm SN}}\right)^2 \tag{3.41}$$

Using the shock injected spectrum as in Eq. (3.31), one finally derives for the confined distribution function

$$f_{\text{conf}}(t, r, p) = A(t_0) p^{-4} \left(\frac{R_s(t_0)}{R_s(t)} \right)^3 \theta \left[p_{\max,0} - p \left(\frac{R_s(t)}{R_s(t_0)} \right)^{3/4} \right] \tag{3.42}$$

Defining the *adiabatic compression factor* $\lambda(t, r)$ from Eq. (3.38) as

$$\lambda(t, r) = \left(\frac{R_s(t)}{R_s(t_0)} \right)^{3/4} \tag{3.43}$$

then the confined density function can be written as

$$f_{\text{conf}}(t, r, p) = \frac{A(t_0)}{\lambda^4(t, r)} p^{-4} \theta \left[p_{\max,0} - p\lambda(t, r) \right] \tag{3.44}$$

Rewriting this equation as a function of all variables at time t, one finally gets

$$f_{\text{conf}}(t, r, p) = A(t, r) p^{-4} \theta \left[\lambda(t, r)(p_{\max}(t, r) - p) \right] \tag{3.45}$$

which shows that the effect of adiabatic losses on the the maximum momentum is such that its temporal and spatial evolution reads as

$$p_{\max}(t, r) = p_{\max,0} \lambda^{-1}(t, r) \tag{3.46}$$

while the normalization of the spectrum scales as

$$A(t, r) = A(t_0) \lambda^{-4}(t, r) \tag{3.47}$$

A visual representation of the radial dependence of the maximum momentum at a fixed observation time is given in Fig. 3.4. Particles located in the inner part of the remnant have been shocked during the ED stage, where the maximum momentum was an increasing function of time. As the ST stage onsets, a break in the maximum momentum function appears, since the maximum momentum starts to decrease with time. Here adiabatic losses are accounted for, hence at the observation time no particles can achieve p_M. Finally, the decrease in time of the maximum momentum during the ST stage appears more pronounced for larger values of δ.

Equation (3.45) can also be used to easily derive the radial dependence of the CR density in the SNR interior: comparing such equation with the expressions in Eqs. (3.31) and (3.42), one gets

$$f_{\text{conf}}(t, r, p) = f_0(p, t) \frac{A(t_0)}{A(t)} \left(\frac{R_s(t_0)}{R_s(t)} \right)^3 \tag{3.48}$$

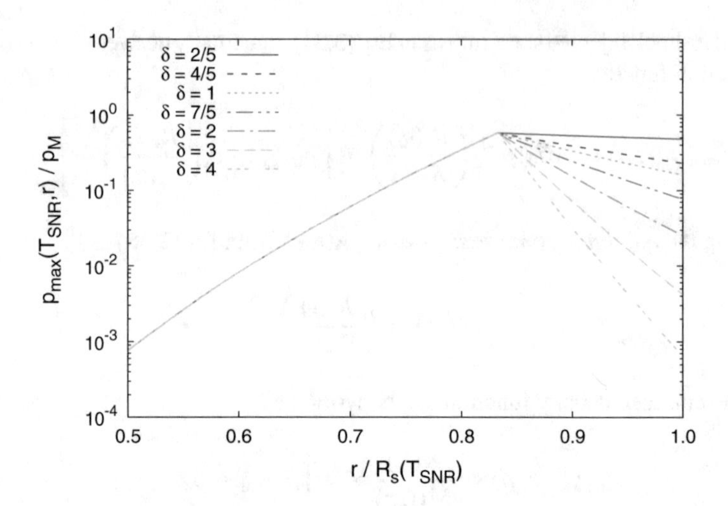

Fig. 3.4 Radial dependence of the maximum momentum, when the observation time is fixed to the SNR age $T_{SNR} = 10^4$ yr. Different values of δ are explored, regulating the time-dependency of the maximum momentum at the shock position as $p_{max,0}(t) \propto t^{-\delta}$ for $t > t_{Sed}$ (see Eq. (3.8)). The acceleration spectrum here assumed has spectral index $\alpha = 4$

Note that, since

$$\frac{A(t_0)}{A(t)} \propto \frac{v_s^2(t_0)}{v_s^2(t)} \propto \left(\frac{t_0}{t}\right)^{-6/5} \tag{3.49}$$

and

$$\left(\frac{R_s(t_0)}{R_s(t)}\right)^3 \propto \left(\frac{t_0}{t}\right)^{6/5} \tag{3.50}$$

as shown in Eq. (3.35), then the confined distribution function in Eq. (3.48) results constant with r (whose dependence was given through t_0). This peculiar result is valid when $\alpha = 4$, and it is a consequence of the fact that adiabatic expansion acting on plasma elements shocked in the past exactly balances the decrease of acceleration efficiency $\propto v_s^2$ in the ST phase. In the next section, it will be shown instead that a radial dependence appears for $\alpha \neq 4$. These results will be used in the following for the computation of the non-confined particle density function.

3.3.2 Normalization at the Shock

A caveat to the derivation of the confined distribution function is mandatory at this point. Indeed, the fact that $A(t) \propto v_s^2(t)$ derives from the assumption that a certain fraction ξ_{CR} of the ram pressure is instantaneously converted into CR pressure at the shock. In this case, the injected spectrum reads as

$$f_0(t,p) = \frac{3\xi_{CR}\rho_{ISM}v_s^2(t)}{4\pi c(m_p c)^{4-\alpha}\Lambda(p_{max}(t))} p^{-\alpha}\theta\left[p_{max}(t) - p\right] \tag{3.51}$$

The pressure into CRs is defined as

$$P_{CR} = \xi_{CR}\rho_{ISM}v_s^2(t) = \frac{4}{3}\pi \int_{p_{min}}^{\infty} p^3 v(p) f_0(p) dp \tag{3.52}$$

where $v(p)$ is the particle speed. Therefore, by substituting Eq. (3.51) into Eq. (3.52), one obtains

$$\xi_{CR}\rho_{ISM}v_s^2(t) = \frac{4}{3}\pi \frac{3\xi_{CR}\rho_{ISM}v_s^2(t)}{4\pi c(m_p c)^{4-\alpha}\Lambda(p_{max}(t))} \int_{p_{min}}^{\infty} p^3\beta(p)cp^{-\alpha}\theta\left[p_{max}(t) - p\right] dp \tag{3.53}$$

from which one can derive

$$\Lambda(p_{max}(t)) = \left(\frac{p_{max}}{m_p c}\right)^{4-\alpha} \int_{p_{min}}^{p_{max}(t)} \frac{1}{\sqrt{p^2 + m_p^2 c^2}} \left(\frac{p}{p_{max}}\right)^{4-\alpha} dp \tag{3.54}$$

where p_{min} is the minimum momentum provided by the acceleration mechanism. The main dependency of such a function on the particle momentum comes from the $p_{max}^{4-\alpha}$ term, rather than from the integral. It is however interesting to evaluate the integral in the limits of high and low momenta, by approximating the integrand function. If $p_{max} \gg m_p c$, it results that:

- if $\alpha > 4$, then $\Lambda(p) \propto \frac{1}{4-\alpha}\left(\frac{p_{min}}{m_p c}\right)^{4-\alpha}$;
- if $\alpha < 4$, then $\Lambda(p) \propto \frac{1}{4-\alpha}\left(\frac{p}{m_p c}\right)^{4-\alpha}$.

On the other hand, if $p_{max} \ll m_p c$ it holds that:

- if $\alpha > 5$, then $\Lambda(p) \propto \frac{1}{5-\alpha}\left(\frac{p_{min}}{m_p c}\right)^{5-\alpha}$;
- if $\alpha < 5$, then $\Lambda(p) \propto \frac{1}{5-\alpha}\left(\frac{p}{m_p c}\right)^{5-\alpha}$.

Such approximations will be considered in Sect. 3.4.

3.3.3 The Distribution Function of Non-confined Particles

After the escape time, particles with $p > p_{max}(t)$ are no more confined by the shock: as a consequence, advection and adiabatic compression can be neglected. Therefore, the transport equation will be simplified as

$$\frac{\partial f}{\partial t} = \frac{1}{r^2}\frac{\partial}{\partial r}\left[r^2 D(r)\frac{\partial f}{\partial r}\right] \tag{3.55}$$

A note of caution needs to be added concerning the assumption of spherical symmetry. While this choice is almost always justified for particles confined inside the SNR, because of the highly turbulent region where they propagate in the shock downstream, escaping particles are expected to diffuse along the magnetic field lines of the CSM. Hence, if the turbulence remains limited ($\delta B \ll B_0$), the geometry should be more similar to a cylindrical flux tube. Nevertheless, matching a spherical geometry inside the remnant with a cylindrical geometry outside it is not a trivial task. For this reason, in the following, a spherical symmetry will be assumed everywhere in the space. Despite this simplified assumption, it can be noted that diffusion in a flux tube is generally more effective, hence the spherical symmetry provides a lower limit to the confinement effect (see e.g. [93]).

Assuming $t = t_{esc}(p)$ as the starting time for particle escape, then the initial condition for the density of non-confined particles reads as

$$f_{esc}(r, t = t_{esc}) = f_{conf}(r, t_{esc}) \tag{3.56}$$

For simplicity, a uniform diffusion coefficient is assumed in the following, so that D is the same inside and outside of the remnant, namely $D = D_{in} = D_{out}$.

The general transport equation with constant diffusion (see Eq. (3.55)) can be expanded as

$$\frac{1}{D}\frac{\partial f}{\partial t} = \frac{\partial^2 f}{\partial r^2} + \frac{2}{r}\frac{\partial f}{\partial r} \tag{3.57}$$

Introducing $u(r, t) = rf(r, t)$, then Eq. (3.57) becomes

$$\frac{1}{D}\frac{\partial u}{\partial t} = \frac{\partial^2 u}{\partial r^2} \tag{3.58}$$

This Equation can be solved with the boundary condition of Eq. (3.56) by means of a Laplace transformation. The full derivation is provided in Appendix C, while the final result on the escape solution reads as

$$\frac{f_{esc}(r, t, p)}{f_{conf}(t_{esc}, p)} = \frac{1}{2}\left[\mathrm{Erf}\left[\frac{R_+}{R_d}\right] + \mathrm{Erf}\left[\frac{R_-}{R_d}\right] + \frac{R_d}{\sqrt{\pi} r}\left(e^{-\left(\frac{R_+}{R_d}\right)^2} - e^{-\left(\frac{R_-}{R_d}\right)^2}\right)\right]\theta[t - t_{esc}(p)] \tag{3.59}$$

where $R_+ = (R_{esc} + r)$, $R_- = (R_{esc} - r)$ and $R_d(t, p) = 2\sqrt{D(p)(t - t_{esc}(p))}$. One can easily check that such solution satisfies the boundary conditions: indeed, if $r = 0$ then $f(0) = 0$, while if $r = \infty$ then $f(\infty) = 0$. Figure 3.5 shows the radial dependence of the non-confined particle distribution function at different times after the escape time for protons of different momenta. As already pointed out, the flat trend

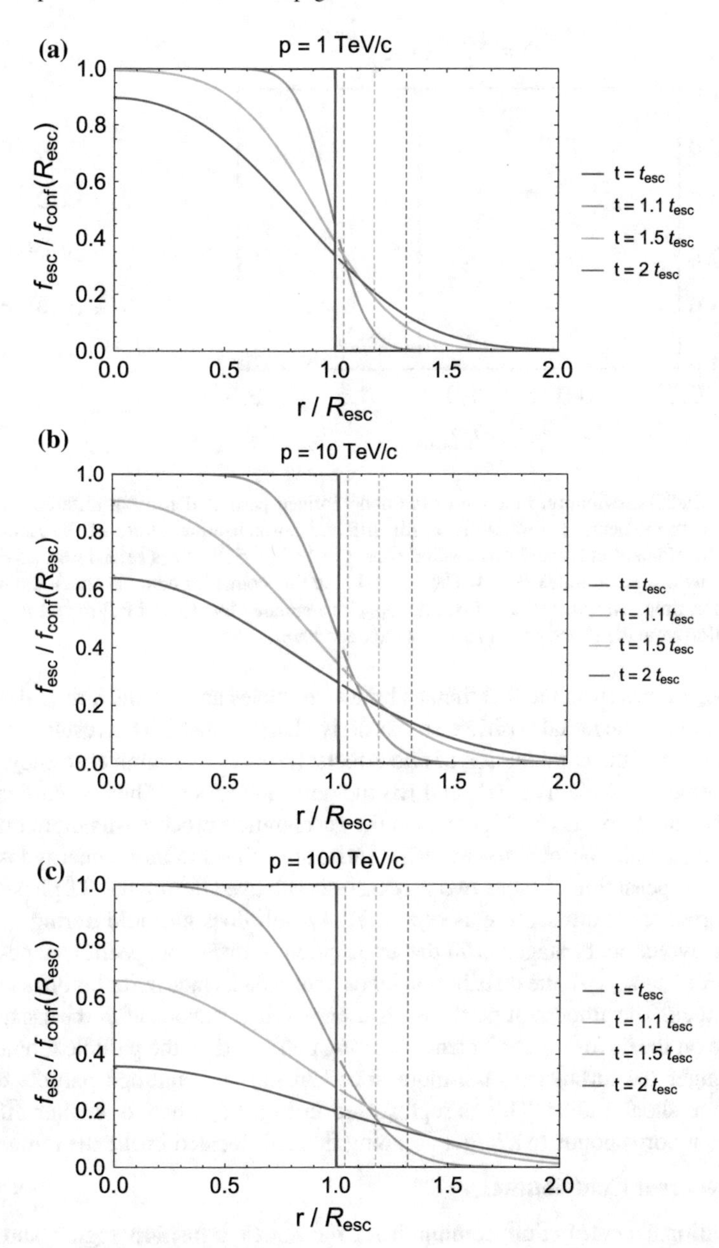

Fig. 3.5 Radial distribution function of the non-confined particle density at different times, for particles with momentum $p = 1\,\mathrm{TeV/c}$ (top), $p = 10\,\mathrm{TeV/c}$ (middle) and $p = 100\,\mathrm{TeV/c}$ (bottom). Solid lines refer to an acceleration spectrum with spectral index $\alpha = 4$. Vertical dashed lines represent the shock position at different times: the blue dashed line coincides with the escape radius. Note however that particles of different energies have a different escape radius. The model here assumes $p_{\max}(t)$ regulated by $p_M = 1\,\mathrm{PeV/c}$ and $\delta = 4$. The diffusion coefficient is Kolmogorov-like, normalized to $D_0 = 10^{26}\,\mathrm{cm^2\,s^{-1}}$. The contribution from the shock precursor is not included here. Figure from Celli et al. [81], reproduced by permission of the Royal Astronomical Society

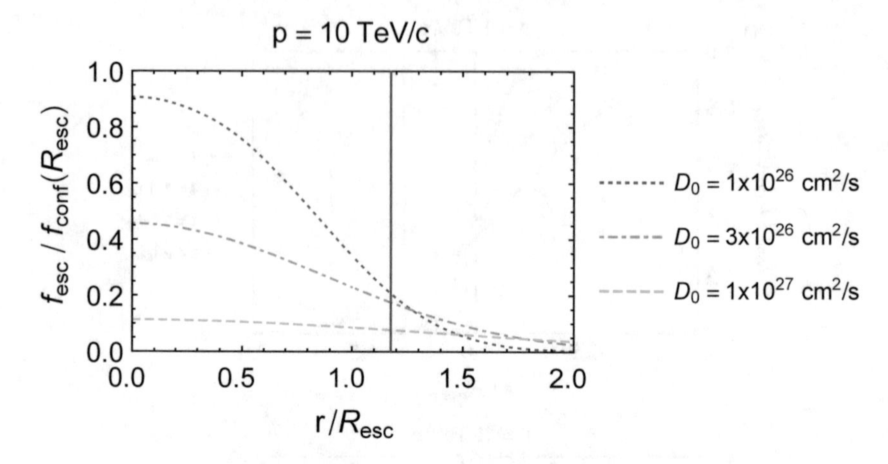

Fig. 3.6 Radial distribution function of the non-confined particle density at different times, for particles with momentum $p = 10\,\mathrm{TeV/c}$, for different normalization values of the Kolmogorov diffusion coefficient at a fixed observation time $t = 1.5t_{\rm esc}$. Solid lines refer to an acceleration spectrum with spectral index $\alpha = 4$. The vertical blue line coincides with the shock radius at the observation time. The model here assumes $p_{\rm max}(t)$ regulated by $p_{\rm M} = 1\,\mathrm{PeV/c}$ and $\delta = 4$. The contribution from the shock precursor is not included here

at $t = t_{\rm esc}$ comes from the fact that, as long as particles are confined, their density is independent of the radial position, as was derived in Eq. (3.48). This result comes as a consequence of the combination of two effects: (i) the acceleration efficiency, which was assumed to scale as $v_{\rm s}^2(t)$, and (ii) the adiabatic losses. Thus, at earlier times when the shock speed was higher, also the acceleration process was more efficient. However, particles accelerated at earlier times are subject to larger energy losses. A perfect compensation of these two processes is achieved during the ST phase for an acceleration spectrum scaling as $\propto p^{-4}$. This result does not hold during different remnant evolutionary stages or in the assumption of different spectral shapes at the shock. For $t > t_{\rm esc}(p)$, the distribution tends to broaden, since particles diffuse ahead of the shock. The amount of particles residing within the shock after the escape time depends on the diffusion coefficient operating there, and on the particle momentum: the stronger the turbulence, the more abundant the non-confined particle density within the shock radius. This is represented in Fig. 3.6, where a smaller diffusion coefficient corresponds to a larger escaping flux still located inside the remnant.

The Precursor Contribution

An additional contribution coming from the shock transition region should be included in the CR escaping density function (see [205]). In fact, at the escape time, some particles are being accelerated up to $p_{\rm max}$ and therefore they can contribute to the escape solution. The CR precursor (obtained from the steady state transport equation in the plane shock approximation) reads as

$$f_{\rm p}(p, r, t) = f_0(p, t) \exp\left[-\frac{v_{\rm s}(t)}{D_{\rm p}(p)}(r - R_{\rm s})\right] \tag{3.60}$$

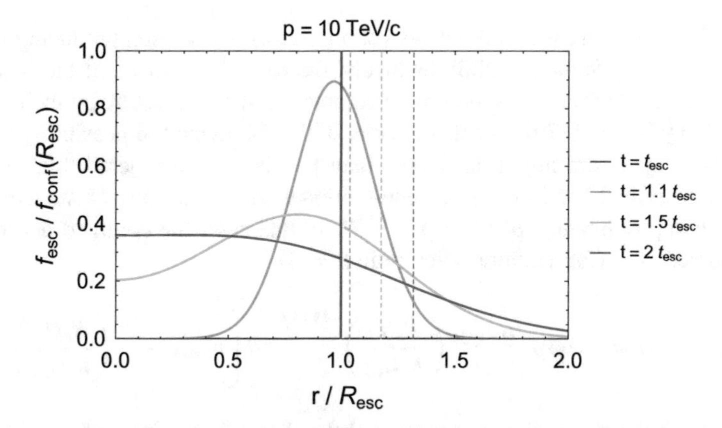

Fig. 3.7 Radial distribution function of the non-confined particle density released by the precursor at different times (solid lines), for particles with momentum $p = 10\,\text{TeV/c}$. Note that the initial condition for particle escape is a δ-function located at the escape radius. Vertical dashed lines represent the shock position at different times. The model here assumes $p_{\text{max}}(t)$ regulated by $p_{\text{M}} = 1\,\text{PeV/c}$ and $\delta = 4$. The diffusion coefficient is Kolmogorov-like, normalized to $D_0 = 10^{26}\,\text{cm}^2\,\text{s}^{-1}$. Figure from Celli et al. [81], reproduced by permission of the Royal Astronomical Society

where the diffusion coefficient $D_{\text{p}}(p)$ is that within the precursor. In order to simplify the computation of the escaping particle density function from the shock transition region, the total number of particles contained in the precursor is assumed to be located at the shock position, namely in $r = R_{\text{s}}$: this assumption is considered in that the analytical expression adopted for the precursor-confined particle density is

$$f_{\text{p,conf}}(p, r, t) = f_0(p, t)\frac{D_{\text{p}}(p)}{v_{\text{s}}(t)}\delta(r - R_{\text{s}}) \tag{3.61}$$

Therefore, these particles start diffusing from the shock and the same procedure as above can be applied. The complete computation is carried out in Appendix C, while here it is simply reported the result of that computation, namely the density function of particles escaping from the shock region, that reads as

$$\frac{f_{\text{p,esc}}(r, t, p)}{f_0(p, t_{\text{esc}})} = \frac{1}{\sqrt{\pi}}\frac{R_{\text{esc}}}{R_{\text{d}}}\frac{D_{\text{p}}(p)}{v_{\text{s}}(t_{\text{esc}})r}\left[e^{-\left(\frac{R_-}{R_{\text{d}}}\right)^2} - e^{-\left(\frac{R_+}{R_{\text{d}}}\right)^2}\right]\theta[t - t_{\text{esc}}(p)] \tag{3.62}$$

The diffusive behavior of the escaped particles from inside the precursor is shown in Fig. 3.7 as a function of the radial position, at four different times. While at $t = t_{\text{esc}}(p)$ the distribution function is schematically shown as a δ-function (see Eq. (3.61)), at later times particles released by the precursor start to diffuse in all the directions. As they get overtaken by the shock itself, they are potentially able to contribute to the distribution function of non-confined particles still located within the remnant after the escape time.

The Case of a Steeper Acceleration Spectrum

Some young SNRs observed in gamma rays (e.g. Tycho and CasA) show steep spectra that, if produced in hadronic interactions, would imply a parent proton spectrum

$\propto p^{-4.3}$. To date, there is no consensus yet on the physical reason producing spectra steeper than p^{-4}. Some possibilities invoke the role of the speed of the scattering centers [180] or the modification produced onto the shock structure by the presence of neutral hydrogen [179]. For this reason, in the following the possibility that the acceleration spectrum might be steeper than p^{-4} is also considered. Interestingly, an exact analytical solution for the non-confined particle density exists within the following approximation $p^{-4.3} \simeq p^{-(4+1/3)}$. In this case, the confined distribution function becomes (for a compression ratio $\sigma = 4$)

$$f_{\rm conf}(t, r, p) = A(t_0) p^{-(4+1/3)} \left(\frac{R_{\rm s}(t)}{R_{\rm s}(t_0)} \right)^{-(3+1/4)} \theta \left[p_{\rm max,0} - p \left(\frac{R_{\rm s}(t)}{R_{\rm s}(t_0)} \right)^{3/4} \right] \tag{3.63}$$

Comparing it to the distribution function at the shock for particles of momentum p at time t

$$f_0(t, p) = A(t) p^{-(4+1/3)} \theta \left[p_{\rm max}(t) - p \right] \tag{3.64}$$

one obtains

$$f_{\rm conf}(t, r, p) = f_0(t, p) \frac{A(t_0)}{A(t)} \left(\frac{R_{\rm s}(t_0)}{R_{\rm s}(t)} \right)^{(3+1/4)} \tag{3.65}$$

Because of Eq. (3.49) and

$$\left(\frac{R_{\rm s}(t_0)}{R_{\rm s}(t)} \right)^{(3+1/4)} = \left(\frac{R_{\rm s}(t_0)}{R_{\rm s}(t)} \right)^{13/4} \propto \left(\frac{t_0}{t} \right)^{13/10} \tag{3.66}$$

the confined distribution function in Eq. (3.65) now results to be

$$f_{\rm conf}(t, r, p) = \frac{3 \xi_{\rm CR} \rho_{\rm ISM}}{25 \pi c (m_{\rm p} c)^{4-\alpha} \Lambda(p_{\rm max}(t_0))} \left(\frac{\xi_0 E_{\rm SN}}{\rho_{\rm ISM}} \right)^{2/5} t_0^{1/10} t^{-13/10} p^{-\alpha} = k(t) r p^{-\alpha} \tag{3.67}$$

given $t_0(t, r)$ as in Eq. (3.41), where the function $k(t)$ is

$$k(t) = \frac{3 \xi_{\rm CR} \rho_{\rm ISM}}{25 \pi c (m_{\rm p} c)^{4-\alpha} \Lambda} \left(\frac{\xi_0 E_{\rm SN}}{\rho_{\rm ISM}} \right)^{1/5} t^{-8/5} \tag{3.68}$$

The linear dependence on the radial distance, shown in Eq. (3.67), leads to a different solution in the escape distribution function with respect to what has been obtained before for an acceleration spectrum $\propto p^{-4}$. The complete derivation is provided in Appendix C, the final result being

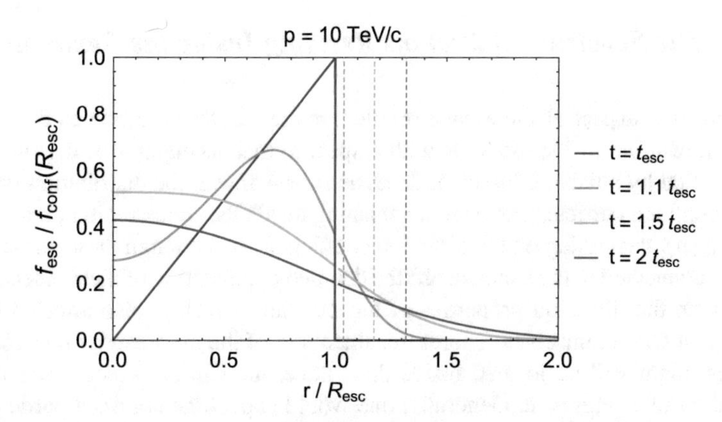

Fig. 3.8 Radial distribution function of the non-confined particle density at different times (solid lines), for particles with momentum $p = 10\,\text{TeV/c}$. Solid lines refer to an acceleration spectrum with spectral index $\alpha = 4 + 1/3$. Vertical dashed lines represent the shock position: the blue dashed line coincides with the escape radius. The model here assumes $p_{\max}(t)$ regulated by $p_{\mathrm{M}} = 1\,\text{PeV/c}$ and $\delta = 4$. The diffusion coefficient is Kolmogorov-like, normalized to $D_0 = 10^{26}\,\text{cm}^2\,\text{s}^{-1}$. The contribution from the precursor is not included here. Figure from Celli et al. [81], reproduced by permission of the Royal Astronomical Society

$$\begin{aligned}\frac{f_{\rm esc}(t,r,p)}{k(t_{\rm esc})} = &\left\{\frac{R_{\rm d}}{\sqrt{\pi}}e^{-\left(\frac{r}{R_{\rm d}}\right)^2} + \frac{R_{\rm d}}{2\sqrt{\pi}}\left(\frac{R_-}{r}\right)e^{-\left(\frac{R_+}{R_{\rm d}}\right)^2} - \frac{R_{\rm d}}{2\sqrt{\pi}}\left(\frac{R_+}{r}\right)e^{-\left(\frac{R_-}{R_{\rm d}}\right)^2} + \right.\\ &+\left(r+\frac{R_{\rm d}^2}{2r}\right)\mathrm{Erf}\left[\frac{r}{R_{\rm d}}\right] + \frac{1}{2}\left(r+\frac{R_{\rm d}^2}{2r}\right)\mathrm{Erfc}\left[\frac{R_+}{R_{\rm d}}\right] + \\ &\left.-\left(1-\mathrm{Erf}\left[\frac{R_-}{R_{\rm d}}\right]\right)\left(\frac{r}{2}+\frac{R_d^2}{4r}\right)\right\}\cdot\theta[t-t_{\rm esc}(p)]\end{aligned} \tag{3.69}$$

This solution can be compared with that presented in Eq. (3.59). A visual representation of the density function reported in Eq. (3.69) at different radial positions is provided in Fig. 3.8, where 10 TeV protons are considered (see Fig. 3.5b for comparison with the p^{-4} case). This figure shows that, at a fixed observation time after the escape time, the contribution from the density of non-confined particles located within the shock is always lower than that expected in the case of a p^{-4} acceleration spectrum. In fact, in the p^{-4} case, the confined density function is radially flat, hence the broadening due to diffusion mainly tends to deprive the inner part of the remnant in order to fill the external part with accelerated particles. On the other hand, for a steeper spectrum as in the $p^{-(4+1/3)}$ case, where the confined density function linearly increases with the radius, the broadening first tends to fill the inner part of the remnant with accelerated particles, and only at later times the same region will be emptied.

3.3.4 The Spectrum of Protons Residing Inside the Remnant

An interesting aspect of the escape problem is the fact that it modifies the particle spectrum residing inside the SNR with respect to that accelerated at the shock and then affected by adiabatic losses. Indeed, as shown above, the distribution function of non-confined particles can be non vanishing inside the shock region up to several times $t_{esc}(p)$, depending on the diffusion coefficient, even though these particles are no more connected to the shock itself: the detailed computation of the residence time depends on the diffusion properties of the medium, which is also affected by the presence of CRs themselves. Therefore, since part of the high-energy non-confined particles might still be located inside the source, the particle spectrum should be affected by their presence. Generally, one would expect the confined particle density to show an exponential suppression at the end of the acceleration spectrum, dictated by the escape process. The additional contribution from the non-confined particles, which however are still diffusing inside the remnant, might possibly produce a spectrum that is closer to a broken power-law. The break in momentum would be regulated by the maximum momentum at the remnant age $p_{max}(T_{SNR})$. Thus, while the low-energy part of the spectrum would simply reflect the acceleration spectrum, non-confined particles would contribute to the high-energy part. Now, since the more energetic particles escape the shock at earlier times and since the diffusion length is generally an increasing function of momentum, these are expected to be less abundant than low-energy particles. Hence, the steepening at high momenta ($p > p_{max}(T_{SNR})$) strongly depends on the temporal dependence of $p_{max}(t)$, i.e. on the δ index.

The proton spectrum resulting from all the particles contained inside the remnant, including both confined and non-confined ones, as well as the contribution from non-confined particles released by the precursor, is computed through

$$J_p(t, p) = \frac{4\pi}{V_{SNR}} \int_0^{R_{SNR}} r^2 \left[f_{esc}(r, t, p) + f_{p,esc}(r, t, p) + f_{conf}(r, t, p) \right] dr \quad (3.70)$$

This is shown in Fig. 3.9 for an acceleration spectrum $\propto p^{-4}$ and $p_M = 1$ PeV/c: different possible slopes δ are explored as well as different absolute normalizations of the diffusion coefficient. Since escape is effective during the ST phase, only middle-aged SNRs are considered in the following. In order to show the effects of this phenomenon, typical parameters are set: namely a SN explosion energy of $E_{SN} = 10^{51}$ erg, a density of the ISM equal to $\rho_{ISM} = 1\ (m_p/\text{g})\ (\text{g/cm}^3)$, a mass in the ejecta of $M_{ej} = 10\ M_\odot$, a remnant age of $T_{SNR} = 10^4$ yr and an efficiency conversion of the bulk energy into CRs $\xi_{CR} = 10\%$. The same values are adopted in the rest of this chapter, unless explicitly mentioned. With this set of values, the Sedov time is equal to $t_{Sed} = 1.6 \times 10^3$ yr. In order to define univocally the escape solution, an effective diffusion coefficient for CR scattering has to be fixed. A focus on the spatial region close to the accelerator, where a stronger turbulence than the average Galactic is expected [93, 156], is provided in the following: thus, a Kolmogorov-like spectrum is set

$$D(p) = D_0 \left(\frac{p}{10\,\text{GeV/c}} \right)^{1/3} \quad (3.71)$$

Table 3.1 Maximum momentum achieved at the shock at $T_{SNR} = 10^4$ yr in the parametrization of Eq. (3.8). The acceleration spectrum is set $\propto p^{-4}$, the upstream density is $n_{up} = 1\,\mathrm{cm}^{-3}$ (it affects the onset of the ST stage), while $p_M = 1$ PeV/c at the ST time. Table from Celli et al. [81], reproduced by permission of the Royal Astronomical Society

δ	$p_{max}(T_{SNR})$ (GeV/c)
1	1.6×10^5
2	2.5×10^4
3	4.1×10^3
4	6.5×10^2

where the normalization is reduced by a factor of 10 with respect to the average Galactic one, i.e. $D_0 = 10^{27}\,\mathrm{cm}^2\,\mathrm{s}^{-1}$. As discussed above, non-confined particles mostly contribute above the energy break, whose value depends on the remnant age and on the slope δ of the maximum momentum temporal dependency. On the other hand, particles connected to the shock contribute to the spectrum in the energy region below the break, which is flat in $p^{\alpha} f(p)$ (for an acceleration spectrum $\propto p^{-\alpha}$). The energy break in the spectrum is due to the maximum momentum achieved at the shock position at the remnant age, its exact value depending on the parameter δ which regulates how fast the maximum momentum is decreasing with time because of particle escape. For a more quantitative estimate, Table 3.1 provides the values of $p_{max}(T_{SNR})$ under different assumptions concerning δ. Furthermore, the spectral trend above the break strongly depends on the energy dependence of the diffusion coefficient assumed: a sort of flattening is visible at high energies, resulting from the sum among non-confined particles and particles escaped from the precursor, both contributions being located within the shock radius but decoupled from it. By increasing the value of the diffusion coefficient, a reduced amount of non-confined particles would be still located inside the remnant shock. Hence it is clear that, if the CRs are able to self-amplify the turbulent magnetic field needed for their efficient scattering, the confinement time within and around the source would increase. In order to confirm this hypothesis, that the self-generated diffusion coefficient in the source vicinity is reduced with respect to the average Galactic value, Sect. 3.3.5 will be devoted to the evaluation of the amount of turbulence generated by the resonant streaming of CRs.

On the other hand, by adopting the approach by [75] for the maximum momentum estimate, the maximum momentum at the shock position in the standard DSA scenario (from Eqs. (3.12) and (3.13)) results higher than in the case of a softer acceleration spectrum (from Eqs. (3.14) and Eq. (3.15)). With the standard set of aforementioned parameters, a remnant expanding into a uniform medium of density $n = 1\,\mathrm{cm}^{-3}$ and observed at a time $T_{SNR} = 10^4$ yr would accelerate particles up to $p_{max}(T_{SNR}) \simeq 5.9 \times 10^3$ GeV/c if $\alpha = 4$ or to $p_{max}(T_{SNR}) \simeq 1.3 \times 10^3$ GeV/c if $\alpha = 4 + 1/3$. The proton spectrum obtained integrating the whole distribution function within the remnant radius is shown in Fig. 3.10, for both the cases of an acceleration spectrum with $\alpha = 4$ and $\alpha = 4 + 1/3$. Here, analogous features to Fig. 3.9 can be noted.

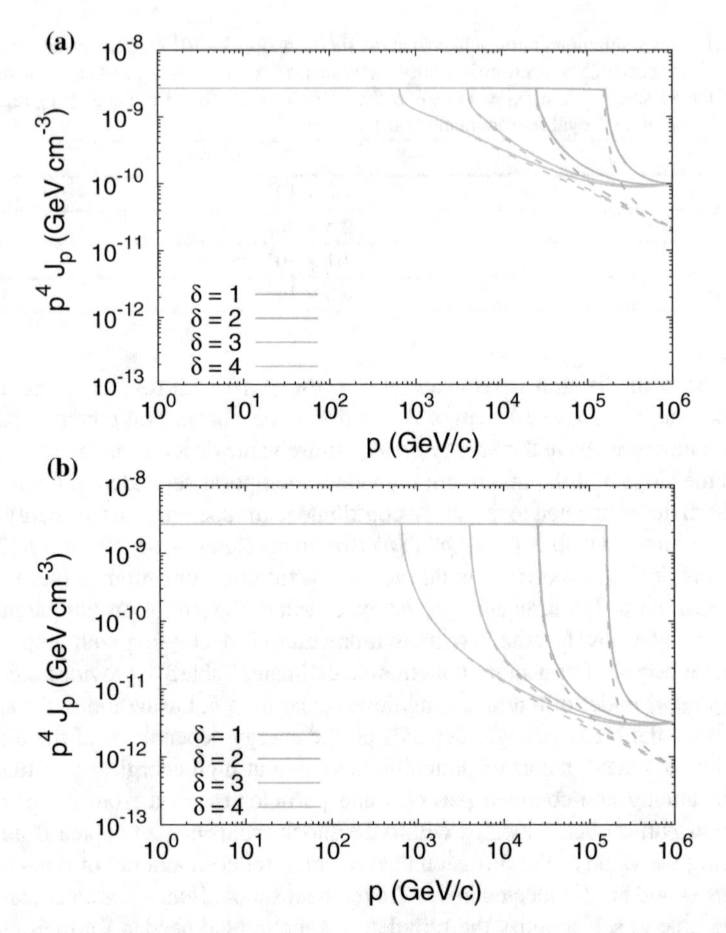

Fig. 3.9 Proton spectrum of confined and non-confined particles still residing inside the remnant at $T_{SNR} = 10^4$ yr for different values of the index δ, which regulates the time-dependence of the maximum momentum at the shock. The acceleration spectrum is set to $\alpha = 4$, the upstream numerical density here assumed is $n = 1\,\mathrm{cm}^{-3}$, the maximum momentum at the Sedov time is set to $p_M = 1$ PeV/c. Diffusion coefficient is Kolmogorov-like, normalized to: *Top*: $D_0 = 1 \times 10^{27}\,\mathrm{cm}^2\,\mathrm{s}^{-1}$, *Bottom*: $D_0 = 1 \times 10^{28}\,\mathrm{cm}^2\,\mathrm{s}^{-1}$. The particle distributions without the shock precursor are always shown as dashed lines. Figure from Celli et al. [81], reproduced by permission of the Royal Astronomical Society

3.3.5 Self-generated Turbulence

It has been previously shown that, in order for the non-confined particles to contribute sizably to the proton spectrum inside the SNR, a reduced value of the diffusion coefficient with respect to the average Galactic one is required. This effect might be ascribed to the presence of spatial gradients in the particle density function. In fact, as discussed in Chap. 2, the amount of self-generated turbulence due to resonant streaming instability depends on the spatial gradient of the distribution function, as

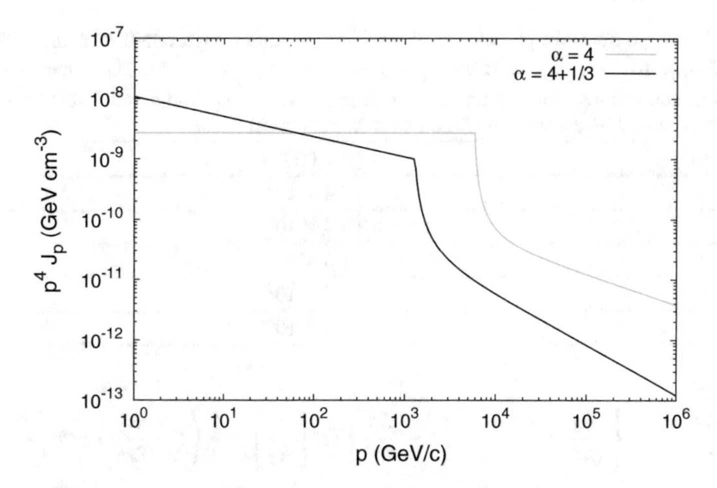

Fig. 3.10 Proton spectrum of confined and non-confined particles inside the remnant. Both lines indicate the modeling according to the maximum momentum time-dependency at the shock taken from [75]: the grey line refers to the case with $\beta = 0$, while the black line refers to $\beta = 1/3$. In both cases, the upstream density is set to $n_{\rm up} = 1\,{\rm cm}^{-3}$, while $D_0 = 1 \times 10^{27}\,{\rm cm}^2\,{\rm s}^{-1}$ and $p_{\rm M} = 1\,{\rm PeV/c}$ were assumed. The source age is fixed to $T_{\rm SNR} = 10^4$ yr

the growth rate of resonant MHD waves $\Gamma_{\rm CR}$ is expressed by Eq. (2.26). On the other hand, if the plasma is completely ionized, a relevant damping mechanism is the non-linear one, expressed by Eq. (2.27). Therefore, by imposing that $\Gamma_{\rm CR} = \Gamma_{\rm D}$, one can evaluate *a posteriori* the amount of turbulence that CRs can self-generate. Note that, for this computation, the analytical expression for the radial gradient of the distribution function can be derived in both the cases of a p^{-4} and $p^{-(4+1/3)}$ acceleration spectra. These are given in the following: for the p^{-4} case the gradient of the non-confined particle density function reads as

$$\frac{1}{f_{\rm conf}}\frac{\partial f_{\rm esc}(r,t,p)}{\partial r} = \frac{1}{\sqrt{\pi} r}\left\{\left[-\frac{R_{\rm d}}{2r} + \frac{r}{R_{\rm d}} - \frac{R_+}{R_{\rm d}}\right] e^{-\left(\frac{R_+}{R_{\rm d}}\right)^2} + \right.$$
$$\left. + \left[\frac{R_{\rm d}}{2r} - \frac{r}{R_{\rm d}} + \frac{R_-}{R_{\rm d}}\right] e^{-\left(\frac{R_-}{R_{\rm d}}\right)^2}\right\} \theta[t - t_{\rm esc}(p)] \tag{3.72}$$

On the other hand, for the $p^{-(4+1/3)}$ case, the gradient of the non-confined particle density function is obtained as

Table 3.2 Escape times for particles of different momentum from an SNR evolving with $M_{ej} = 10 M_{\odot}$, $E_{SN} = 10^{51}$ erg, $\xi_{CR} = 10\%$, $n_{up} = 1\,\mathrm{cm}^{-3}$ and $T_{SNR} = 10^4$ yr. The parametrization of escape time adopted here follows Eq. (3.9), with $p_M = 10^6$ GeV/c and $\delta = 3$. Table from Celli et al. [81], reproduced by permission of the Royal Astronomical Society

p (GeV/c)	t_{esc} (yr)
10	7.4×10^4
10^2	3.4×10^4
10^3	1.6×10^4
10^4	7.4×10^3
10^5	3.4×10^3

$$\frac{\partial f_{esc}(r,t,p)}{\partial r} = \left\{ \frac{R_d}{\sqrt{\pi} r} e^{-\left(\frac{r}{R_d}\right)^2} + \left(1 - \frac{R_d^2}{2r^2}\right) \mathrm{Erf}\left[\frac{r}{R_d}\right] + \frac{1}{2}\left(1 - \frac{R_d^2}{2r^2}\right) \mathrm{Erfc}\left[\frac{R_+}{R_d}\right] + \right.$$
$$- \left(\frac{1}{2} - \frac{R_d^2}{4r^2}\right)\left(1 - \mathrm{Erf}\left[\frac{R_-}{R_d}\right]\right) - \frac{1}{\sqrt{\pi}} e^{-\left(\frac{R_+}{R_d}\right)^2} \left(\frac{R_d R_{esc}}{2r^2} + \frac{R_{esc}^2}{r R_d} + \frac{R_d}{2r}\right) +$$
$$\left. + \frac{1}{\sqrt{\pi}} e^{-\left(\frac{R_-}{R_d}\right)^2} \left(\frac{R_d R_{esc}}{2r^2} - \frac{R_{esc}^2}{r R_d} - \frac{R_d}{2r}\right) \right\} k(t_{esc}) \theta[t - t_{esc}(p)] \tag{3.73}$$

where $k(t_{esc})$ was defined at the end of Sect. 3.3.3. Furthermore, when computing the CR self-generated turbulence in the $p^{-(4+1/3)}$ case, one should account for the non vanishing radial gradient of the confined particle distribution function. Finally, also the precursor particle density function has a non null radial gradient.

Hence, by using the spatial CR gradient obtained with a given assumption on D_{out}, it is possible to calculate the level of self-generated turbulence due to streaming instability (namely the power in turbulent magnetic waves as given in Eq. (2.28)) and consequently to derive the diffusion coefficient $\hat{D} \propto \mathcal{F}^{-1}$ effectively operating within or outside the shock at the remnant age. Even if such a calculation is not a self-consistent one, it can show whether or not the streaming instability can be responsible for the reduction of D_{out}. It is worth stressing that one should account for the duration of the wave amplification process: on a general ground, one can expect that a suppression of the diffusion coefficient with respect to the average Galactic value is achieved within few escape times, but later on, when the CR density diminishes, also the amplification of the magnetic turbulence fades. In other words, the streaming instability cannot be responsible for the needed magnetic turbulence during the whole remnant lifetime, but only as long as a consistent number of particles are being accelerated. In order to facilitate the comparison among the remnant age and the escape time of particles at different energies, Table 3.2 reports the expected escape time, computed according to Eq. (3.9).

Assuming the same benchmark values as in Sect. 3.3.4 with a background magnetic field $B_0 \simeq 3\mu$G and $\alpha = 4$, the ratio $D_{out}/\hat{D}$ was calculated for $D_0 = 1 \times 10^{27}\,\mathrm{cm}^2\,\mathrm{s}^{-1}$ and $D_0 = 1 \times 10^{26}\,\mathrm{cm}^2\,\mathrm{s}^{-1}$. Results are shown in the top panel of Fig. 3.11. As visible, in both cases, the level of self-generated turbulence is such that

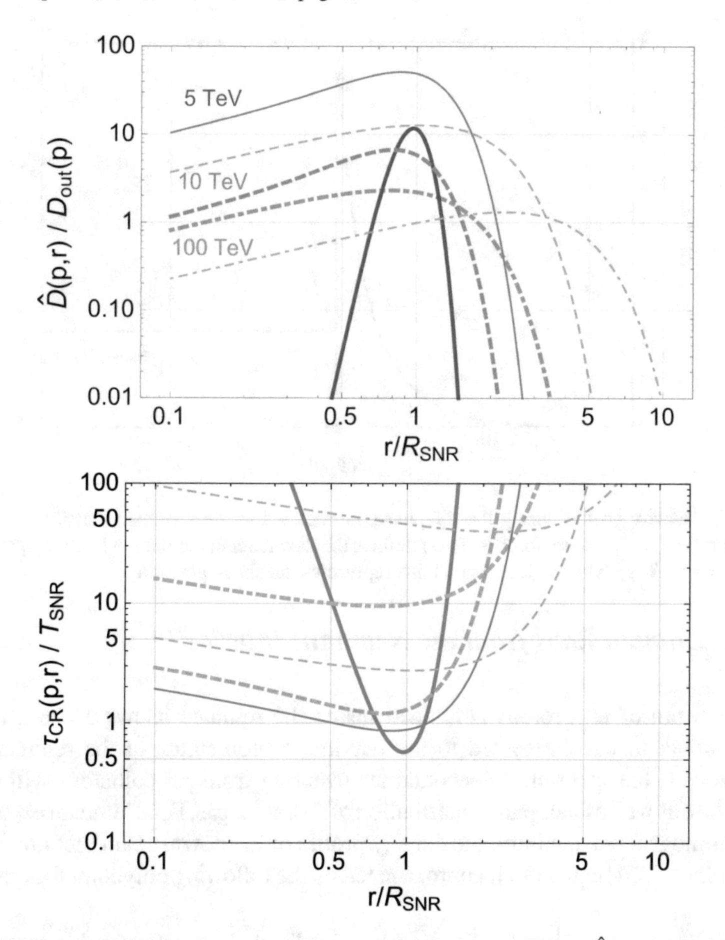

Fig. 3.11 *Top*: spatial dependence of self-generated diffusion coefficient $\hat{D}(p, r)$, normalized to $D_{out}(p)$ for $D_0 = 1 \times 10^{27}\,\mathrm{cm^2\,s^{-1}}$ (thin lines) and $D_0 = 1 \times 10^{26}\,\mathrm{cm^2\,s^{-1}}$ (thick lines), calculated by setting an acceleration spectrum with slope $\alpha = 4$. The parametrization of escape time adopted here follows Eq. (3.9), with $\delta = 3$. The three sets of lines correspond to three different particle energies: 1 TeV (solid), 10 TeV (dashed) and 100 TeV (dot-dashed). *Bottom*: corresponding excitation time for the streaming instability in unit of the SNR age ($T_{SNR} = 10^4$ yr) and for the same energy values as the top panel. Figure from Celli et al. [81], reproduced by permission of the Royal Astronomical Society

$\hat{D} \lesssim D_{out}$ for $pc \lesssim 10$ TeV in a region of about the size of the SNR. On the other hand, the timescale to excite the instability $\tau_{CR} = 1/\Gamma_{CR}$, reported in the bottom panel of the same figure, is smaller, or comparable, to the SNR age only for energy lower than ~1 TeV, meaning that only below such energy the resonant streaming instability is able to reduce the diffusion coefficient at least by a factor of 10 with respect to the average Galactic D_{Gal}, thus increasing the particle residence time around the remnant.

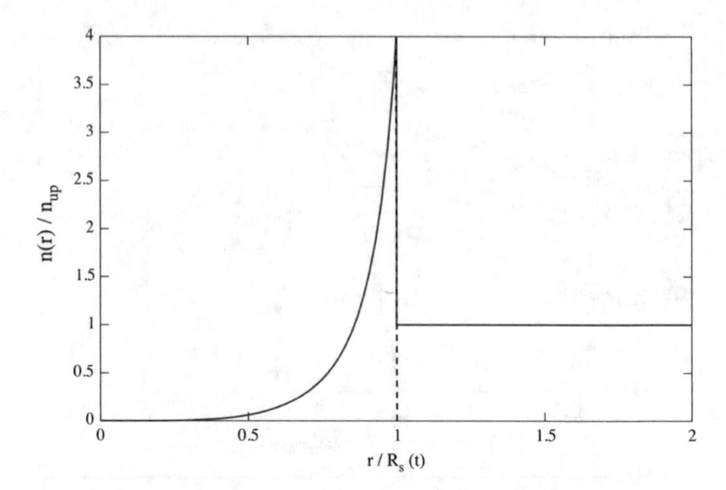

Fig. 3.12 Density profile during the ST phase, normalized to the upstream density, for a compression ratio $\sigma = 4$. The region $r \leq R_s(t)$ defines the downstream medium, while the upstream is located in $r > R_s(t)$. The vertical dashed line represents the shock position

3.3.6 Gamma Rays from the Remnant Interior

The spectrum of all protons contained inside the remnant shows a characteristic energy break that is connected to the maximum momentum at the remnant age. Analogously, the spectrum of secondaries resulting from pp collisions will reflect this feature as well as the spatial distribution of the target gas. For a remnant expanding into a homogeneous medium, the density profile of the downstream medium during the ST phase [214] can be well approximated by the following polynomial expression

$$n_{\text{down}}(r,t) = n_{\text{up}}\sigma \left[a_1 \left(\frac{r}{R_s(t)} \right)^{\alpha_1} + a_2 \left(\frac{r}{R_s(t)} \right)^{\alpha_2} + a_3 \left(\frac{r}{R_s(t)} \right)^{\alpha_3} \right] \quad (3.74)$$

where σ is the compression ratio and the upstream density n_{up} is assumed to be constant. The parameters in Eq. (3.74) have been derived by fitting the radial profile of the shocked medium, as presented in [214], resulting into $a_1 = 0.353$, $a_2 = 0.204$, $a_3 = 0.443$, $\alpha_1 = 4.536$, $\alpha_2 = 24.18$ and $\alpha_3 = 12.29$. A visual representation of such a density profile is given in Fig. 3.12: it is worth noting here that the plasma compression ratio σ is only achieved at the shock position.

Convolving the differential energy spectrum of protons residing in the remnant interior with the density profile of Eq. (3.74), one obtains

$$I(t, T_p) = \left(\frac{d^3 p}{dT_p} \right) 4\pi \int_0^{R_{\text{SNR}}} r^2 \left[f_{\text{esc}}(r,t,p) + f_{\text{p,esc}}(r,t,p) + f_{\text{conf}}(r,t,p) \right] n_{\text{down}}(r,t) dr \quad (3.75)$$

and considering the differential cross-section for pp-interactions through

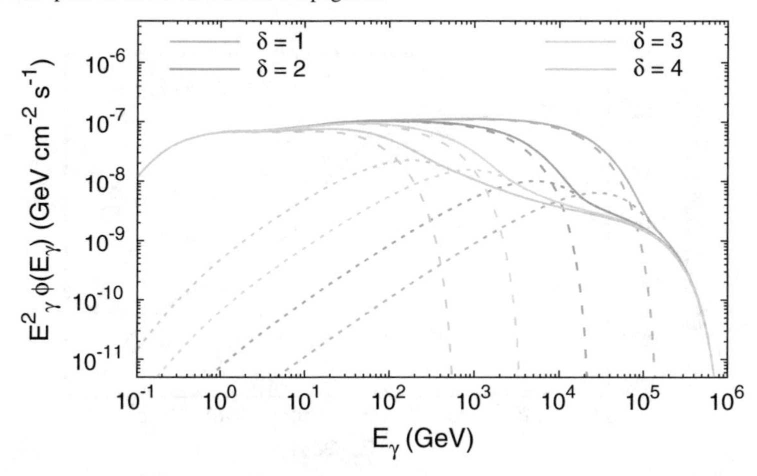

Fig. 3.13 Gamma-ray flux from pp collisions of confined and non-confined particles located inside the remnant. Same simulation parameters as the top panel of Fig. 3.9. Dashed lines for the confined distribution function, dotted for the non-confined one and solid lines for their sum. The source distance here is set to $d = 1$ kpc. Figure from Celli et al. [81], reproduced by permission of the Royal Astronomical Society

$$\epsilon_\gamma(t, E_\gamma) = 4\pi \int dT_\mathrm{p} \frac{d\sigma_{pp}}{dE_\gamma}(T_\mathrm{p}, E_\gamma) I(t, T_\mathrm{p}) \tag{3.76}$$

the gamma-ray differential energy flux at every time and energy reads as

$$\phi_\gamma(t, E_\gamma) = \frac{c}{4\pi d^2} \epsilon_\gamma(t, E_\gamma) \tag{3.77}$$

The gamma-ray flux expected from the interactions of the accelerated protons with the target gas inside the shell of an SNR located at a distance of $d = 1$ kpc is represented in Fig. 3.13. Here, the same set of parameters as the top panel of Fig. 3.9 was assumed. As was already visible in the spectrum of protons contained inside the remnant, the gamma-ray energy flux at the highest energies is essentially due to the non-confined particles, whose onset strongly depends on the parametrization of the maximum momentum time-dependence (namely on the index δ).

Analogously, the gamma-ray flux obtained with the maximum-momentum evolutionary parametrization by [75] can be derived. The resulting emission (corresponding to the proton spectrum shown in Fig. 3.10) is shown in Fig. 3.14. At this point, it is worth checking whether the modeling of $p_\mathrm{max}(t)$ from [75], or alternatively a simple power-law, is able to adequately reproduce the high-energy radiation observed from middle-aged SNRs, as will be discussed in Sect. 3.5.

Given the large number of parameters involved in the model, in order to disentangle their effects, each of them will be varied independently in the following. In particular, Figs. 3.15, 3.16 and 3.17 show the resulting gamma-ray spectrum when the upstream density, the diffusion coefficient and remnant age are varied respectively. In general, the higher is the upstream density and the earlier will the Sedov time be achieved: this implies that more particles will be able to contribute to the

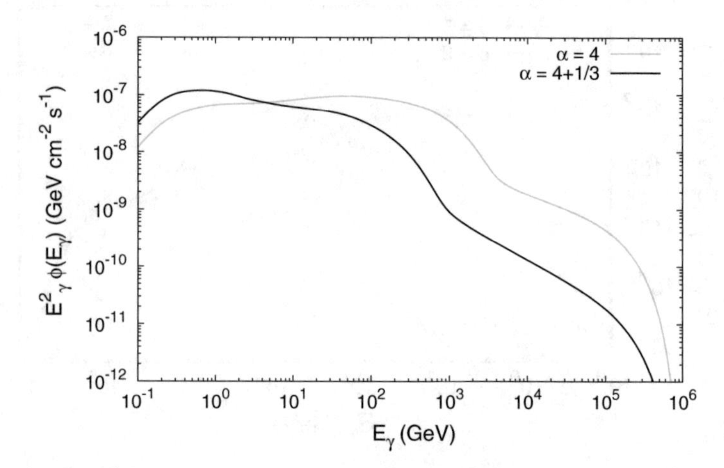

Fig. 3.14 Gamma-ray flux from *pp* collisions of confined and non-confined particles inside the remnant. Here the time-dependence of the maximum momentum at the shock is taken from [75] with $\beta = 0$ for the grey line and $\beta = 1/3$ for the black one. Same simulation parameters as in Fig. 3.10. The source distance here is set to $d = 1$ kpc

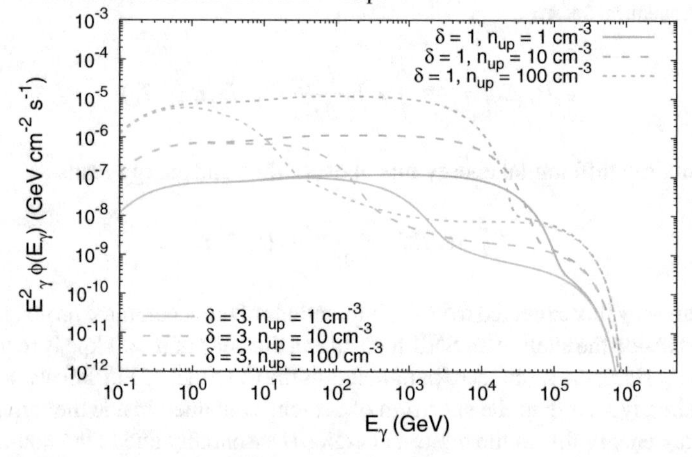

Fig. 3.15 Gamma-ray flux from *pp* collisions of confined and non-confined particles located inside the remnant at $T_{\rm SNR} = 10^4$ yr for different values of the slope δ and upstream density. The acceleration spectrum has index $\alpha = 4$, the diffusion coefficient energy-dependence here is Kolmogorov-like, normalized to $D_0 = 3 \times 10^{27}$ cm^2 s^{-1}, $p_M = 1$ PeV/c, while the downstream radial profile assumed is the Sedov solution [214]. The source distance here is set to $d = 1$ kpc

non-confined particle density function, once the age of the remnant is fixed. Moreover, the larger is the upstream density and the higher is the expected gamma-ray emissivity, as in hadronic interactions the photon intensity scales linearly with the target density. On the other hand, the smaller the diffusion coefficient is, the larger is the confinement time $\tau_{\rm diff} = R^2_{\rm SNR}/6D(p)$, implying that a larger amount of particles will be still residing within the remnant at a fixed age. Finally, for younger remnants, the lower energy particles contribute less to the non-confined particle density.

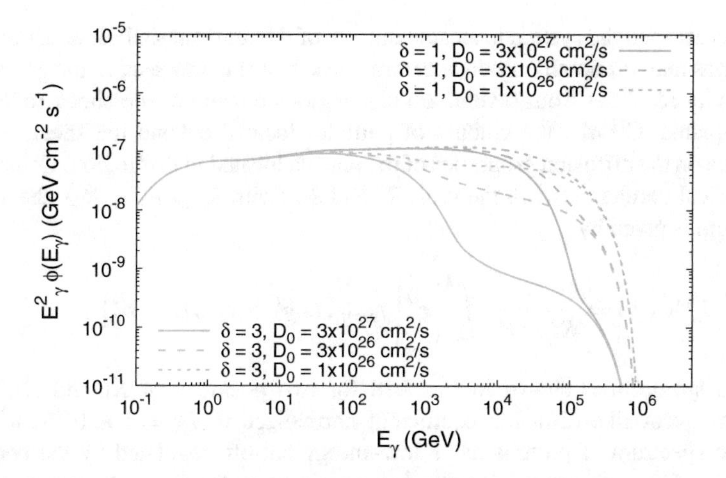

Fig. 3.16 Gamma-ray flux from *pp* collisions of confined and non-confined particles located inside the remnant at $T_{\rm SNR} = 10^4$ yr for different values of the slope δ and D_0. The acceleration spectrum has index $\alpha = 4$, the upstream numerical density here assumed is $n = 1\,{\rm cm}^{-3}$, $p_{\rm M} = 1$ PeV/c, while the downstream radial profile assumed is the Sedov solution [214]. The source distance here is set to $d = 1$ kpc

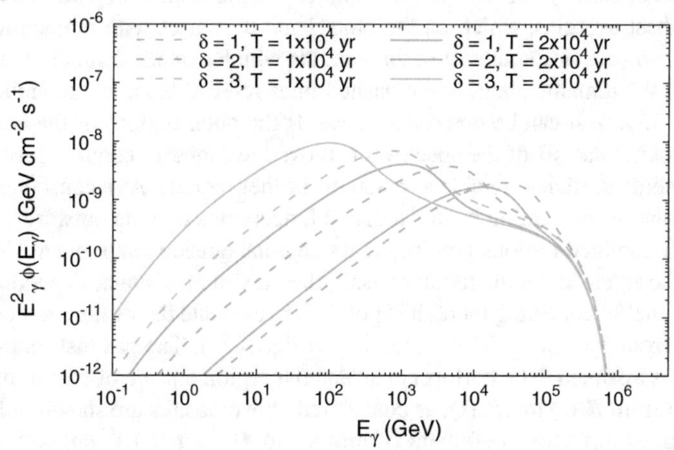

Fig. 3.17 Gamma-ray flux from *pp* collisions of non-confined particles located inside the remnant at different times and for different values of the slope δ. The acceleration spectrum has index $\alpha = 4$, the diffusion coefficient here is Kolmogorov-like, normalized to $D_0 = 3 \times 10^{27}\,{\rm cm}^2\,{\rm s}^{-1}$, $p_{\rm M} = 1$ PeV/c, the upstream numerical density here assumed is $n = 1\,{\rm cm}^{-3}$, while the downstream radial profile assumed is the Sedov solution [214]. The source distance here is set to $d = 1$ kpc

3.3.7 Protons and Gamma Rays Produced Outside of the Remnant

In the escape scenario, high-energy particles are able to leave the shock at earlier times and freely diffuse in the CSM. Thus, one would expect that part of the accelerated flux that has escaped the system is located outside of the remnant shell. Interactions with target gas may eventually produce gamma rays, hence in this section the gamma-ray emissivity resulting from several annular regions outside of the shock radius will be evaluated. Clearly, the amount of particles located beyond the shock strongly depends on the diffusion properties of the plasma located in that region. Considering a spherical corona between the radii R_1 and R_2 (with $R_s \leq R_1 < R_2$), the average spectrum is given by

$$J_p^{out}(t,p) = \frac{3}{R_2^3 - R_1^3} \int_{R_1}^{R_2} r^2 \left[f_{esc}(r,t,p) + f_{p,esc}(r,t,p) \right] dr \tag{3.78}$$

Such a spectrum is shown in Fig. 3.18 for two positions of R_1 and R_2, where a Kolmogorov-like diffusion coefficient normalized to $D_0 = 1 \times 10^{27}\,\mathrm{cm^2\,s^{-1}}$ is set: the spectrum of protons has a low-energy cut-off, regulated by the condition $p = p_{max}(T_{SNR})$, due to the fact that low-energy particles did not have enough time to achieve the location of the region considered. The contribution from the precursor is well visible at the highest energies, where the spectrum flattens: such a behavior results from the balance between the number of particles contained in the precursor, which is an increasing function of momentum, and the number of particles contained at the highest energies, which on the other hand decreases with momentum. Solid lines refer to particles enclosed in the annulus which extends from $R_s(t)$ to $2R_s(t)$ outside of the remnant shell, while dashed lines refer to those in the annulus from $2R_s(t)$ to $3R_s(t)$. It can be noted that, towards the outer regions of the accelerator, the low-energy cut-off of the spectrum is moved to highest energies since only the highest energy particles are able to reach the farther regions. As a consequence, also the spectrum normalization is affected, and it decreases moving outwards.

As non-confined protons possibly reach regions outside the remnant shell as far as twice the extension of the remnant itself, the maximum distance depending on the diffusion coefficient acting there, it is possible to evaluate the gamma-ray emissivity resulting from pp interactions, according to Eq. (3.77). The gas distribution in the upstream is assumed to be uniform in an annular region outside of the remnant shell, extending from $R_s(t)$ to $2R_s(t)$, as considered above: results are shown in Fig. 3.19, for a reduced diffusion coefficient of protons to $D_0 = 1 \times 10^{27}\,\mathrm{cm^2\,s^{-1}}$. Again, a bump-like feature appears, showing similar features concerning the low-energy cut-off as in Fig. 3.18. However, the gamma-ray flux is shifted towards lower energies, as a consequence of the secondary production, and it appears smoothed in the low-energy part. Next-generation gamma-ray instruments, as CTA, would possibly investigate the presence of such features in bright emitters: however, a correct evaluation of the instrument performances requires to account for the spatial extent of the region under investigation. For instance, a middle-aged remnant ($T_{SNR} = 10^4$ yr) at a distance of 1 kpc would cover an angular area of radius ~0.8 deg, resulting into an even more extended halo of escaping particles. The large amount of background coincident with such an extended angular search window tends to degrade the instrument sensitivity level, as more quantitative estimated in Chap. 5.

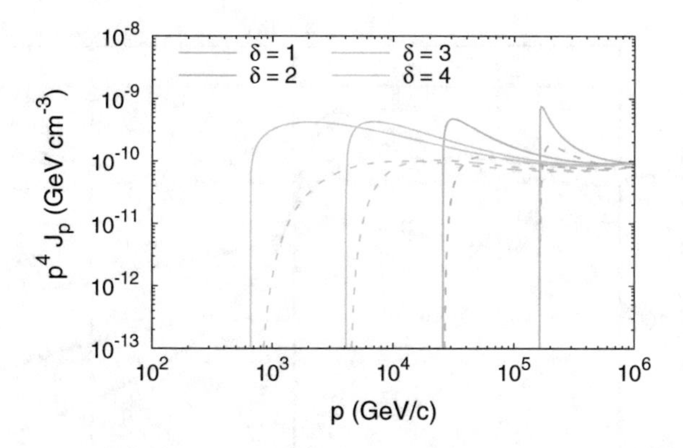

Fig. 3.18 Non-confined proton spectrum at $T_{SNR} = 10^4$ yr located outside of the remnant shell for different values of the slope δ, which regulates the time-dependence of the maximum momentum at the shock. The acceleration spectrum has index $\alpha = 4$, the upstream numerical density here assumed is $n = 1\,\text{cm}^{-3}$, while the diffusion coefficient is Kolmogorov-like, normalized to $D_0 = 1 \times 10^{27}\,\text{cm}^2\,\text{s}^{-1}$. The maximum momentum at the Sedov time is set to $p_M = 1\,\text{PeV/c}$. Solid lines for the particle spectrum within an annulus extending from R_{SNR} to $2R_{SNR}$ outside of the remnant, dashed lines for those contained in an annulus extending from $2R_{SNR}$ to $3R_{SNR}$ outside of the remnant. Figure from Celli et al. [81], reproduced by permission of the Royal Astronomical Society

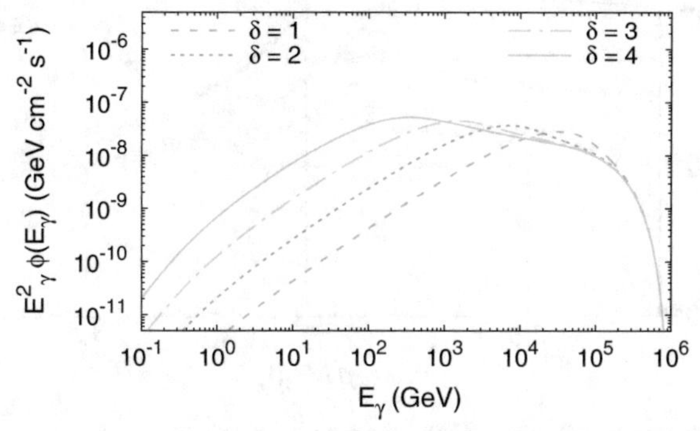

Fig. 3.19 Gamma-ray flux from pp collisions in an annulus extending outside of a remnant from R_{SNR} to $2R_{SNR}$, with an age of $T_{SNR} = 10^4$ yr. The upstream density assumed here is $n_{up} = 1\,\text{cm}^{-3}$ and the acceleration spectrum has index $\alpha = 4$. The diffusion coefficient is Kolmogorov-like, normalized to $D_0 = 1 \times 10^{27}\,\text{cm}^2\,\text{s}^{-1}$, $p_M = 1\,\text{PeV/c}$, while the source distance is $d = 1\,\text{kpc}$. Figure from Celli et al. [81], reproduced by permission of the Royal Astronomical Society

3.3.8 The Gamma-Ray Radial Profile

The volume-integrated emission is not always the best quantity to compare with the observations if the object under investigation is extended and diffuse. In this case, precious information can be derived from the remnant morphology, especially from the radial profile of the emissivity. In order to compare the observed radial profiles with the model predictions, the emission projected along the line of sight has to be

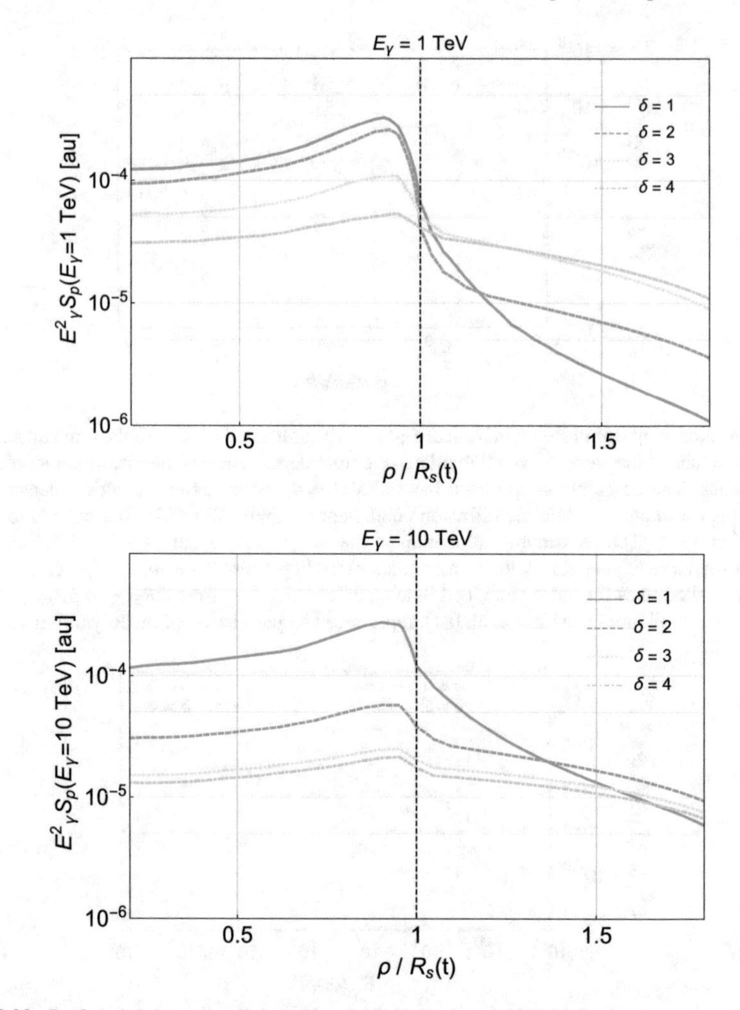

Fig. 3.20 Surface brightness radial profile of 1 TeV (top) and 10 TeV (bottom) gamma rays for a remnant age of $T_{\rm SNR} = 10^4$ yr located at a distance $d = 1$ kpc. The upstream density assumed here is $n_{\rm up} = 1\,{\rm cm}^{-3}$ and the acceleration spectrum has index $\alpha = 4$. The diffusion coefficient energy dependence is Kolmogorov-like, normalized to $D_0 = 1 \times 10^{27}\,{\rm cm}^2\,{\rm s}^{-1}$, while $p_{\rm M} = 1$ PeV/c. The CR acceleration efficiency here assumed is $\xi_{\rm CR} = 10\%$. Figure from Celli et al. [81], reproduced by permission of the Royal Astronomical Society

computed. Under the assumption of spherical symmetry, the gamma-ray emissivity is a function of the radius alone and the projected emission is simply computed by integration along the line of sight l:

$$S_{\rm p}(E_\gamma, t, \rho) = 2 \int_0^{\sqrt{R^2_{\rm max}-\rho^2}} S\left(E_\gamma, t, r = \sqrt{\rho^2 + l^2}\right) dl \qquad (3.79)$$

where $S(E_\gamma, t, r)$ represents the surface brightness and $R_{\max}$ defines the extension of the region considered in the projection. A sharp drop in the emission profile is expected between the position of the contact discontinuity and that of the shock for a shell-like SNR. Figure 3.20 provides a representation of the expected gamma-ray surface brightness profile arising from pp interactions at different photon energies and for different slopes δ. Here, the instrumental performances are also accounted for, in that a Gaussian smearing of the angular resolution is applied to the profile model of Eq. (3.79). The PSF considered is that of CTA-South, with values of $\sigma_{\rm CTA}(E_\gamma = 1~{\rm TeV}) = 0.051°$ and $\sigma_{\rm CTA}(E_\gamma = 10~{\rm TeV}) = 0.037°$.[2] As visible from the gamma-ray profile obtained, the smaller is δ and the larger are the fluctuations expected between the region inside and that outside the shell. The drop however shrinks with increasing energy, as parent particles are able to reach larger distances. As the emission profile drop ranges from about one to two orders of magnitude, it appears likely that the next-generation instruments will achieve the sensitivity level necessary for detecting such an emission from outside of the shell of bright emitters.

3.4 The Cosmic-Ray Spectrum Injected in the Galaxy

Once the distribution function of the particles accelerated at the shock is known at any given time and position, it is possible to compute the spectrum of particles released in the Galaxy by an individual SNR. In order to compute the number of particles (per unit energy) injected with momentum p, a spatial integration of the particle distribution function at the escape time has to be performed as

$$\begin{aligned} f_{\rm inj}(p, t_{\rm esc}) &= 4\pi \left[\int_0^{R_{\rm esc}(p)} r^2 f_{\rm conf}(t_{\rm esc}, r, p) dr + \int_{R_{\rm esc}(p)}^{\infty} r^2 f_{\rm p}(t_{\rm esc}, r, p) dr \right] \\ &\equiv f_1(p) + f_2(p) \end{aligned} \tag{3.80}$$

where the first contribution comes from the particles confined within the escape radius and the second one derives from particles enclosed in the precursor. In the following, these contributions are evaluated individually: in order to perform the integration in Eq. (3.80), the radial dependency of the particle distribution function has to be clearly explicited.

The distribution function of confined particles was derived in Sect. 3.3.1, including adiabatic losses, for two different shock-acceleration spectra (with $\alpha = 4$ and $\alpha = 4 + 1/3$). For a generic acceleration spectrum with spectral index α and compression ratio σ, this function can be written as

[2] https://www.cta-observatory.org/science/cta-performance/.

$$f_{\rm conf}(t, r, p) = \frac{3\xi_{\rm CR}\rho_{\rm ISM} v_s^2(t)}{4\pi c(m_p c)^{4-\alpha}\Lambda(p_{\rm max}(t))} p^{-\alpha}\theta\left[p_{\rm max}(t) - p\right]\left(\frac{R_s(t)}{R_s(t_0)}\right)^{-\alpha(\sigma-1)/\sigma} \quad (3.81)$$

During the ST phase one obtains

$$f_{\rm conf}(t, r, p) = \frac{3\xi_{\rm CR}\rho_{\rm ISM} v_s^2(t)}{4\pi c(m_p c)^{4-\alpha}\Lambda(p_{\rm max}(t))} p^{-\alpha}\theta\left[p_{\rm max}(t) - p\right]\left(\frac{t}{t_0}\right)^{-2\alpha(\sigma-1)/5\sigma} \quad (3.82)$$

In the following, the parameter ϵ is defined as

$$\epsilon = \frac{2}{5}\alpha\left(\frac{\sigma-1}{\sigma}\right) \quad (3.83)$$

This, in the case of the test-particle DSA theory with $\alpha = 4$ and $\sigma = 4$, assumes the value $\epsilon = 6/5$, while it increases for steeper spectra at the shock (for instance, for $\alpha = 4 + 1/3$ it becomes $\epsilon = 13/10$). In this way, the confined distribution function at the escape time reads as

$$f_{\rm conf}(t_{\rm esc}, r, p) = \frac{3\xi_{\rm CR}\rho_{\rm ISM} v_s^2(t_{\rm esc})}{4\pi c(m_p c)^{4-\alpha}\Lambda(p_{\rm max}(t_{\rm esc}))} p^{-\alpha}\theta\left[p_{\rm max}(t_{\rm esc}) - p\right]\left(\frac{t_{\rm esc}}{t_0(t_{\rm esc}, r)}\right)^{-\epsilon} \quad (3.84)$$

In order to explicitly express the radial dependence of $f_{\rm conf}$, one should recall the definition of $t_0(t, r)$, as in Eq. (3.41). Hence, the following relation is derived

$$\frac{t_{\rm esc}}{t_0(t_{\rm esc}, r)} = \frac{t_{\rm esc}^4}{r^{10}}\left(\frac{\xi_0 E_{\rm SN}}{\rho_{\rm ISM}}\right)^2 \Longrightarrow \left(\frac{t_{\rm esc}}{t_0(t_{\rm esc}, r)}\right)^{-\epsilon} = \left(\frac{r}{R_{\rm esc}}\right)^{10\epsilon} \quad (3.85)$$

through which the confined function at the escape time can be expressed as

$$f_{\rm conf}(t_{\rm esc}, r, p) = f_0(p)\left(\frac{r}{R_{\rm esc}}\right)^{10\epsilon} \quad (3.86)$$

Introducing this expression into Eq. (3.80), one obtains

$$f_1(p) = 4\pi f_0(p) R_{\rm esc}^{-10\epsilon}\int_0^{R_{\rm esc}} r^{2+10\epsilon} dr = 4\pi f_0(p)\frac{R_{\rm esc}^3(p)}{3 + 10\epsilon} \quad (3.87)$$

On the other hand, the contribution from particles accelerated within the shock precursor is simply described through Eq. (3.61). Therefore the spatial integration of the second term simply reads as

$$f_2(p) = 4\pi f_0(p)\frac{D_p(p)}{v_s(t_{\rm esc})}\int_{R_{\rm esc}}^{\infty} r^2\delta(r - R_{\rm esc}) dr = 4\pi f_0(p)\frac{D_p(p)}{v_s(t_{\rm esc})} R_{\rm esc}^2(p) \quad (3.88)$$

In principle, $D_{\rm p}$ depends also on time: however, a non-linear approach is beyond the scope of this work. The total spectrum of particles with momentum between p and $p + dp$ injected in the Galaxy at the escape time thus amounts to

$$f_{\rm inj}(p, t_{\rm esc}(p)) = 4\pi f_0(p) R_{\rm esc}^3(p) \left[\frac{1}{3 + 10\epsilon} + \frac{D_{\rm p}(p)}{R_{\rm esc}(p) v_{\rm esc}(p)} \right] \tag{3.89}$$

where the two terms within square brackets represent respectively the contribution from particles within the escape radius and that enclosed in the shock precursor. As explained in Sect. 3.3.2, the acceleration spectrum at every time is assumed to scale as a fixed fraction of the ram pressure, namely $f_0(p) \propto v_{\rm esc}^2(p) p^{-\alpha}/\Lambda(p)$, and therefore

$$f_{\rm inj}(p, t_{\rm esc}(p)) \propto \frac{v_{\rm esc}^2(p) R_{\rm esc}^3(p)}{\Lambda(p)} p^{-\alpha} \left[\frac{1}{3 + 10\epsilon} + \frac{D_{\rm p}(p)}{R_{\rm esc}(p) v_{\rm esc}(p)} \right] \tag{3.90}$$

In the assumption that $t_{\rm esc}(p) \propto p^{-1/\delta}$ (see Eq. (3.9)) and that the remnant is undergoing the ST phase, then $R_{\rm esc}(p) \propto p^{-2/5\delta}$ and $v_{\rm esc}(p) \propto p^{3/5\delta}$. Hence, the term $v_{\rm esc}^2(p)$ perfectly balances $R_{\rm esc}^3(p)$, and the momentum dependence of the spectrum injected in the Galaxy as provided by particles contained into the escape radius is

$$f_{\rm inj}(p) \propto \frac{p^{-\alpha}}{\Lambda(p)} \tag{3.91}$$

Here, the contribution from the precursor has been omitted, as will be justified at the end of this section. Neglecting the dependency on particle momentum provided by $\Lambda(p)$, one would derive that the the spectrum injected in the Galaxy coincides with the acceleration spectrum. On the other hand, in order to account for this term as well, the approximate expression for the integral in $\Lambda(p)$ presented in Sect. 3.3.2 will be considered in the following. In the limiting case of relativistic particles ($p \gg m_{\rm p}c$), the approximation of $\Lambda(p)$ reads as:

- if $\alpha > 4$, then $\Lambda(p) \propto (p_{\rm min}/m_{\rm p}c)^{4-\alpha} \implies f_{\rm inj}(p) \propto p^{-\alpha}$: if the acceleration spectrum is steeper than p^{-4}, the spectrum injected in the Galaxy will show the same steepness, thus coinciding with the acceleration spectrum;
- if $\alpha < 4$, then $\Lambda(p) \propto (p/m_{\rm p}c)^{4-\alpha} \implies f_{\rm inj}(p) \propto p^{-4}$: if the acceleration spectrum is flatter than p^{-4}, the spectrum injected in the Galaxy will be a p^{-4} power law, regardless of the acceleration spectrum.

This result can be summarized as: for particles with $p \gg m_{\rm p}c$ it holds that

$$f_{\rm inj}(p) \propto \begin{cases} p^{-\alpha} & \alpha > 4 \\ p^{-4} & \alpha < 4 \end{cases} \tag{3.92}$$

while for particles with $p \ll m_{\rm p}c$ it rather holds

$$f_{\rm inj}(p) \propto \begin{cases} p^{-\alpha} & \alpha > 5 \\ p^{-5} & \alpha < 5 \end{cases} \tag{3.93}$$

Few words of caution are mandatory at this stage: the spatial integration performed in Eq. (3.80) on the confined density function extends until $R_{\rm esc}(p)$, in the assumption that the remnant evolution proceeds entirely through the ST stage. Formally, one should consider whether the shock is still able to accelerate particles even when the lowest energy particles are escaping. In fact, it might either be that the remnant is evolving in the snowplow phase or alternatively that the acceleration process has stopped since the shock speed has become comparable to the sound speed of the CSM. For simplicity reasons, here the injection spectrum was considered as released by a remnant evolving purely in the ST phase.

At this point it is worthwhile stressing that the result obtained in Eq. (3.92) coincides with past calculations by [75, 189, 212], obtained under the same assumption that a fixed fraction of the shock energy is transferred to CRs. However, the definition of escaping particles adopted here is different from what has been assumed in the cited works. In fact, in [75, 189, 212], the escaping spectrum at time t is modeled as a δ-function in energy which carries a fixed fraction of the kinetic energy that the shock has at the same moment t, namely $E_{\rm esc} \propto \rho_0 v_{\rm s}^2(t)$. On the contrary, in the model presented here, the escaping flux at each fixed time t includes particles that have been accelerated in the past when the shock speed was faster than $v_{\rm s}(t)$, and have also suffered adiabatic losses. In other words, the energy carried by the particles escaping at time t is not a fixed fraction of $\rho_0 v_{\rm s}^2(t)$. The definition used by [75, 189, 212] is probably more suitable to describe the escaping process during the initial phase of the remnant life, when particles at the maximum energy are located only at the shock, while in the interior of the remnant there are only particles with smaller energies. Nonetheless, the results obtained in the relativistic regime are consistent with each other.

The result for non-relativistic energies ($p \ll m_{\rm p}c$) predicts a spectral steepening if $\alpha < 5$. This result is at odd with the CR spectrum observed by Voyager, where a hardening is rather observed. The disagreement is not surprising, in that two strong assumptions were set, which likely are not realized in reality: i) the shock keeps accelerating particles always maintaining the same efficiency, and ii) the remnant evolution proceeds all the way through the ST stage. For instance, by releasing the latter assumption, one can derive the effects produced by the end of the acceleration spectrum on the injected spectrum. By assuming that he acceleration suddenly stops at the beginning of the snowplow phase, which is reached at a time $t_{\rm sp}$ when the temperature of the shocked gas drops below 10^6 K (at a remnant age of $t_{\rm sp} \simeq 50$ kyr for the standard set of parameters considered here), it is possible to show that all those particles still located inside the SNR (namely with $p < p_{\rm max,0}(t_{\rm sp}) \simeq 40$ GeV/c) are instantaneously released into the ISM without suffering further adiabatic losses. As a consequence, the spectrum below 40 GeV/c is $\propto p^{-\alpha}$ and, interestingly, if $\alpha < 4$ a break in the injected spectrum appears right at this energy. In summary, the spectrum injected into the Galaxy is a feature-less power law under two conditions: i) the acceleration spectrum has to be steeper than p^{-4}, and ii) the acceleration should stop when the maximum energy is still in the relativistic domain. The latter condition also translates into an upper bound for δ which, for the parameter values adopted here, has to be ≤ 4.

In the following it is shown that the contribution to $f_{\rm inj}$ from the shock precursor, namely f_2, is always negligible with respect to f_1 (as defined in Eq. (3.80)). In fact, the relative weight between particles injected by the inner shock and those injected by the shock precursor is:

$$\frac{f_2(p)}{f_1(p)} = \frac{D_{\rm p}(p)}{R_{\rm esc}(p)v_{\rm esc}(p)}(3+10\epsilon) \tag{3.94}$$

The contribution from particles in the shock precursor is thus negligible if $D_{\rm p}(p) \ll D^*(p)$, where

$$D^*(p) = \frac{R_{\rm esc}(p)v_{\rm esc}(p)}{3+10\epsilon} = \frac{2}{5(3+10\epsilon)}t_{\rm esc}^{-1/5}(p)\left(\frac{\xi_0 E_{\rm SN}}{\rho_{\rm ISM}}\right)^{2/5} \tag{3.95}$$

In the assumptions of Eq. (3.9), the critical diffusion coefficient is

$$D^*(p) = \frac{2}{5(3+10\epsilon)}\left(\frac{\xi_0 E_{\rm SN}}{\rho_{\rm ISM}}\right)^{2/5} t_{\rm Sed}^{-1/5}\left(\frac{p}{p_{\rm M}}\right)^{1/5\delta} \tag{3.96}$$

Using the definition for the Sedov time as provided in Eq. (3.3), the final expression for $D^*(p)$ is

$$D^*(p) = \frac{2}{5(3+10\epsilon)}\xi_0^{2/5}E_{\rm SN}^{1/2}\rho_{\rm ISM}^{-1/3}M_{\rm ej}^{-1/6}\left(\frac{p}{p_{\rm M}}\right)^{1/5\delta} \tag{3.97}$$

For a p^{-4} acceleration spectrum, assuming $E_{\rm SN} = 10^{51}$ erg, $M_{\rm ej} = 10M_\odot$, $\rho_{\rm ISM} = 1\,(m_{\rm p}/{\rm g})\,({\rm g/cm^3})$ and $p_{\rm M} = 1\,{\rm PeV/c}$, the critical value reads as

$$D^*(p) = \frac{3\times 10^{26}}{10^{1/\delta}}\left(\frac{p}{10\,{\rm GeV/c}}\right)^{1/5\delta}\ {\rm cm^2\,s^{-1}} \tag{3.98}$$

Considering for $D_{\rm p}(p)$ a Bohm diffusion coefficient within a few μG magnetic field, the condition $D^*(p) \gg D_{\rm p}(p)$ is always verified below $p \simeq 10\,{\rm TeV/c}$ for $\delta > 2/5$. Hence, the number of particles escaping from the shock precursor is generally negligible with respect to that of particles escaping from within the shock radius. It is worth recalling that this conclusion holds for remnants expanding in homogenous media. Possibly, the particles from the shock precursor become relevant in the highest energy domain. However, if the amplified turbulence is driven by the resonant CR streaming, the Bohm diffusion regime is likely to occur: it leads to a maximum energy of the order of $p_{\rm max} \simeq 100\,{\rm TeV/c}$ and thus the condition that $D^*(p) \gg D_{\rm p}(p)$

will generally be verified at all the interesting energy scales. If, on the other hand, the amplified turbulence is driven by the non-resonant CR streaming, a more efficient acceleration will take place, which highly reduces the diffusion coefficient with respect to the Bohm value, and leads to $p_{\max} \simeq 1\,\mathrm{PeV/c}$. However, if $D_{\mathrm{p}}(p)$ is reduced, then the escaping particles contained in the shock precursor become less important for the spectrum of particles injected in the Galaxy. As a consequence, the spectrum injected by middle-aged SNRs in the Galaxy can safely be assumed to coincide with that defined by Eqs. (3.92) and (3.93).

3.5 Application to Some Middle-Aged SNRs

The particle propagation model presented so far allows the description of the CR density distribution (i) within a source, as due to both particles confined by the shock and particles which have effectively escaped the system, but are still diffusing within the shock radius, and (ii) outside the source, as due to non-confined particles only and discussed in Sect. 3.3.7. In this section the former point is considered, and the gamma-ray flux emerging from CR interactions with a target density is evaluated. Such a VHE radiation can in fact be used to test the particle escape modeling presented and to define the values of unknown parameters which currently characterize it. Clearly, this approach only represents an attempt to show the possible applications of the model: in fact, realistic SNRs strongly differ from the spherically symmetric approximation adopted here, and also their evolution is significantly affected by the non-homogeneous density distribution of the surrounding medium. With these caveats, the model predictions are applied to some interesting SNRs, namely middle-aged remnants which are embedded in molecular clouds [129], in that they constitute bright sources of gamma rays. In particular, IC 443, W 51C and W 28N are considered in the following. Note that all of them belong to the SNOB catalog [175]. As already mentioned, the treatment developed so far for the description of the particle escape process from SNRs only applies to middle-aged sources. In fact, specific assumptions were set regarding both the evolutionary stage of remnant (see Eq. (3.5)) and the density profile of compressed matter (Eq. 3.74), which are valid for times larger than the Sedov phase onset. Thus, a different solution should be adopted in the case of young remnants.

3.5.1 *The Case of IC 443*

IC 443 is a middle-aged (~3–30 kyr) shell type SNR, probably originated by a core-collapse SN, located at a distance of 1.5 kpc[3] from Earth in the direction of the Galactic anticenter. It is well known for its radio, optical, X-ray and gamma-

[3] http://tevcat.uchicago.edu/?mode=1;id=120.

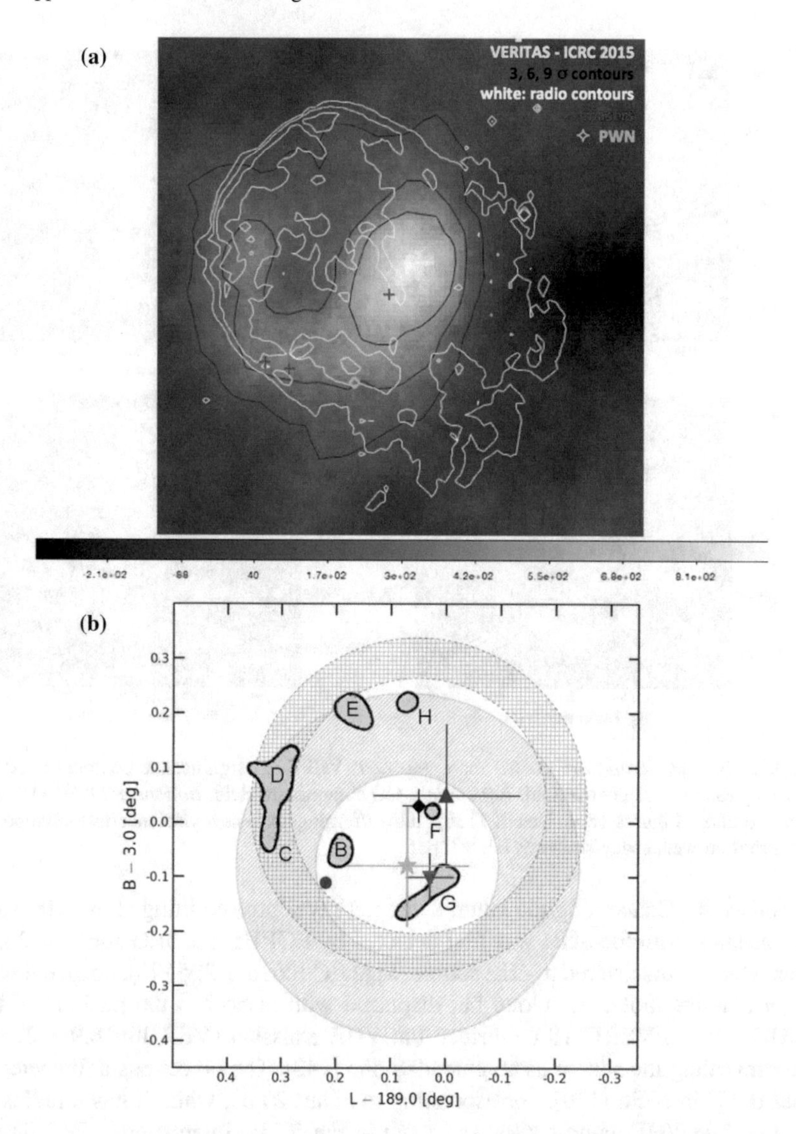

Fig. 3.21 IC 443. *Top*: VERITAS excess map, with white contours indicating the radio shell and black contours for the significance of the VERITAS observations at the 3, 6, and 9 σ levels. Figure from [142] under the CC BY LICENSE. *Bottom*: Locations of the gamma-ray sources: EGRET centroid (blue up triangle); MAGIC centroid (red down triangle); VERITAS centroid (green star) and Fermi-LAT centroid (black diamond). Best-fit spatial extensions of the Fermi (cross-hatched band) and VERITAS (striped green band) sources are drawn as rings: the central sources are located 0.12° apart, corresponding to 1.5 times the VERITAS localization error (68% C.L.). The PWN location is shown as a dot. Figure from Abdo et al. [13] ©AAS. Reproduced with permission

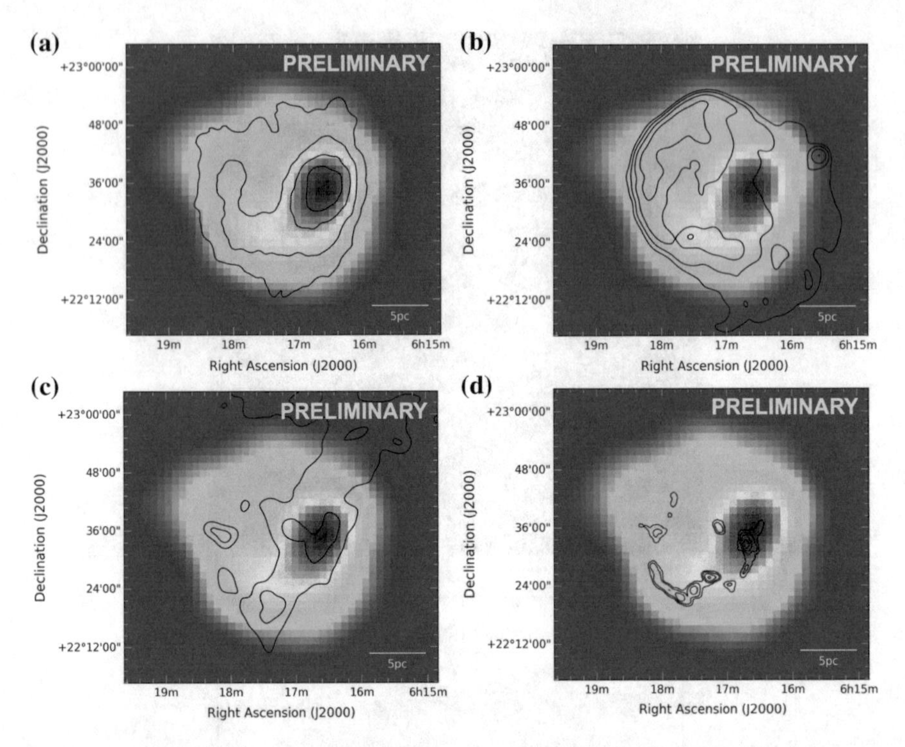

Fig. 3.22 IC 443 Fermi-LAT counts map. *Top Left*: VERITAS significance contours overlaid. *Top right*: Radio contours overlaid. *Bottom left*: CO contours overlaid. *Bottom right*: HCO^+ contours overlaid. Figures from Hewitt [136], https://fermi.gsfc.nasa.gov/science/mtgs/symposia/2015/program/wednesday/session9/JHewitt.pdf

ray emissions. Concerning the latter, a gamma-ray source emitting above 100 MeV and cospatial with the SNR was first detected by EGRET, i.e. 3EG J0617 + 2238. Later, MAGIC discovered a VHE source, MAGIC J0616 + 225 [48], cospatial with a very massive molecular cloud but displaced with respect to the position of the EGRET source. VERITAS confirmed the VHE emission (VER J0616.9 + 2230), while revealing the source as an extended one [142]. The object has a diameter of about 0.75° in radio [129], corresponding to about 20 pc, while it has a radius of about 0.3° in VHE gamma rays, as shown in Fig. 3.21a. Interestingly, Fermi-LAT detected a morphology above 5 GeV similar to that in the VHE domain and revealed the pion bump, distinctive feature of hadronic processes [23]. However, the GeV and the TeV emission centroids appear significantly displaced from each other and from the known PWN CXOU J061705.3 + 222127 (by ~10–20 arcmin), which suggests that the PWN is not the major emitter in the VHE energy band, as shown in Fig. 3.21b. The displacement might be connected with a non-homogeneous density profile of the surrounding medium. Furthermore, the TeV shell appears thicker than the GeV one, as one would expect in an escape scenario. The gamma-ray emission results brighter in a region where a maser is located, supporting the interpretation that the shock would be responsible for the stimulated emission. The remnant is well

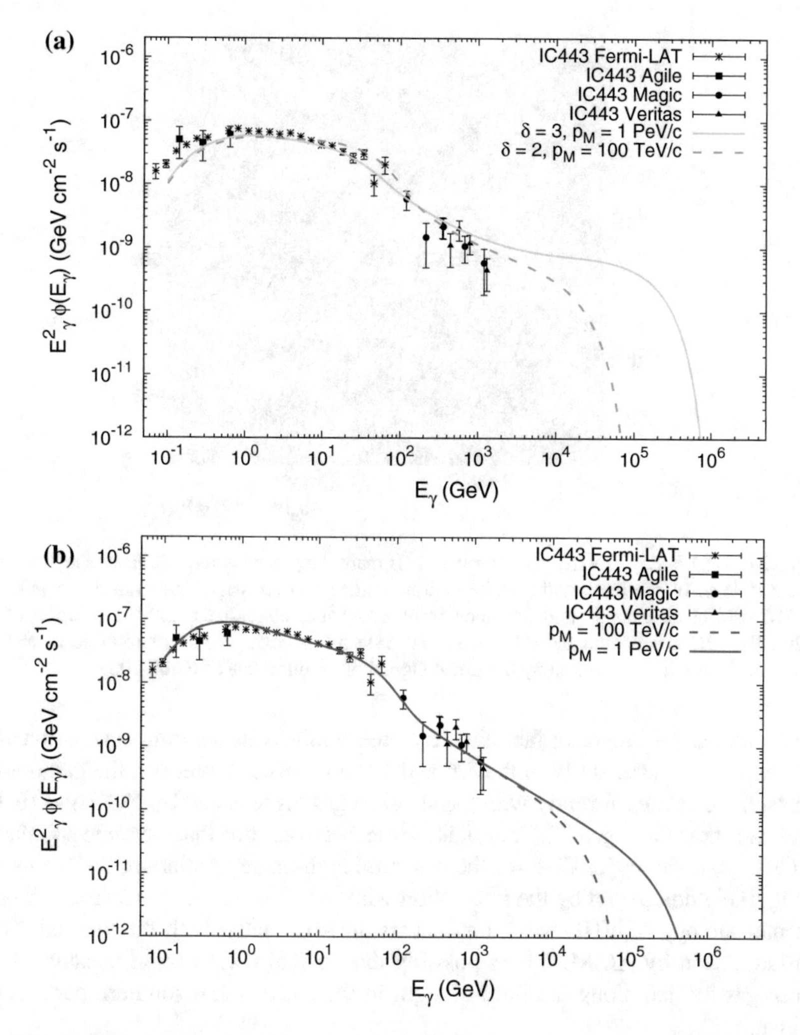

Fig. 3.23 Modeling of the GeV-TeV emission of IC 443 for confined and non-confined protons interacting with the target gas. The remnant age is set to $T_{SNR} = 1.5 \times 10^4$ yr, and the target numerical density is assumed equal to $n_{up} = 10\,\mathrm{cm}^{-3}$. *Top*: The escape model assumed here features a maximum momentum time-dependence in the form of Eq. (3.8) with δ and p_M as indicated in the legend, an acceleration spectrum is set to p^{-4}, a Kolmogorov-like diffusion coefficient within the remnant normalized to $D_0 = 1 \times 10^{27}\,\mathrm{cm}^2\,\mathrm{s}^{-1}$, resulting into a CR acceleration efficiency of $\xi_{CR} = 2\%$. *Bottom*: The escape model assumed here follows the description by [75] with p_M as indicated in the legend, an acceleration spectrum is set to $p^{-(4+1/3)}$, a Kolmogorov-like diffusion coefficient within the remnant normalized to $D_0 = 3 \times 10^{26}\,\mathrm{cm}^2\,\mathrm{s}^{-1}$, resulting into a CR acceleration efficiency results of $\xi_{CR} = 2\%$. Spectral data points in both panels from [20, 23, 48, 226]

known also from multi-wavelength observations, in that it is located into a complex environment and appears to be interacting with a molecular cloud, as detected in radio and illustrated in Fig. 3.22b. While the EGRET source is located in the center of the SNR, the VHE gamma-ray source is displaced to the south, in direct correlation

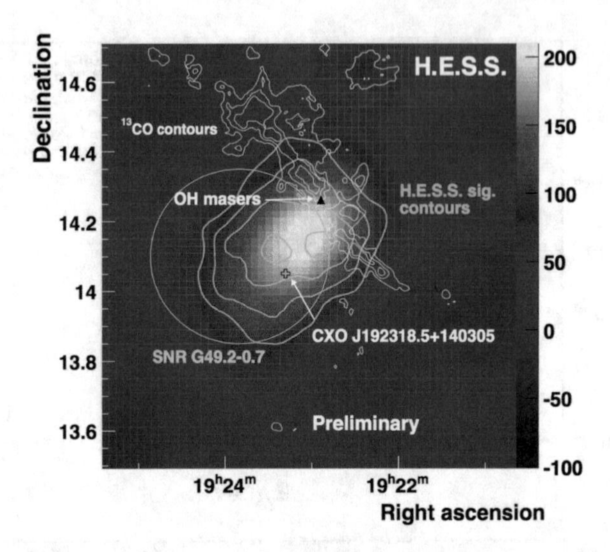

Fig. 3.24 W51 H.E.S.S. VHE gamma-ray excess map: the green contours indicate the 3–6 σ significance level of the excess, the white contours show CO emission, the cyan circle indicates W 51C, the black triangle and the open black cross indicate respectively the position of the 1720 MHz OH masers and the PWN CXO J192318.5+140305. Figure reproduced from Fiasson et al. [109] with permission by the Local Organizing Committee of ICRC 2009

with a molecular cloud. In fact, the TeV emission reveals a strong correlation with dust emission, particularly in the 4.6 and 12 μm ranges. Moreover, the gamma-ray emission correlates strongly with the shocked gas, as revealed by ^{12}CO and HCO$^+$ lines and shown in Fig. 3.22. Thus, it is reasonable to assume that the same population of CRs might be responsible for the observed high-energy radiation, their emission being likely dominated by the interaction with gas close to the shock front. Finally, the morphology of VHE radiation appears anti-correlated with the thermal X-ray emission seen by XMM: this is possibly due to X-ray absorption because of the dense gas located along the lines of sight in the western and southern parts of the remnant.

The VHE data show a quite steep spectrum $\phi(E) \propto E^{-\Gamma}$: MAGIC reported a spectral index $\Gamma = 3.1 \pm 0.3$ [48], and VERITAS data show for the entire emitting region $\Gamma = 2.99 \pm 0.38$ [20]. Fixing the remnant age to $T_{\rm SNR} = 1.5 \times 10^4$ yr and the density of the upstream medium to $n_{\rm up} = 10\,{\rm cm}^{-3}$, the remnant radial extent would amount to 0.38°. With these parameters, it is possible to reproduce the GeV-TeV data within the context of a maximum momentum temporal dependence in the form of a power-law trend, as visible from Fig. 3.23a, provided that a diffusion coefficient around the remnant of $D_0 = 10^{27}\,{\rm cm^2\,s^{-1}}$ is achieved and the acceleration spectrum has slope $\alpha = 4$. The spectral modeling can be achieved by either setting $\delta = 3$ and $p_{\rm M} = 1\,{\rm PeV/c}$ or $\delta = 2$ and $p_{\rm M} = 100\,{\rm TeV/c}$, therefore constraining the CR acceleration efficiency to $\xi_{\rm CR} \simeq 2\%$. It is worth to note that a clear spectral hardening is expected at the highest energies in the case of $p_{\rm M} = 1$ PeV/c, as due to the precursor contribution. This feature will possibly be revealed by next-generation instruments, and it will allow to distinguish between the different scenarios presented here. Alternatively, a satisfactory description of the data can also be achieved in the context of the model by [75], as shown in Fig. 3.23b, when assuming $\alpha = 4 + 1/3$ coupled with a diffusion coefficient of $D_0 = 3 \times 10^{26}\,\mathrm{cm^2\,s^{-1}}$. Here, the same CR acceleration efficiency as the former model would be required. However, with current data, it is not possible to discriminate between the scenario featuring $p_\mathrm{M} = 100\,\mathrm{TeV/c}$ or the one with $p_\mathrm{M} = 1\,\mathrm{PeV/c}$.

3.5.2 The Case of W 51C

W 51 is a radio complex composed of two HII regions, W 51A and W 51B. These are enclosed in a giant molecular cloud (GMC), that hosts several star forming regions. A third component is visible in the radio complex: the SNR W 51C (also called G 49.2 − 0.7), appearing as a shell with ∼30′ radius in radio continuum, likely located at a distance of 5.4 kpc.[4] Two 1720 MHz OH masers have been detected towards the North of the SNR shell, proving that the remnant is interacting with the GMC. The SNR W 51C has also been detected in X rays, where a composite structure appears composed by: (i) a thermal emission detected from most parts of the shell, plus (ii) a non-thermal component arising from the PWN CXO J192318.5 + 140505. Gamma-ray emission from the same region was discovered by H.E.S.S. (>1 TeV) [109], with an integrated flux above 1 TeV equivalent to 3% of the Crab Nebula. Subsequently it was also identified in Fermi-LAT data (between 0.2 and 50 GeV) [14] and later on confirmed by MAGIC (between 50 GeV and few TeV) [49]. The VHE emissions are shown in Fig. 3.24 for the H.E.S.S. detection and in Fig. 3.25 for the MAGIC one. The HE and VHE emission radially extends for about 0.2°. Interestingly, the MAGIC flux map between 300 and 1000 GeV shows an overall shape which is elongated towards the south-east, where the PWN is located, while the maximum of the emission coincides with the region of the shocked gas. To date, it is not possible to exclude underlying structures, namely to clearly discriminate between an extended source of excess or two individual sources. As a consequence, the following modeling the VHE emission takes into account the whole emission as connected to a unique source: as estimated in [49], the maximum contribution of the PWN to the VHE emission would amount to about 20%.

The spectral index measured by MAGIC in the VHE domain amounts to $\Gamma = 2.58 \pm 0.07$ [49]. On the other hand, data from Fermi-LAT show an energy break at $E_b \simeq 2.7\,\mathrm{GeV}$, with a spectral slope above the break of $\Gamma = 2.52 \pm 0.07$ [144]. The very steep spectrum at multi-GeV requires an acceleration spectrum as $\propto p^{-(4+1/3)}$. A scenario with escaping CRs from a remnant as old as $T_\mathrm{SNR} = 3 \times 10^4$ yr, and interacting with a target numerical density of $n_\mathrm{up} = 10\,\mathrm{cm^{-3}}$, well reproduces the HE and VHE data: the remnant radial extension at this age would amount to 0.2°. The spectral modeling can either be realized within the power-law like maximum momentum

[4] http://tevcat.uchicago.edu/?mode=1;id=178.

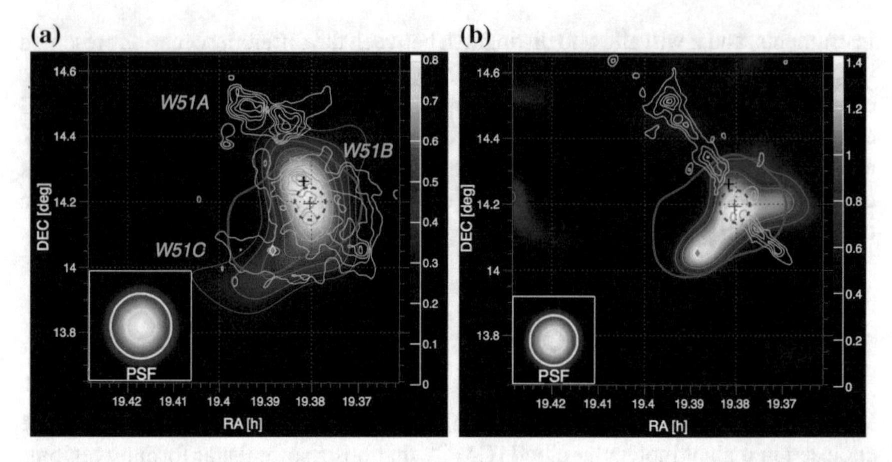

Fig. 3.25 W 51 MAGIC flux map: the blue diamond represents the position of CXO J192318.5+140305, the black cross the position of the OH maser emission, the red dashed ellipse represents the region of shocked atomic and molecular gas, while pink contour displays counts above 1 GeV determined by Fermi-LAT. *Left*: Energies from 300 GeV to 1 TeV, the green contours represent the 21 cm radio continuum emission. *Right*: Energies above 1 TeV, the green contours represent the ^{13}CO (J= 1 − 0) intensity map. Credit: Aleksic et al. [49] reproduced with permission ©ESO

time-dependence parametrization or within the [75] scenario. The first case is shown in Fig. 3.26: it is possible to reproduce the data by setting $\delta = 2$ either coupled with (i) $\alpha = 4$, $p_M = 100$ TeV/c and $D_0 = 1 \times 10^{27}\,\mathrm{cm^2\,s^{-1}}$, resulting into $\xi_{CR} = 20\%$; or (ii) $\alpha = 4 + 1/3$, $p_M = 100$ TeV/c (or 1 PeV/c) and $D_0 = 3 \times 10^{26}\,\mathrm{cm^2\,s^{-1}}$, resulting into $\xi_{CR} = 15\%$ (or $\xi_{CR} = 12\%$). The second case is illustrated in Fig. 3.27: here a diffusion coefficient normalization equal to $D_0 = 1 \times 10^{26}\,\mathrm{cm^2\,s^{-1}}$ is rather required, with $\alpha = 4 + 1/3$ and $\xi_{CR} = 11\%$. All the aforementioned values for the CR acceleration efficiency result well consistent with the standard assumptions in the context of the SNR paradigm for the origin of CRs.

3.5.3 The Case of W 28N

W 28 (G 6.4 − 0.1) is a mixed-morphology SNR, composed by a radio shell enclosing a center-filled thermal X-ray emission, located within a complex star-forming region in the Galactic Plane, towards large HII regions (M 8 and M 20) and young clusters (NGC 6530), at a distance of ~2 kpc.[5] The overall shape of the SNR is elliptical with dimensions 50′ × 45′. The high concentration of 1720 MHz OH masers and the high density ($n_H > 10^3\,\mathrm{cm^{-3}}$) shocked gas reveal the interactions of the SNR with molecular clouds along its northern and northeastern boundaries. The estimated age of the system is largely uncertain, likely ranging between 3×10^4 and 1.5×10^5 yr,

[5]http://tevcat.uchicago.edu/?mode=1;id=162.

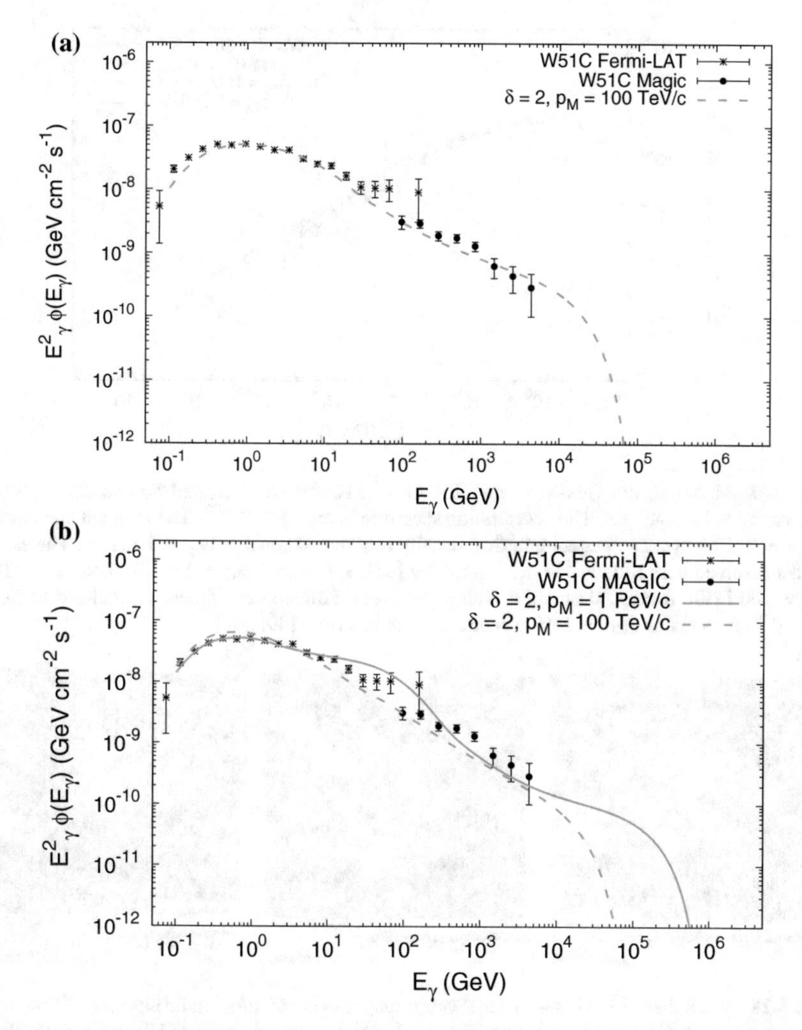

Fig. 3.26 Modeling the GeV-TeV emission of W 51C for confined and non-confined protons interacting with target gas. The remnant age is set to $T_{\mathrm{SNR}} = 3 \times 10^4$ yr. The target numerical density is assumed equal to $n_{\mathrm{up}} = 10\,\mathrm{cm}^{-3}$. *Top*: The escape model assumed here features a maximum momentum time-dependence in the form of Eq. (3.8) with $\delta = 2$ and $p_{\mathrm{M}} = 100\,\mathrm{TeV/c}$ (green dashed), a Kolmogorov-like diffusion coefficient normalized to $D_0 = 1 \times 10^{27}\,\mathrm{cm}^2\,\mathrm{s}^{-1}$, an acceleration spectrum with slope $\alpha = 4$ and a CR acceleration efficiency of $\xi_{\mathrm{CR}} = 20\%$. *Bottom*: The escape model assumed here features a maximum momentum time-dependence in the form of Eq. (3.8) with $\delta = 2$ and either (i) $p_{\mathrm{M}} = 100\,\mathrm{TeV/c}$ (green dashed) and $\xi_{\mathrm{CR}} = 15\%$ or (ii) $p_{\mathrm{M}} = 1\,\mathrm{PeV/c}$ and $\xi_{\mathrm{CR}} = 12\%$. A Kolmogorov-like diffusion coefficient normalized to $D_0 = 3 \times 10^{26}\,\mathrm{cm}^2\,\mathrm{s}^{-1}$ was set, as well as an acceleration spectrum with slope $\alpha = 4 + 1/3$. Spectral data points in both panels from [49, 144]

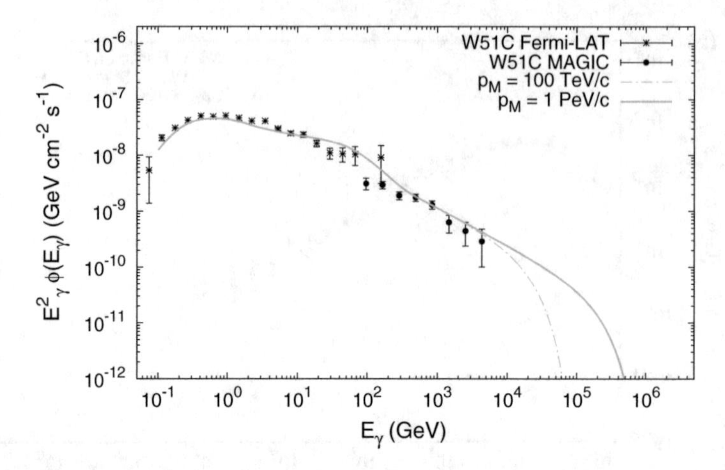

Fig. 3.27 Modeling the GeV-TeV emission of W 51C for confined and non-confined protons interacting with target gas. The acceleration spectrum is set $\propto p^{-(4+1/3)}$. The remnant age is set to $T_{\rm SNR} = 3 \times 10^4$ yr. The target numerical density is assumed equal to $n_{\rm up} = 10\,{\rm cm}^{-3}$. The escape model assumed here follows the description by [75], with either $p_{\rm M} = 1\,$PeV/c (orange solid) or $p_{\rm M} = 100\,$TeV/c (orange dashed), a Kolmogorov-like diffusion coefficient normalized to $D_0 = 1 \times 10^{26}\,{\rm cm}^2\,{\rm s}^{-1}$ and $\xi_{\rm CR} = 11\%$. Spectral data points from [49, 144]

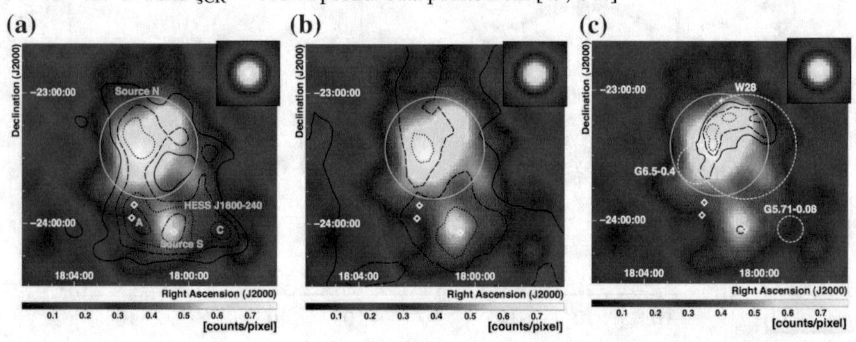

Fig. 3.28 W 28. Fermi-LAT 2 – 10 GeV count map: the green circle indicates size of Source N, the green cross indicates the position of Source S, white diamonds indicate HII regions (W 28A2, G 6.1 – 0.6 and G6.225 – 0.569), the diamond on the right is W 28A2. *Left*: Black contours represent the H.E.S.S. significance map for TeV gamma rays at 20, 40, 60 and 80% of the peak value, while bright TeV spots in the south are HESS J1800 – 240A, B and C. *Middle*: Black contours give CO (J= 1 – 0) line intensity taken by NANTEN at 25, 50, 75% levels. *Right*: Black contours indicate the VLA 90cm image at 25, 50, 75% of the peak intensity. Outer boundaries of the SNR, as determined by the radio images, are drawn as white dashed circles. Figures from Abdo et al. [15] ©AAS. Reproduced with permission

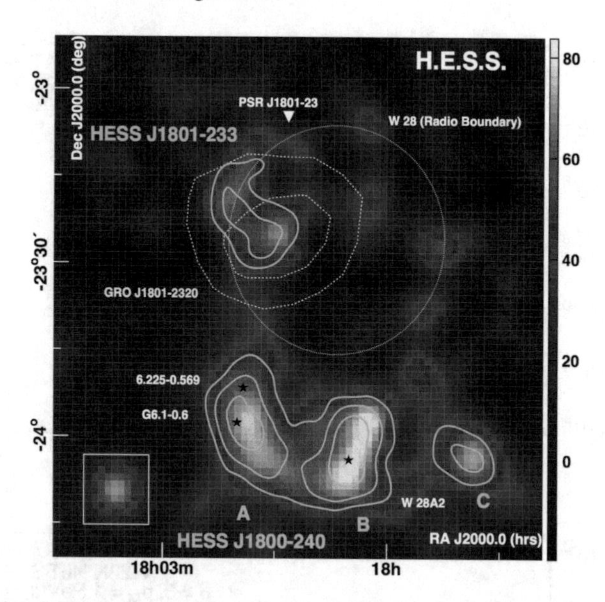

Fig. 3.29 W 28. VHE gamma-ray excess counts as revealed by H.E.S.S.: the thin-dashed circle depicts the approximate radio boundary of the SNR W 28, guided predominantly by the bright northern emission. Identified here are VHE source regions HESS J1801 − 233 to the northeast, and a complex of sources HESS J1800 − 240 (A, B & C) to the south of W 28. Also indicated are: HII regions (black stars); W 28A2, G 6.1 − 0.6, G 6.225 − 0.569, the 68% and 95% location contours (thick-dashed yellow lines) of the $E > 100$ MeV EGRET source GRO J1801 − 2320, and the pulsar PSR J1801 − 23 (white triangle). Credit: Aharonian et al. [41] reproduced with permission ©ESO

where the upper limit is set by its enhanced optical emission, suggesting that the SNR has entered its radiative phase of evolution [163]. Nonetheless, since pronounced Balmer lines have been observed only in some parts of the remnant, and given the intrinsic complexity of the non uniform SNR dynamical evolution (whose modeling is beyond the scope of this work), W 28 will be considered in the following as if it were completely evolving through its adiabatic phase.

H.E.S.S. observations of the W 28 field have revealed four TeV gamma-ray sources positionally coincident with molecular clouds [41]: HESS J1801 − 233, located along the northeastern boundary of W 28, and a complex of sources, HESS J1800 − 240A, B and C, located ~30′ south of SNR W 28. GeV data from Fermi-LAT [15] have revealed the presence of two sources in the vicinity of W 28, as visible from Fig. 3.28: 1FGL J1801.3 − 2322c (also called Source N, a disc-like source of radial extension 0.39° coincident with HESS J1801 − 233) and 1FGL J1800.5 − 2359c (also called Source S, a point-like source spatially coinciding with HESS J1800 − 240B). On the other hand, no significant GeV counterpart was found at the positions of HESS J1800 − 230*A* and HESS J1800 − 230C. The spectrum of Source N smoothly connects with the TeV spectrum, suggesting a physical relationship between the two. In particular, the measured spectral index by Fermi-LAT above the energy break at $E_b = 1$ GeV amounts to $\Gamma = 2.74 \pm 0.06$ [15]; consistently, H.E.S.S. data indicate $\Gamma = 2.66 \pm 0.27$ [41]. A modeling of the

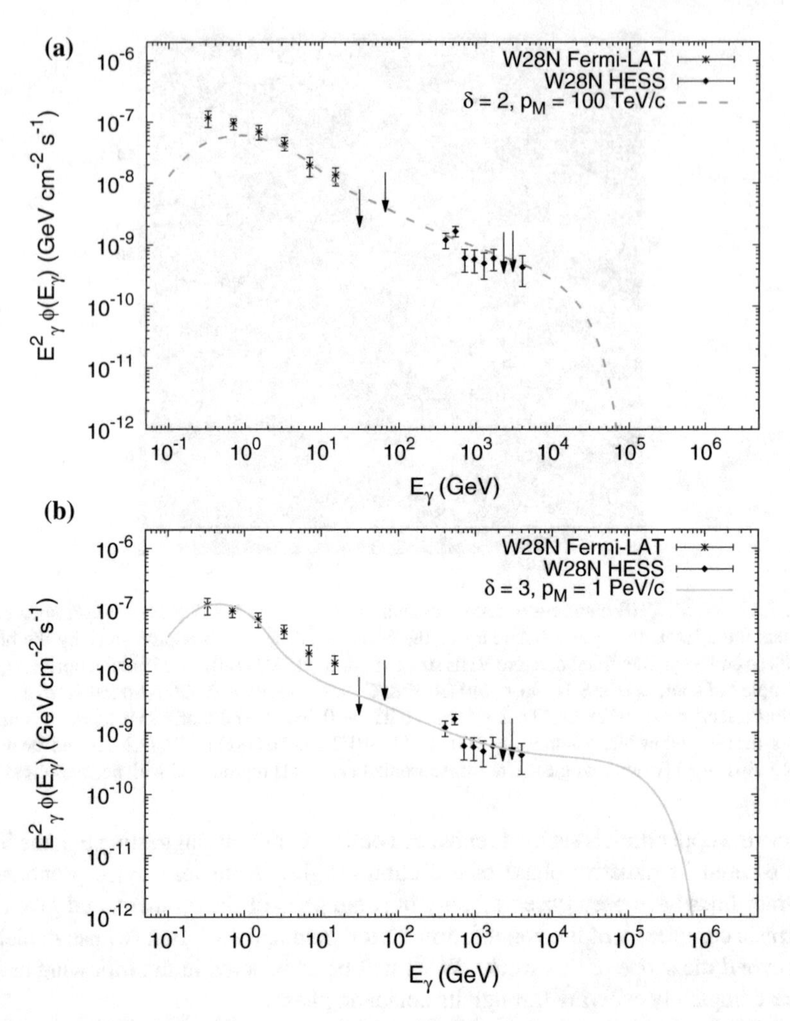

Fig. 3.30 Modeling of the GeV-TeV emission of W 28N for confined and non-confined protons interacting with target gas. The escape model assumed here features a maximum momentum time-dependence in the form of Eq. (3.8). The remnant age is set to $T_{\rm SNR} = 4 \times 10^4$ yr, while the target numerical density is assumed equal to $n_{\rm up} = 10\,{\rm cm}^{-3}$ and the acceleration spectrum is set $\propto p^{-4}$. *Top*: The $p_{\rm max}(t)$ is defined by $\delta = 2$ and $p_{\rm M} = 100\,{\rm TeV/c}$ (green dashed), the Kolmogorov-like diffusion coefficient around the remnant is normalized to $D_0 = 1 \times 10^{27}\,{\rm cm}^2\,{\rm s}^{-1}$ and the CR acceleration efficiency results in $\xi_{\rm CR} = 4\%$. *Bottom*: The $p_{\rm max}(t)$ is defined by $\delta = 3$ and $p_{\rm M} = 1\,{\rm PeV/c}$ (pink solid), the Kolmogorov-like diffusion coefficient is normalized to $D_0 = 3 \times 10^{27}\,{\rm cm}^2\,{\rm s}^{-1}$, while $\xi_{\rm CR} = 15\%$. In both panels, Fermi-LAT data points from [15], black diamonds for HESS J1801 − 233 from [41]

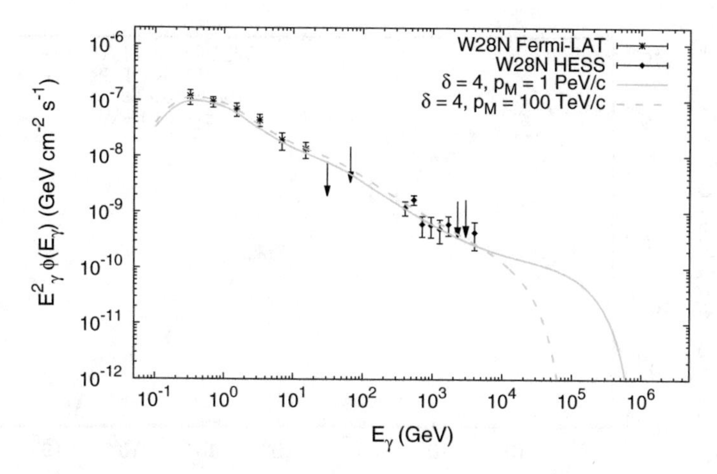

Fig. 3.31 Modeling of the GeV-TeV emission of W 28N for confined and non-confined protons interacting with target gas. The escape model assumed here features a maximum momentum time-dependence in the form of Eq. (3.8). The remnant age is set to $T_{\mathrm{SNR}} = 4 \times 10^4$ yr, while the target numerical density is assumed equal to $n_{\mathrm{up}} = 10\,\mathrm{cm}^{-3}$ and the acceleration spectrum is set $\propto p^{-(4+1/3)}$. The Kolmogorov-like diffusion coefficient around the remnant is normalized to $D_0 = 1 \times 10^{27}\,\mathrm{cm}^2\,\mathrm{s}^{-1}$. The $p_{\mathrm{max}}(t)$ is defined by $\delta = 4$ and either (i) $p_{\mathrm{M}} = 100\,\mathrm{TeV/c}$ (cyan dashed), implying $\xi_{\mathrm{CR}} = 20\%$, or (ii) $p_{\mathrm{M}} = 1\,\mathrm{PeV/c}$ (cyan dashed), implying $\xi_{\mathrm{CR}} = 15\%$. Fermi-LAT data points from [15], black diamonds for HESS J1801 − 233 from [41]

GeV-TeV emission of Source N is here attempted. Several scenarios reproduce the data: in the context of a maximum momentum power-law time-dependence, both a standard p^{-4} acceleration spectrum and a steeper $p^{-(4+1/3)}$ can reproduce the spectral shape and intensity of the gamma-ray data, as shown in Figs. 3.30 and 3.31. Here, the remnant age was set to $T_{\mathrm{SNR}} = 4 \times 10^4$ yr with a target density of $n_{\mathrm{up}} = 10\,\mathrm{cm}^{-3}$, implying a radial extension of 0.42°. In the former case, a reasonable data description is achieved when assuming a Kolmogorov-like diffusion coefficient normalized to $D_0 = 1 \times 10^{27}\,\mathrm{cm}^2\,\mathrm{s}^{-1}$, $p_{\mathrm{M}} = 100\,\mathrm{TeV/c}$ and $\delta = 2$, implying a CR acceleration efficiency of $\xi_{\mathrm{CR}} = 4\%$. Alternatively, normalizing the diffusion coefficient to $D_0 = 3 \times 10^{27}\,\mathrm{cm}^2\,\mathrm{s}^{-1}$, data can also by $p_{\mathrm{M}} = 1\,\mathrm{PeV/c}$ and $\delta = 3$, coupled with $\xi_{\mathrm{CR}} = 15\%$. In the latter case, when a steeper acceleration spectrum is set, the diffusion coefficient that allows to reproduce the data is $D_0 = 1 \times 10^{27}\,\mathrm{cm}^2\,\mathrm{s}^{-1}$ together with $\delta = 4$ and either $p_{\mathrm{M}} = 100\,\mathrm{TeV/c}$ (implying $\xi_{\mathrm{CR}} = 20\%$) or $p_{\mathrm{M}} = 1\,\mathrm{PeV/c}$ (implying $\xi_{\mathrm{CR}} = 15\%$). On the other hand, in the $p_{\mathrm{max}}(t)$ formulation by [75], only an acceleration spectrum as steep as $p^{-(4+1/3)}$ can describe the data, as shown in Fig. 3.32. In this case, either $D_0 = 1 \times 10^{26}\,\mathrm{cm}^2\,\mathrm{s}^{-1}$ or $D_0 = 3 \times 10^{26}\,\mathrm{cm}^2\,\mathrm{s}^{-1}$, the CR acceleration efficiency amounts to $\xi_{\mathrm{CR}} = 2\%$. Note that the suppression of the diffusion coefficient in the source region is consistent with previous estimates [122] (Fig. 3.29).

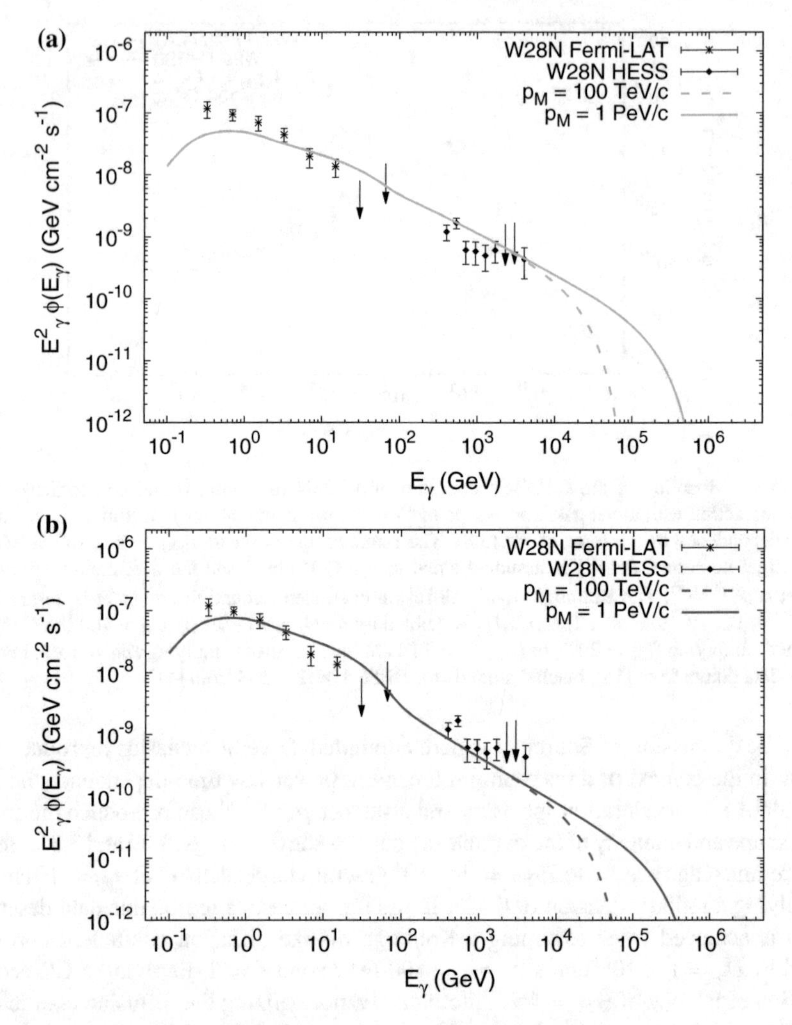

Fig. 3.32 Modeling of the GeV-TeV emission of W 28N for confined and non-confined protons interacting with target gas. The escape model assumed here follows the description by [75]. The acceleration spectrum is set $\propto p^{-(4+1/3)}$. The remnant age is set to $T_{SNR} = 4 \times 10^4$ yr. The target numerical density is assumed equal to $n_{up} = 10\,\mathrm{cm}^{-3}$. *Top*: The Kolmogorov-like diffusion coefficient is normalized to $D_0 = 1 \times 10^{26}\,\mathrm{cm}^2\,\mathrm{s}^{-1}$. Both lines refer to a CR acceleration efficiency of $\xi_{CR} = 2\%$. *Bottom*: The Kolmogorov-like diffusion coefficient is normalized to $D_0 = 3 \times 10^{26}\,\mathrm{cm}^2\,\mathrm{s}^{-1}$. Both lines refer to a CR acceleration efficiency of $\xi_{CR} = 2\%$. In both panels Fermi-LAT data points from [15], black diamonds for HESS J1801 − 233 from [41]

Chapter 4
The Galactic Center Region

The supermassive black-hole in the center of the Milky Way, located in the radio source Sgr A*, is one of the most interesting astronomical objects (see [126] for an extensive review). It is now in a state of relative inactivity [201] but it might well be a non-stationary source. For instance, there are interesting hints of a much stronger emission few hundreds years ago [154]; on the timescale of 4×10^4 years, major variability episodes are expected [111] and the Fermi bubbles [223] could be the visible manifestations [92] of such intense activity. Therefore, it is reasonable to expect that a past emission from the Galactic Center might lead to observable effects, as it was recently considered in [112].

The recent observations by the H.E.S.S. observatory [19] show that various regions around Sgr A* emit gamma rays up to many tens of TeV, offering new reasons to investigate this object. In particular, the gamma-ray spectrum from the region in the closest vicinity of Sgr A* is different from the one coming from its outskirts, the latter extending up to very high energies ($\sim$35 TeV) without a perceivable cut-off. Most likely, the gamma rays seen by H.E.S.S. can be attributed to CR collisions [19], even though a conclusive proof of this statement requires neutrino telescopes. Hence, it is essential to derive reliable predictions concerning the neutrino signal expected from Sgr A* and its surroundings. In this regard, H.E.S.S. measurements constitute very valuable information, in that these refer to an energy range similar to that where neutrino telescopes operate. Remarkably, the possibility that the Galactic Centre is a significant neutrino source is discussed since the first works [253]: in fact, Sgr A* is one of the main targets in the point-like source search of current neutrino observatories, namely ANTARES [26] and IceCube [6]. In few words, the hypothesis of a PeVatron in the Galactic Center motivates the search for neutrinos.

Part of this chapter has already been published in Palladino A. & Vissani F., 'Neutrinos and gamma rays from the Galactic Center Region after H.E.S.S. multi-TeV measurements', European Physics Journal C, Vol. 77, Issue 3, Id 66 (2017).

S. Celli, *Gamma-ray and Neutrino Signatures of Galactic Cosmic-ray Accelerators*, Springer Theses, https://doi.org/10.1007/978-3-030-33124-5_4

In this chapter, the implications of the H.E.S.S. results are discussed: a brief review of the known VHE sources within the inner Galactic Center region is provided in Sect. 4.1. Then, in Sect. 4.2, the very central sources are considered in order to study the effect of gamma-ray absorption. At the currently probed energies, the known background radiation fields have a minor impact, whereas it is not possible to exclude larger effects due to an additional infrared radiation field near the very Center, as examined in details in Sect. 4.3. The expected signal in neutrino telescopes, evaluated at the best of the present knowledge, is obtained in Sect. 4.4, while precise upper limits on neutrino fluxes are quantified in Sect. 4.4.1, where the underlying hypotheses are discussed. The expected number of events for ANTARES, IceCube and KM3NeT, based on the H.E.S.S. measurements, are calculated. It is shown that km^3-class telescopes in the Northern Hemisphere have the potential of observing high-energy neutrinos from this important astronomical object and can hence prove the existence of a hadronic PeV Galactic accelerator. It is thus argued that the PeVatron hypothesis makes the case for a cubic kilometer class neutrino telescope, located in the Northern Hemisphere, more compelling than ever.

4.1 The Morphology of the Inner Galactic Center Region

Detailed morphological and spectroscopic studies of the VHE gamma-ray emission in the central 200 pc of the Galaxy, the so-called *Galactic Ridge*, have been conducted with the H.E.S.S. observatory [11]. Several spatial components have been identified as contributors to its total emission. After accounting for the estimated charged particle background in the region, the Galactic large scale unresolved emission and the CR induced gamma-ray emission in the Ridge, two bright sources can be distinguished, as shown in the top panel of Fig. 4.1: HESS J1745-290 [42], coincident with Sgr A*, and G 0.9+0.1, a composite SNR whose gamma-ray emission appears to originate in its plerionic core [38]. Investigating the residual map, a fainter diffuse emission is visible, where the massive molecular complexes Sgr B2, Sgr C and Sgr D are clearly resolved (see the bottom panel of Fig. 4.1). Moreover, the extended source HESS J1745-303 and the recently identified source HESS J1746-285, at the edge of the so-called *GC radio Arch*, are visible. A schematic table of all the observed emission components is given in Table 4.1. In the context of hadronic models for the interpretation of this VHE radiation, the massive cloud systems provide an adequate target for *pp* interactions. It is natural hence to expect also VHE neutrinos from the same spatial regions and it appears timely to investigate the potential of current and future generation large-volume neutrino telescopes. A focus on the very central part of the Galactic Ridge is provided in this chapter, while the following chapter will present a study dedicated to the whole extended region of the Ridge.

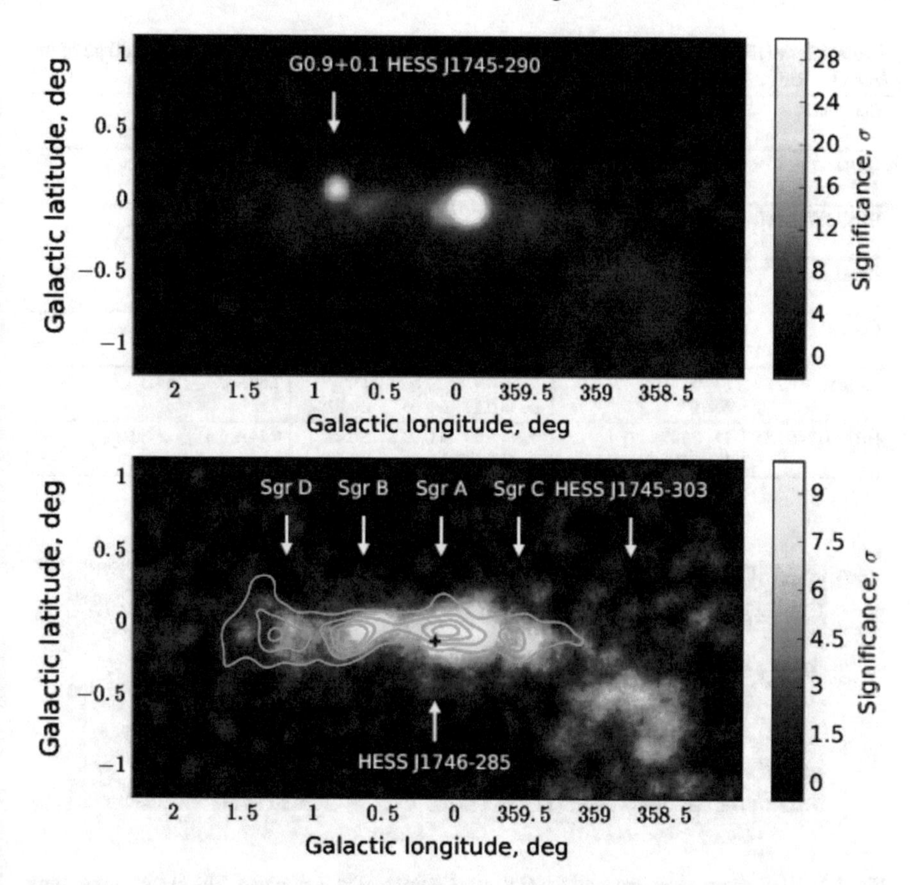

Fig. 4.1 VHE gamma-ray images of the GC region in Galactic coordinates, smoothed with the H.E.S.S. PSF. *Top panel*: Gamma-ray significance map. *Bottom panel*: Residual significance map after subtraction of the two point-like sources G $0.9 + 0.1$ and HESS J1745 − 290. The cyan contours indicate the density of molecular gas as traced by the Cesium, while the black cross marks the position of HESS J1746 − 285, coincident with the GC radio arch. Credit: Abdalla et al. [11] reproduced with permission ©ESO

4.2 The Gamma-Ray Spectra from the Galactic Center Region

The excess of VHE gamma rays recently reported by the H.E.S.S. Collaboration [19] can be ascribed to two regions around the Galactic Center, as represented in Fig. 4.2:

(i) the point-like source HESS J 1745 − 290, identified by a circular region centered on the radio source Sgr A* with a radius of 0.1°;
(ii) a diffuse emission, coming from an annular region located between 0.15° and 0.45° (corresponding to 20 − 63 pc) from the very center.

Table 4.1 VHE gamma-ray emission components identified by H.E.S.S. in the inner 200 pc of the Galactic Ridge. Credit: Abdalla et al. [11] reproduced with permission ©ESO

Component	Position (Galactic coordinates)	Extension (°)	Flux (10^{-12} TeV^{-1} cm^{-2} s^{-1})
G0.9+0.1	$l = 0.86°$ $b = 0.069°$	–	$0.88 \pm 0.04_{stat} \pm 0.25_{sys}$
HESS J1745-290	$l = 359.94°$ $b = -0.05°$	–	$2.9 \pm 0.4_{stat} \pm 0.8_{sys}$
Dense gas	$l = 0°$ $b = 0°$	$\sigma = 1.11° \pm 0.17°_{stat} \pm 0.17°_{sys}$	$4.3 \pm 0.9_{stat} \pm 1.5_{sys}$
Central	$l = 0°$ $b = 0°$	$\sigma = 0.11° \pm 0.01°_{stat} \pm 0.02°_{sys}$	$1.03 \pm 0.05_{stat} \pm 0.25_{sys}$
Large scale	$l = 0°$ $b = 0°$	$\sigma_x = 0.97°^{+0.04°}_{-0.02°} \pm 0.13°_{sys}$ $\sigma_y = 0.22° \pm 0.06°_{stat} \pm 0.07°_{sys}$	$2.68 \pm 0.6_{stat} \pm 1.3_{sys}$
HESS J1746-285	$l = 0.14°$ $b = -0.11°$	$\sigma_x = 0.03° \pm 0.03°_{stat} \pm 0.03°_{sys}$ $\sigma_y = 0.02° \pm 0.02°_{stat} \pm 0.03°_{sys}$	$0.24 \pm 0.03_{stat} \pm 0.07_{sys}$

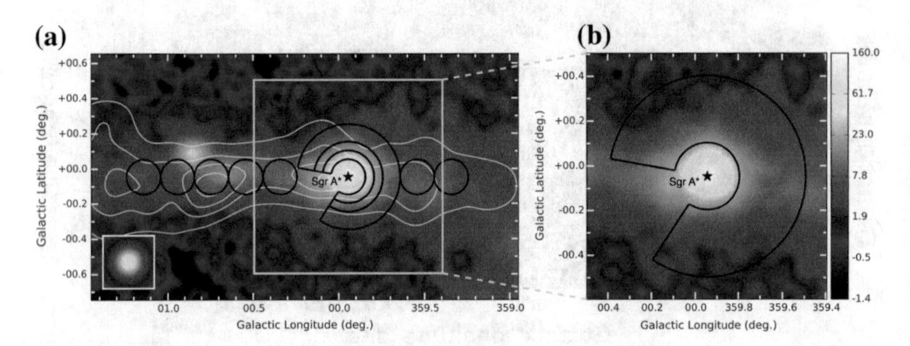

Fig. 4.2 VHE gamma-ray image of the Galactic Centre region. *Left panel*: The black lines outline the regions used to calculate the CR energy density throughout the CMZ, as shown in Fig. 1.3. White contour lines trace the density distribution of molecular gas, as defined by its Cesium line emission. *Right panel*: Zoomed view of the inner ~70 pc and the contour of the region used to extract the spectrum of the diffuse emission. Figure from Abramowski et al. [19], reprinted by permission from Springer Nature

The observed spectrum from the point-like source is described by an exponentially suppressed power-law distribution, as

$$\phi_\gamma(E) = \phi_0 \left(\frac{E}{1\,\mathrm{TeV}}\right)^{-\Gamma} \exp\left(-\frac{E}{E_{cut}^{\gamma}}\right) \tag{4.1}$$

while in the case of diffuse emission an unbroken power law is preferred. The H.E.S.S. Collaboration has summarized its observations by means of the following parameter sets:

- Best fit of the Point-like Source (PS) region:

$$\begin{cases} \Gamma = 2.14 \pm 0.10 \\ \phi_0 = (2.55 \pm 0.37) \times 10^{-12}\,\mathrm{TeV}^{-1}\,\mathrm{cm}^{-2}\,\mathrm{s}^{-1} \\ E^{\gamma}_{\mathrm{cut}} = 10.7 \pm 2.9\,\mathrm{TeV} \end{cases} \quad (4.2)$$

- Best fit of the Diffuse (D) region:

$$\begin{cases} \Gamma = 2.32 \pm 0.12 \\ \phi_0 = (1.92 \pm 0.29) \times 10^{-12}\,\mathrm{TeV}^{-1}\,\mathrm{cm}^{-2}\,\mathrm{s}^{-1} \end{cases} \quad (4.3)$$

These fits are shown in Fig. 4.3. In the diffuse case, however, also exponentially suppressed power-law fits are compatible with H.E.S.S. data: in particular, assuming

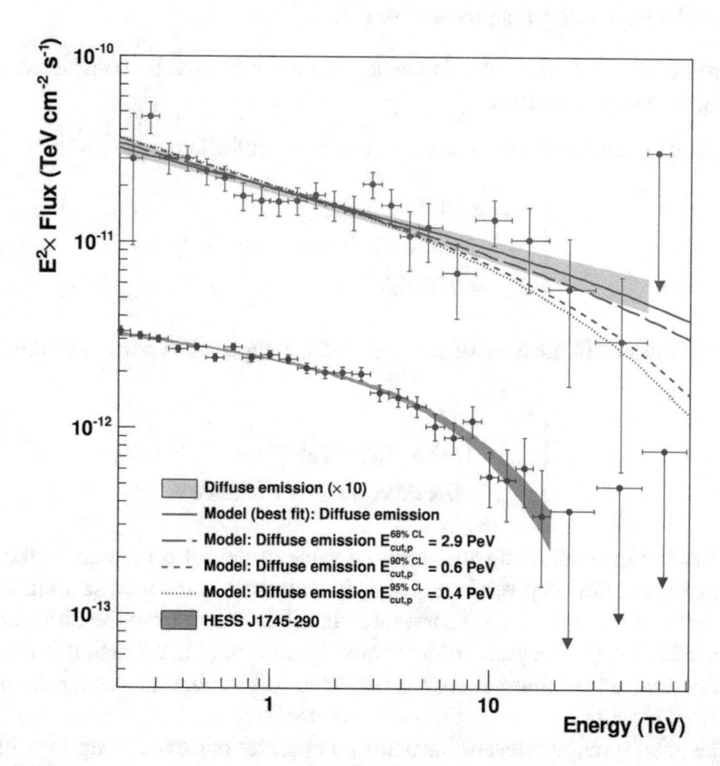

Fig. 4.3 VHE gamma-ray spectra of the diffuse emission and HESS J1745 − 290. Vertical and horizontal error bars show the 1σ statistical error and bin size, respectively. Arrows represent 2σ flux upper limits. The 1σ confidence bands of the best-fit spectra of the diffuse and HESS J1745 − 290 are shown in red and blue shaded areas, respectively. The red lines are the results of numerical computations assuming hadronic gamma-ray production. The fluxes of the diffuse emission spectrum and models are multiplied by 10. Figure from Abramowski et al. [19], reprinted by permission from Springer Nature

a cut-off in the parent proton spectrum from the diffuse region, the resulting secondary gamma-ray spectrum would deviate from the H.E.S.S. data at 68%, 90% and 95% confidence levels respectively for cut-off energies located at 2.9, 0.6 and 0.4 PeV.

These functional forms correspond to the gamma-ray spectrum *observed* at Earth. However, in the case of a significant absorption of gamma rays in their travel towards the Earth, the observed spectrum does not coincide with the *emission* spectrum, namely the spectrum at the source. On the other hand, as neutrinos are not affected by the same absorption processes involving photons, in order to correctly predict the neutrino spectrum at Earth, the gamma-ray spectrum at the source is needed. In this section, the implications of the following assumptions are discussed, namely whether:

(i) the emitted gamma-ray spectra coincide with the observed gamma-ray spectra, as described by the functional forms reported in Eqs. (4.2) and (4.3);
(ii) the gamma-ray emission at the source is described by different functional forms or model parameters than those observed.

In regards to the last issue, the following models will also be considered for the emitted gamma-ray spectrum:

- Point Source emission with a larger value of the cut-off energy (PS*):

$$\begin{cases} \Gamma = 2.14 \\ \phi_0 = 2.55 \times 10^{-12}\,\mathrm{TeV}^{-1}\,\mathrm{cm}^{-2}\,\mathrm{s}^{-1} \\ E^{\gamma}_{\mathrm{cut}} = 100\,\mathrm{TeV} \end{cases} \tag{4.4}$$

- Diffuse emission in the form of an exponentially suppressed power law (DC) with:

$$\begin{cases} \Gamma = 2.32 \\ \phi_0 = 1.92 \times 10^{-12}\,\mathrm{TeV}^{-1}\,\mathrm{cm}^{-2}\,\mathrm{s}^{-1} \\ E^{\gamma}_{\mathrm{cut}} = 0.4\,\mathrm{PeV},\, 0.6\,\mathrm{PeV}\,\mathrm{or}\,2.9\,\mathrm{PeV} \end{cases} \tag{4.5}$$

The interest in considering an *increased* value of the cut-off energy in the point-like source spectrum, as given in Eq. (4.4) is motivated in the next section. Instead, the inclusion of an exponential suppression in the emission from the diffuse region, as indicated in Eq. (4.5), agrees with the observations of H.E.S.S. and it is motivated by the fact that a maximum energy is expected to be achieved during the particle acceleration process.

As the energy range relevant for neutrino observations extends up to ~100 TeV (as clear e.g. from Figs. 2 and 3 of [88] and Fig. 1 of [243]) while the H.E.S.S. observations extend up to 20–40 TeV, the gamma-ray data currently available only cover the lower energy interval where the number of neutrinos is expected to be significant. In other words, it should be kept in mind that until gamma-ray observations up to few 100 TeV will become available, thanks to future measurements by HAWC [220] and CTA [237], the expectations for neutrinos will rely in part on extrapolation and/or on theoretical modeling. In this work, unless otherwise stated, the 'minimal extrapolation' is considered, namely the assumption that the functional forms in Eqs. (4.2)–(4.5) of the gamma-ray spectrum are valid for the emission spectrum.

A precise upper limit on the expected neutrino flux can be determined from the H.E.S.S. measurement, by assuming a hadronic origin of the observed gamma rays. Note that the presence of a significant leptonic component in the measured gamma rays would imply a smaller neutrino flux. On the other hand, several effects might in principle increase the expected flux of neutrinos. For instance, hadronic interactions in other regions close the the Galactic Center, but not probed by H.E.S.S., could produce high-energy gamma rays and neutrino radiation, leading to an interesting signal. In relation to this, the annulus reported in the H.E.S.S. analysis resembles more a region selected for observational purposes rather than an object with an evident physical meaning.[1] Another reason to expect an increased neutrino flux is that the ice-based neutrino telescope IceCube integrates on an angular extension of about 1°, which is 5 times larger than the angular region covered in [19]. In view of these motivations, the theoretical upper limit on the neutrino flux that will be derived represents the *minimum* that is justified by the current gamma-ray data. Moreover, another specific phenomenon that potentially increases the expected neutrino flux, as derived from the gamma-ray flux currently measured by H.E.S.S, is the absorption of gamma rays from non-standard radiation fields, as discussed in the next section.

4.3 Absorption of Gamma Rays

During their propagation within the background radiation fields of the Milky Way, high-energy photons are subject to absorption due to the pair production process. If absorption is effective, the *observed* gamma-ray spectrum would be suppressed with respect to the *emitted* one. However, the neutrino spectrum would not suffer any significant absorption, as it reflects the emission spectrum, and thus it would result larger than the one obtained by converting the observed gamma-ray spectrum. In order to derive the source emission spectrum, hence, it is necessary to model and then to remove the effect of the absorption (de-absorption) from the observed spectrum. Note that the idea that gamma rays could suffer significant absorption in the energy range probed by H.E.S.S. was put forward in [19]; in the following, this idea is examined in details.

The existence of the Cosmic Microwave Background, that uniformly pervades the whole space, leads to absorption of gamma rays of very high energies, around PeV. In addition, the model by [202], adopted e.g. in the GALPROP simulation program,[2] can be conveniently used to describe gamma-ray absorption due to other

---

[1]Because of this consideration, and also in view of the fact that the angular resolution of the next-generation water-based neutrino telescopes matches the physical size of the two regions, the predictions for the point-like source and the diffuse region will be presented separately.

[2]http://galprop.stanford.edu/.

background radiation fields that populate the Galaxy, as the infrared (IR) and starlight (SL) (see e.g., [86, 105, 162, 183]), whose absorption effect is relevant for lower energy photons. It is convenient to group these three radiation fields (CMB, IR and SL) as 'known' radiation fields. However, it is not possible to exclude that additional, and possibly more intense, radiation fields exist in the vicinity of the Galactic Center. The formal description of the absorption effects can be simplified without significant loss of accuracy if the background radiation field is effectively parameterized as the combination of thermal and quasi-thermal distributions, where the latter ones are just proportional to a thermal distribution with the same temperature. In fact, for a perfect blackbody (BB) of temperature T emitting photons of energy ϵ, the differential number density of emitted photons is given by

$$n_{\rm BB}(\epsilon, T) = \frac{\epsilon^2}{\pi^2}\left[\frac{1}{\exp^{\epsilon/T} - 1}\right] \tag{4.6}$$

where natural units ($c = k = 1$) have been set. On the other hand, a grey body (GB) of the same temperature is simply rescaled as $n_{\rm GB}(\epsilon, T) = \xi n_{\rm BB}(\epsilon, T)$, where the *non-thermal parameter* ξ has been introduced. As the total radiation field is generally composed by several components, the i-th one is defined by two parameters: the temperature T_i and its coefficient of proportionality to the thermal distribution ξ_i. In the following, the formalism adopted for the computation of the optical depth due to the total radiation field is presented.

Formalism A photon with energy E_γ emitted from an astrophysical source can interact during its travel to the Earth with ambient photons, producing electron-positron pairs. The probability that it will reach the Earth is

$$P(E_\gamma) = \exp\left[-\tau(E_\gamma)\right] \tag{4.7}$$

where τ is the opacity. For a uniform ambient radiation field with density n (as the CMB for instance), the optical depth for photons of energy E_γ coming from a source at a distance L can be calculated as

$$\tau(E_\gamma) = L \int\int \sigma_{\gamma\gamma}(E_\gamma, \epsilon) n(\epsilon) \left[\frac{1 - \cos\theta}{2}\right] \sin\theta d\epsilon d\theta \tag{4.8}$$

where θ is the angle between the momenta of the two interacting photons and $\sigma_{\gamma\gamma}$ is the pair production cross-section. For non uniform radiation fields, the optical depth due to pair production can be computed similarly to Eq. (4.8), with the additional complication regarding the integration along the line of sight, since the field density depends on the position in the space.

In the ISM, different radiation fields can offer a target to astrophysical photons, causing their absorption: the total opacity is therefore the sum of various contributions, namely

$$\tau = \sum_i \tau_i \tag{4.9}$$

where the index i indicates the components of the background radiation field that contributes to the absorption process. These include the CMB, as well as the IR and SL backgrounds, and possibly additional ones, present near the region of the Galactic Center. For the i-th component of a thermal distribution (or of a distribution proportional to it), the integrals in Eq. (4.8) provide an optical depth simply given by

$$\tau_i(E_\gamma) = 1.315 \times \frac{L_i}{L_0} \times \frac{n_{\gamma,i}}{n_{\gamma,\text{CMB}}} \times f\left(\frac{E_i}{E_\gamma}\right) \tag{4.10}$$

where the quantities chosen for the normalization are $L_0 = 10$ kpc (a typical Galactic distance) and $n_{\gamma,\text{CMB}} = 410.7\ \text{cm}^{-3}$. The multiplicative factor in Eq. (4.10) is obtained by combining several constants (gathered from the CMB density, the interaction cross section and L_0) into

$$\frac{\pi}{2\,\zeta(3)} \times L_0\, r_e^2 \times n_{\gamma,\,\text{CMB}} = 1.315 \tag{4.11}$$

where $r_e = e^2/(m_e c^2) \approx 2.818 \times 10^{-13}$ cm is the classical electron radius and $\zeta(3) \approx 1.20206$ the Riemann's zeta function. The various quantities appearing in Eq. (4.10) are now discussed.

(i) The parameter L_i is the effective size of the region where the background radiation field is present. In the case of the CMB, this is just the distance between the Galactic Center and the detector (i.e., 8.3 kpc), as the CMB is uniformly distributed throughout the ISM. On the other hand, the density of IR and SL photons follow an approximate exponential distribution, peaked in the Galactic Center [183], namely the density of photon is $n_{\gamma,i}(L) = n_{\gamma,i}\, e^{-L/L_i}$. Thus, the product $n_{\gamma,i} \times L_i$ represents the column density of photons. The values of the scales L_i are given in Table 4.2: these are introduced in order to avoid the spatial integration in the computation of the optical depth at the expenses of schematically considering the i-th radiation field as uniform over the length scale L_i.

(ii) $n_{\gamma,i}$ is the total density of photons of the considered background radiation field. In the case of IR and SL fields, the total density of photons is given by the following expression:

Table 4.2 Values of the parameters adopted for modeling the background radiation fields during the computation of the absorption factor of gamma rays from the Galactic Center: the black body temperature T_i; the non-thermal parameter ξ_i, defined in Eq. (4.12); the typical length L_i, namely the distance from the Galactic Center for the CMB and the exponential scales for the IR and SL radiation fields [183]; the total density of photons $n_{\gamma,i}$, as obtained from Eq. (4.12); the typical energy E_i, as obtained from Eq. (4.13). Table reproduced from Celli et al. [80], under the CC BY LICENSE

Rad. field	T_i (eV)	ξ_i	L_i (kpc)	$n_{\gamma,i}$ (cm^{-3})	E_i (TeV)
CMB	2.35×10^{-4}	1	8.3	410.7	1111
IR	3.10×10^{-3}	1.55×10^{-4}	4.1	146.0	84.23
SL	3.44×10^{-1}	1.47×10^{-11}	2.4	19.0	0.26

$$n_{\gamma,i}(\xi_i, T_i) = \xi_i \times n_{\gamma,\,\text{CMB}} \times \left(\frac{T_i}{T_{\text{CMB}}}\right)^3 \tag{4.12}$$

where $\xi_i \leq 1$ is the non-thermal parameter, chosen to reproduce the observed distribution of ambient photons (the case $\xi_i = 1$ is the thermal one).

(iii) The energy E_i is connected to the energy threshold for pair production, or in other words to the blackbody (or quasi-blackbody for IR and SL fields) temperature T_i, through the relation

$$E_i(T_i) = \frac{m_e^2}{T_i} \tag{4.13}$$

where the assumed distribution is proportional to a thermal distribution with temperature T_i.

(iv) The adimensional function $f(x)$, where $x = E_i/E_\gamma$, describes how the absorption varies with the energy of the gamma ray. This function was first obtained in [183] by using the results of [72]. As it encloses the integral presented in Eq. (4.8), the precise values obtained by numerical integration are provided in Sect. 4.3.1, where furthermore an approximate expression for such a function is discussed, reading as

$$f(x) \simeq -a\,x\,\log[1 - \exp(-b\,x^c)], \qquad \begin{cases} a = 3.68 \\ b = 1.52 \\ c = 0.89 \end{cases} \tag{4.14}$$

which is very easy to use and accurate to within 3%.

The values of T_i, ξ_i and L_i for the known components of the radiation field are found by fitting the energy spectra of radiation as reported in GALPROP: results for the CMB, IR and SL fields are summarized in Table 4.2. The parameters $n_{\gamma,i}$ and E_i are also given in Table 4.2, to allow one to use directly Eq. (4.10) in the computation of the optical depth, by just replacing the appropriate numerical values. As the characteristic temperatures of the IR and SL fields are larger than that of the CMB, their contributions is expected to affect the survival probability of gamma rays at energies smaller than those due to the CMB photons. In this formalism, a population of quasi-thermal background photons is characterized by the parameters T_i, ξ_i, L_i, and it will yield the opacity factor $\tau_i = \tau(E_\gamma, T_i, \xi_i \times L_i)$. Note that ξ_i and L_i appear in Eq. (4.10) only through their product, so that, for each component of the background radiation field (known or hypothetical), the photon absorption is described by *two parameters*.

The reliability of this procedure for the description of the Galactic absorption was already tested in [86, 162]. The formalism can be simplified even further without significant loss of accuracy thanks to the fully analytical (albeit approximate) formula provided in Eq. (4.14), which will be derived and discussed in details in Sect. 4.3.1. The consistency with the other approach–based on [202]—was checked by comparing these results with Fig. 3 of [105].

A few advantages of this procedure are:

(1) the results are exact in the case of the CMB distribution, that is a thermal one;
(2) such procedure allows one to easily vary the parameters of the radiation field discussing the effect of errors and uncertainties;
(3) the very same formalism allows to model the effect of new hypothetical radiation background fields.

4.3.1 *Study of the Function $f(x)$*

The evaluation of the effects of absorption via Eq. (4.10) rests on the estimation of the properties of the background radiation field and on the calculation of the function $f(x)$. Note that, even if the interest here is in the gamma rays emitted from the Galactic Center, the results concerning gamma-ray absorption can be applied to a very large variety of cases and situations.

In this section, a detailed study of $f(x)$ is provided, together with a table of numerical values for this function, and the bases of the approximation given in Eq. (4.14) are discussed. The following material is useful to validate the results and to compare them with what has been obtained in the literature.

The pair creation process [72]

$$\gamma + \gamma_{\rm bkg} \longrightarrow e^+ + e^-$$

in the background of thermal photons with temperature T_i gives the opacity factor [183],

$$\tau_i = \frac{r_e^2}{\pi} L_i\, T_i^3\, f\left(\frac{m_e^2}{T_i\, E_\gamma}\right)$$

that can be rewritten introducing the thermal photon density $n_{\gamma,i} = 2\zeta(3)T_i^3/\pi^2$. The function $f(x)$ is defined as

$$f(x) = x^2 \int_0^1 d\beta\, R(\beta)\, \psi\left(\frac{x}{1-\beta^2}\right) \tag{4.15}$$

Here β is the velocity of the outgoing electron in the center of mass frame, while the two auxiliary functions are

$$R(\beta) = \frac{2\beta}{(1-\beta^2)^2}\left[(3-\beta^4)\log\left(\frac{1+\beta}{1-\beta}\right) - 2\beta(2-\beta^2)\right]$$

$$\psi\left(\tfrac{x}{1-\beta^2}\right) = -\log\left(1 - e^{-\frac{x}{1-\beta^2}}\right)$$

Table 4.3 The values of the function $f(x)$ as given in Eq. (4.15). In bold, the value of x in which the function reaches the maximum and half of the maximum. Table reproduced from Celli et al. [80], under the CC BY LICENSE

x	$f(x)$	x	$f(x)$
10^{-10}	7.32×10^{-9}	10^{-1}	6.32×10^{-1}
10^{-9}	6.57×10^{-8}	**0.503**	**1.076**
10^{-8}	5.81×10^{-7}	1	9.07×10^{-1}
10^{-7}	5.05×10^{-6}	**1.77**	**0.538**
10^{-6}	4.29×10^{-5}	5	3.27×10^{-2}
10^{-5}	3.54×10^{-4}	10	2.92×10^{-4}
10^{-4}	2.78×10^{-3}	20	1.78×10^{-8}
10^{-3}	2.04×10^{-2}	30	9.65×10^{-13}
10^{-2}	1.31×10^{-1}	50	$\simeq 0$
0.0756	**0.538**		

By solving numerically the integral in Eq. (4.15), the values reported in Table 4.3 are obtained.

Firstly, the behavior of the integrand function in β is examined.

The function $R(\beta) \sim 4\beta^2$ if $\beta \to 0$; on the contrary, when $\beta \to 1$, it diverges like $R(\beta) \sim -\log(1-\beta)/(1-\beta^2)^2$. The divergence is compensated by the behavior of the function ψ, that follows from

$$\psi\left(\frac{x}{1-\beta^2}\right) = \sum_{n=1}^{\infty} \frac{1}{n} e^{-n\left(\frac{x}{1-\beta^2}\right)}$$

at high values of $x/(1-\beta^2)$. Finally,

$$\psi\left(\frac{x}{1-\beta^2}\right) \approx -\log\left(\frac{x}{1-\beta^2}\right)$$

at small values of $x/(1-\beta^2)$.

At this point, the behavior of $f(x)$ in x is studied:

- For high x, only the first term of the expansion of ψ is relevant, and hence the function $f(x)$ is well approximated by:

$$f(x) \approx x^2 \times \int_0^1 d\beta\, R(\beta) \exp\left(-\frac{x}{1-\beta^2}\right)$$

within an accuracy of 1% for $x > 3$.
- For small x, the most important contribution to the integral is produced when R diverges and ψ is not exponentially suppressed. This condition is realized when

$\beta < \sqrt{1-x} \approx 1 - x/2$ and in this region

$$\psi\left(\frac{x}{1-\beta^2}\right) \approx -\log\left(\frac{x}{1-\beta^2}\right)$$

Concerning $R(\beta)$, its asymptotic expression can be used, i.e.

$$R(\beta) \simeq \frac{4}{(1-\beta^2)^2} \log\left(\frac{2}{1-\beta}\right)$$

The approximation of the function $f(x)$ is thus given by:

$$f(x) \approx -4x^2 \int_0^{1-x/2} d\beta \frac{\log\left(\frac{2}{1-\beta}\right) \log\left(\frac{x}{1-\beta^2}\right)}{(1-\beta^2)^2}$$

This implies the behavior $f(x) \approx -3.076\, x\, \log(x)$ to within an accuracy of about 3% in the interval $10^{-10} \leq x \leq 10^{-5}$.

A global analytical approximation of the $f(x)$, that respects the behavior for small and large values of x, is given by Eq. (4.14). Its accuracy is $\sim$3% into the interval $10^{-10} \leq x \leq 10$. When $x > 10$ the function rapidly decreases, as can be seen from Table 4.3, where the values obtained by numerical integrations (without any approximation) are provided.

4.3.2 Results

The effects of absorption due to the known radiation fields of Table 4.2, that concern the gamma rays propagating from the Galactic Center to the Earth, are illustrated in the left panel of Fig. 4.4. They become relevant at some hundreds of TeV, thus the gamma rays presently observed by H.E.S.S. from the diffuse region (namely the models D and DC introduced in Eqs. (4.3) and (4.5)) are not significantly influenced by this phenomenon, when considering only the standard radiation fields adopted for instance in GALPROP [202]. Therefore, it is possible to use directly the observed diffuse gamma-ray flux in order to obtain the gamma-ray flux at the source, modulo the caveats concerning the extrapolation at high energy, as explained in Sect. 4.2.

On the other hand, the flux from the point-like source could be affected by the absorption due to a new, non-standard and intense radiation field close to the central black-hole. In fact, the physical conditions in the close vicinity of the supermassive black-hole are very uncertain and in particular the local radiation field is not directly measured. Recall that the gamma-ray spectrum from the point-like source *observed* by H.E.S.S. deviates from a power-law distribution at the highest energies currently probed (see Eq. (4.2)): in the presence of a new radiation field, the observed spectrum would naturally differ from the *emitted* gamma-ray spectrum. In such a case, while the

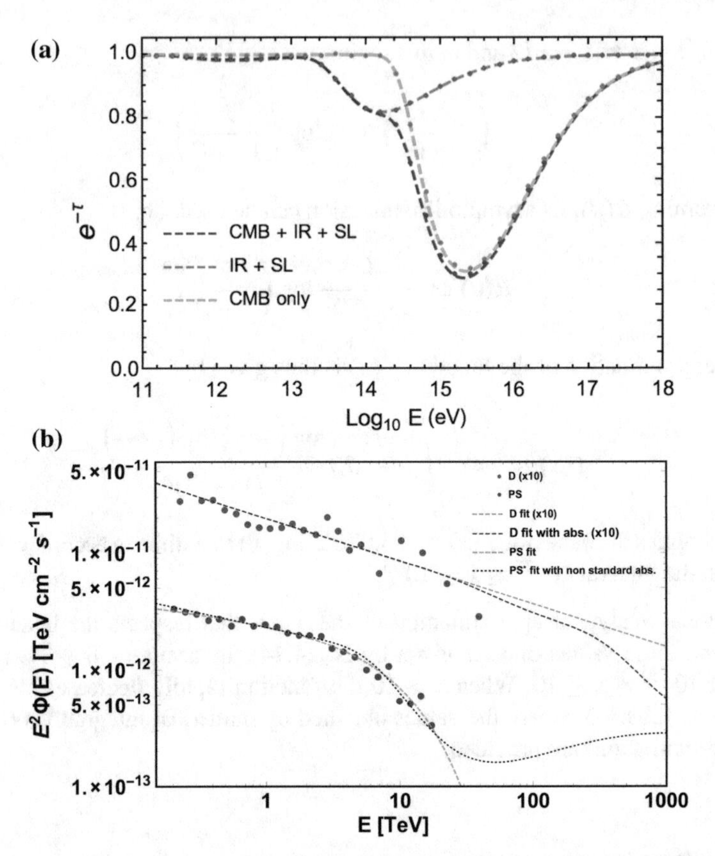

Fig. 4.4 *Top:* Absorption of gamma rays from the Galactic Center at different energies, due to the interaction with CMB, IR and SL. *Bottom*: Fits of the H.E.S.S. data. D is the Diffuse flux, PS is the Point-like Source flux. For D, absorption due to CMB, IR and SL is considered. For PS, an increased absorption, due to a non-standard radiation field, is considered. As visual aid, the central values measured by H.E.S.S. [19] in the case of the Diffuse and the Point-like Source are shown by red and blue dots respectively. Figure reproduced from Celli et al. [80], under the CC BY LICENSE

cut-off in the *observed* spectrum would probe the features of the absorbing radiation field, the cut-off in the emitted spectrum would be connected to the acceleration process, and located at higher energies.[3]

In such a scenario, namely when the absorption by an additional radiation field would be responsible for the observed cut-off in the point-like source at the Galactic Center, the VHE data provide valuable information concerning the intensity and the characteristic energy of such background photons. In fact, the energy where the cut-off is measured is connected to the characteristic energy of the absorb-

[3]Note incidentally that the position of the cut-off in the spectrum of CRs accelerated near the supermassive black-hole is to date unknown.

ing photons, while the amount of the attenuation is determined by the column density of the field along the line of sight. As a result, in order to reproduce the H.E.S.S. observations concerning the high-energy attenuation in the spectrum of the point-like source, as shown in Fig. 4.4b, the background radiation field that would be responsible for the absorption is found to be characterized by a temperature $T = 1.3 \times 10^{-2}$ eV= 155 K: the photon wavelength corresponding to the peak of the emission can be derived through the Wein law as $\lambda_{max} = b/T \simeq 20\,\mu$m (where $b \simeq 2900\,\mu$m K) and would thus be located in the far infrared (FIR) spectral band, corresponding to a frequency of $\nu_{max} = 15$ THz. For a blackbody field, the corresponding column density as extracted by the absorption dip would amount to $L_{min} \times n_\gamma \simeq 5000$ kpc cm^{-3} = 1.5×10^{25} cm^{-2}. In order to match the characteristic length scale of the central source observed by H.E.S.S., one should assume $L_{min} = L_{PS} = 22$ pc, which implies a photon density of the novel radiation field of $n_\gamma = 2.3 \times 10^5$ cm^{-3}. With such a value, the deviation from the perfect blackbody would amount to $\xi = 0.003$ (see Eq. (4.12)). This is illustrated in Fig. 4.4b: the curve called 'PS fit with non standard abs.' represents a power-law spectrum with the same spectral index and normalization of the PS model given in Eq. (4.2), modified by taking into account the above scenario for gamma-ray absorption.

Given the expected intensity of the speculative IR field, it appears extremely relevant to compare it to the sensitivity of current generation instruments. In the thermal case, a maximum radiance of $B_{BB}(\nu_{max}, T) = 4 \times 10^{-10}$ erg cm^{-2} s^{-1} sr^{-1}Hz^{-1} is expected. However, in the more realistic situation of a grey body with the aforementioned non-thermal parameter $\xi = 0.3\%$, one derives a maximum radiance of $B_{GB}(\nu_{max}, T) = \xi B_{BB}(\nu_{max}, T) \simeq 1.3 \times 10^{11}$ Jy/sr. This value is well within the sensitivity of IR instruments, as IRAS,[4] the most recent sky survey operating in the FIR domain (at 12, 25, 60 and 100 μm wavelengths). The telescope has recorded some enhanced activity in the GC region, as illustrated in Fig. 4.5: in particular, the Galactic Nucleus appears extremely bright in the 12 μm map ($\nu_{obs} = 25$ THz). The detected map, with a few arcmin resolution, shows a flux intensity ranging from $\sim 5 \times 10^6$ Jy/sr up to $\sim 2 \times 10^{10}$ Jy/sr [90], which has been attributed to the presence of dust grains with typical temperature of 27 K. The striking similarity between the size of the emission region in the FIR domain and that of the gamma-ray central source might indicate a physical connection among the two.

Some remarks to this scenario are in order:

(1) The non-standard IR radiation field could be produced in the reprocessing of the radiation from the central source, due to collision with CircumNuclear Disc clumps, as reported in [126]. The Galactic Center region is indeed very active in star formation processes and it is also populated with abundant dust grains.
(2) Evidently, the Diffuse component would not be affected by this new IR radiation, because it is far enough from the Galactic Center.
(3) As one understands from Fig. 4.4, in order to observationally determine whether the exponential suppression measured in the point-like source spectrum is intrinsic to the source or is an absorption feature, measurements of gamma rays at ener-

[4]https://lambda.gsfc.nasa.gov/product/iras/.

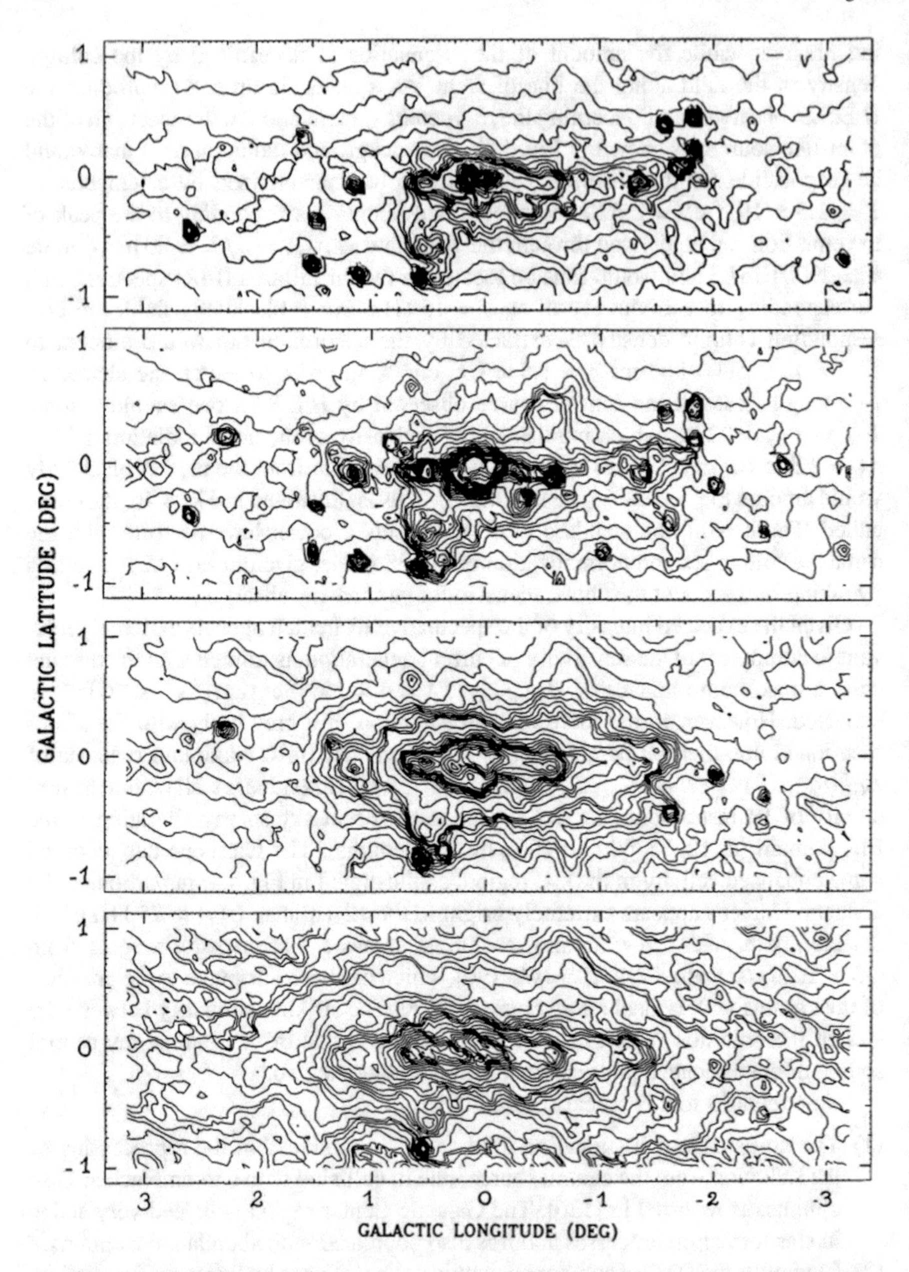

Fig. 4.5 IRAS contour plots of the surface brightness distribution of the Galactic Center over a region $6° \times 3°$ at 12, 25, 60 and 100 μm from top to bottom. The maps have been corrected for the zodiacal emission and for the diffuse emission associated with the Galactic Disc. Stripping effects appearing at 100 μm are artifacts due to the instrument calibration process. Figure from Cox & Laureijs [90], reprinted by permission from Springer Nature

gies above tens of TeV are required: indeed, the absorption results in a peculiar distortion of the power-law spectrum, that is expected to be different from the effect of an exponential cut-off above ~50 TeV. This discrimination might be achieved with CTA [24] or with other future instruments.

Note that, in the hypothesis that the gamma rays close to the black-hole are significantly absorbed, the flux emitted from the point-like source is still compatible with a power-law distribution at the energies observed by H.E.S.S. and above, just as the one due to the diffuse gamma-ray component. This speculative scenario allows to estimate in a reasonable way the maximum effect of absorption due to yet unknown radiation fields. This has a direct implication on the neutrino signal, that is quantified in the next section.

4.4 High-Energy Neutrinos from the Galactic Center Region

If the Galactic Center region hosts a PeVatron (or more than one), neutrinos could be produced in hadronic interactions of PeV protons with the ambient gas: since each neutrino carries about 5% of the energy of the parent proton, they have to be expected in the multi-TeV range, in angular correlation with the high-energy gamma rays emitted in the same region. This scenario is supported by the observed correlation between the gamma-ray emission and the molecular clouds of the CMZ, as described in [19].

To date, IceCube has set the best 90% C.L. upper limit on the $\nu_\mu + \bar{\nu}_\mu$ flux expected from Sgr A* [6], assuming an unbroken E^{-2} spectrum. This limit reads as

$$\phi_{\nu_\mu} + \phi_{\bar{\nu}_\mu} = 7.6 \times 10^{-12} \left(\frac{E}{\text{TeV}}\right)^{-2} \text{TeV}^{-1}\text{cm}^{-2}\text{s}^{-1} \quad (4.16)$$

It corresponds to the absence of a significant event excess over the known background, that in the IceCube analysis [3] amounts to 25.2 background events within 1° from Sgr A*. This limit has been obtained by means of downward-going track-type events, as discussed in Sect. 4.4.1. Presumably, this is the most reliable information currently available on the neutrino emission from Sgr A*, even if it is based on an unrealistic neutrino emitted spectrum. In principle, the assumption of an E^{-2} differential energy spectrum for neutrinos would follow from the first order Fermi acceleration mechanism, though there is no observational evidence that this is a reliable assumption, and moreover, this is not supported by the H.E.S.S. observations of the gamma-ray spectrum.

The model predictions derived in the following are based instead on the current gamma-ray observations and on the assumption that such emission is fully hadronic. In this scenario, assuming a gamma-ray emission spectrum $\phi_\gamma(E)$, the muon neutrino and antineutrino spectrum can be calculated through the precise relations based on the assumption of CR-gas collisions as [239]

$$\phi_{\overset{(-)}{\nu}_\mu}(E) = \alpha_\pi\, \phi_\gamma\left(\frac{E}{1-r_\pi}\right) + \alpha_K\, \phi_\gamma\left(\frac{E}{1-r_K}\right) + \int_0^1 \frac{dx}{x}\, K_{\overset{(-)}{\nu}_\mu}(x)\, \phi_\gamma\left(\frac{E}{x}\right) \tag{4.17}$$

where $x = E/E_\pi$. The multiplicative factors read as: $\alpha_\pi = 0.380$ (0.278) and $\alpha_K = 0.013$ (0.009) for ν_μ and $\bar{\nu}_\mu$ respectively, $r_\pi = (m_\mu/m_{\pi^\pm})^2 = 0.573$ and $r_K = (m_\mu/m_K)^2 = 0.0458$. In this expression, the first two contributions describe neutrinos from the two-body decay of pions and kaons, while the third term accounts for neutrinos from muon decay. The kernels for muon neutrinos $K_{\nu_\mu}(x)$ and for muon antineutrinos $K_{\bar{\nu}_\mu}(x)$, which also account for all flavor oscillations from the source to the Earth (in the full pion decay chain hypothesis), are

$$K_{\nu_\mu}(x) = \begin{cases} x^2(15.34 - 28.93x) & 0 < x \leq r_K \\ 0.0165 + 0.1193x + 3.747x^2 - 3.981x^3 & r_K < x < r_\pi \\ (1-x)^2(-0.6698 + 6.588x) & r_\pi \leq x < 1 \end{cases}$$

$$K_{\bar{\nu}_\mu}(x) = \begin{cases} x^2(18.48 - 25.33x) & 0 < x \leq r_K \\ 0.0251 + 0.0826x + 3.697x^2 - 3.548x^3 & r_K < x < r_\pi \\ (1-x)^2(0.0351 + 5.864x) & r_\pi \leq x < 1 \end{cases}$$

By applying such procedure, the expected (upper limits on the) neutrino spectra are obtained from the gamma-ray spectrum.

Figure 4.6 shows the sum of the muon neutrino and antineutrino fluxes: they are derived by using alternatively the four models of $\phi_\gamma(E)$ introduced in Sect. 4.2,

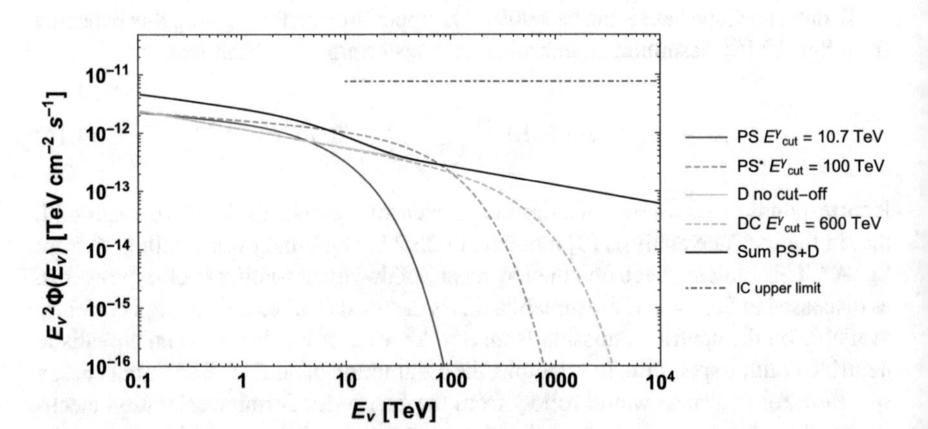

Fig. 4.6 Sum of the muon neutrino and antineutrino fluxes predicted from the Point-like Source best-fit with a cut-off at $E^\gamma_{cut} = 10.7$ TeV, for the Point-like Source with an arbitrary cut-off at $E^\gamma_{cut} = 100$ TeV, for the Diffuse gamma-ray best-fit without a cut-off and with a cut-off at $E^\gamma_{cut} = 600$ TeV. The sum of the total neutrino flux expected in the inner 0.45°, coinciding with the whole region of the H.E.S.S. observations, is indicated by the black solid line (PS + D emissions). The red dot-dashed line represents the 90% C.L. IceCube upper limit, as in Eq. (4.16), in the energy range of the measurement. Figure reproduced from Celli et al. [80], under the CC BY LICENSE

respectively Eqs. (4.2), (4.3), (4.4) and (4.5) (the latter with $E_\gamma = 600\,\text{TeV}$). The resulting neutrino fluxes represent the most precise information within a model-independent approach, as they only rely on the hypothesis that the gamma rays observed by H.E.S.S. are fully hadronic. In this respect, the model prediction derived can also be regarded as a *theoretical upper bound*. As can be deduced from Fig. 4.6, the expected fluxes are compatible with the IceCube non-detection. In fact, the minimum flux that IceCube would be able to measure (see Eq. (4.16)) is much larger and harder than the upper limit on the neutrino flux derived from the gamma-ray observations. It is very important however to clearly distinguish the experimental upper limit of Eq. (4.16) from the theoretical upper limit on the expected neutrino signal, as derived through the observed gamma rays. The latter is more realistic and also much more stringent, but, just as the former, it depends on various theoretical assumptions. In fact, the experimental upper limit showed in Eq. (4.16) is based on the assumption that the neutrino differential energy distribution follows an E^{-2} power law: this is not confirmed by the data measured by H.E.S.S., as the spectral trend appears steeper (see Eqs. (4.2) and (4.3)), though these data cannot exclude a hardening of the $\nu_\mu + \bar{\nu}_\mu$ spectrum to E^{-2} above 20–40 TeV. However, the normalization of this component has to be $\sim$5 times smaller than Eq. (4.16), if the spectrum is a smooth distribution (a continuous function) linked to the observed gamma-ray spectrum. This kind of very speculative scenario, along with other scenarios mentioned in Sect. 4.2, might increase the expected neutrino signal. On the other hand, a conservative approach is provided in this work through the minimal scenario, as defined in Sect. 4.2 and motivated by H.E.S.S. measurements. It will be shown that the theoretical limit on the neutrino flux is smaller than the minimum measurable flux for the current experiments, whereas it could be within the reach of future detectors.

Our results compare reasonably well with the fluxes given in the Extended Data Fig. 3 of [19], that however concern the total flux of neutrinos (i.e., all three flavors). The conclusion stated in [19], based on the observed gamma-ray fluxes and on the criterion stated in [243], is that these fluxes are potentially observable. In the next section, a thorough analysis is provided to the reader in order to clarify the condition for the detectability in the existing detectors and quantify the expected number of signal events that can be detected. Moreover, the dependence of such conclusions upon the detector features is discussed.

4.4.1 Expected Signal in Neutrino Telescopes

The operating neutrino telescopes, like ANTARES [29], IceCube [5] and those under construction, as KM3NeT [27] and Baikal-GVD [55], could be able to detect the neutrinos from the Galactic Center region by looking for track-like events from the direction of this source.

The use of track-like events for the search of point-like sources is desirable because of the relatively good angular resolution, of the order of 1° in ice and several times better in water. This allows the detectors to operate with a manageable rate of background events, due to atmospheric muons and neutrinos. The atmospheric neutrinos are an irreducible source of background events for all detectors. They can be discriminated from the signal of a point-like source due to the fact that they do not have a preferential direction, and moreover they have a softer energy spectrum than the one expected from the Galactic Center region. Part of the atmospheric neutrinos from above can be identified and excluded thanks to the accompanying muons, for neutrino energies above 10 TeV and zenith angles smaller than 60° [211]. This rejection method works in the search for high-energy starting events above 30 TeV in IceCube, since it removes 70% of atmospheric neutrinos from the Southern Hemisphere [2]. Its application in the specific case of the GC region is less effective. The first reason is obvious: Sgr A* is observed at a large zenith angle from the South Pole, $\theta_Z = 90° - 29° = 61°$. The second reason is that an important fraction of the signal is expected below 10 TeV, as discussed later in this section. For what concerns atmospheric muons, one should distinguish, broadly speaking, the cases when the track-type events of interest for the search of the signal are upward-going or downward-going:

(1) The first class of events is not subject to the contamination of atmospheric muons: due to the position of Sgr A*, this kind of events can be observed by detectors located in the Northern Hemisphere.
(2) The second class of events is subject to the contamination of atmospheric muons: due to the position of Sgr A*, this is relevant event class for IceCube.

IceCube has successfully exploited a subset of downward-going track events with the purpose of investigating neutrino emission from Sgr A* [4], by requiring the additional condition that the production vertex of the downward-going tracks is contained in the detector. The fraction of time when the source is below the horizon is given by the expression $f_{\text{below}} = 1 - \text{Re}[\arccos(-\tan\delta\tan\varphi)]/\pi$ [88]. Its value is

$$f_{\text{below}} = 0\%, 64\%, 68\%, 76\% \tag{4.18}$$

for IceCube, KM3NeT-ARCA, ANTARES and GVD respectively, where the declination of the Galactic Center is equal to $\delta_{\text{Sgr A}*} = -29.01°$ and where the latitudes of the various detectors are $\varphi = 90°$ S, 36.27° N, 42.79° N, 51.83° N for IceCube, KM3NeT-ARCA, ANTARES and GVD respectively. In this fraction of time, the atmospheric muon background is suppressed and the search for a signal is more favorable. The search for a signal with upward-going tracks allows to increase the effective volume of neutrino detection to the surrounding region, whenever the muon range allows to reach the detector with sufficient energy to be detected. Moreover, the condition that the vertex is contained in the detector fiducial volume reduces the atmospheric muon background greatly, even in the low-energy region where it is more abundant. The containment condition required by down-going searches, however, does not allow to use the full volume of the detector but only a part of it, which hinders the search for a signal, especially a weak signal as the one here discussed. Another specific circumstance favors the Cherenkov telescopes operated in water, in comparison to those operated in ice, in the search for a neutrino signal at low energies. This is due to the angular resolution $\delta\theta$, that is better in water than in ice. The number of background events N_b, contained in a given search window, decreases as $\delta\theta^2$; the observable signal N_s depends upon $N_s/\sqrt{N_b}$, that scales as $1/\delta\theta$. In any detector, there is a minimum energy below which the search for a signal becomes very challenging, since the number of background events tends to be excessively large. In water-based detectors, this energy is smaller than in the case of ice-based detectors, because the number of background events decreases with angular resolution. For this reason, the neutrino telescopes operated in water are more sensitive than the telescopes operated in ice, as they can afford to use a smaller energy threshold for data taking.

The IceCube upper limit mentioned in Sect. 4.4 is based on downgoing tracks and of course this telescope is operated in ice. As will be shown in the next paragraph, the use of a water-based telescope in the Northern Hemisphere, capable of achieving good performances at low energies, can provide significantly better results and has even the potential to probe the predictions of the gamma-ray absorption model presented here. In principle, IceCube can also look at the Southern sky by exploiting the HESE sample [5], namely high-energy events whose vertex is contained inside the detector. The HESE sample was obtained by adopting a very high energy threshold ($\sim$30 TeV): this warrants a sufficiently clean sample, but requires a rather intense flux to produce an observable signal. However, in the case of Sgr A*, the interest moves to a point-like source at lower energy, so that a high precision on the reconstructed event direction and an energy threshold much lower than 30 TeV are needed. The importance of these considerations is showed in the following by a direct quantitative evaluation of the HESE event rate.

Effective Areas The angular resolution for neutrino telescopes is such that both the PS and D regions are seen as point-like regions. In the case of water-based detector, the current generation of telescopes has been able to achieve an angular resolution for track-like events of $E \geq 10$ TeV lower than 0.3° (and it is expected to further decrease with the adoption of a larger instrumentation), hence it is potentially able to distinguish the PS from the D region. On the other hand, ice-based detectors suffer from a larger uncertainty in the track reconstruction, because of the scattering properties of the ice, thus they rather see the total emission from the GC as the sum among the PS and the D emissions. In the following, the effective areas considered for ANTARES [25] and IceCube [4] are those adopted in the search of point-like sources within the declination range that includes the Galactic Center. Likewise, the effective area of KM3NeT-ARCA[5] refers to the point-like source search, though it is an average over the whole sky: it is applied to the next configuration including two building blocks. These effective areas, used for the following calculation of the event rates, are shown in Fig. 4.7.

The number of events that a detector is able to measure, assuming a certain angular search region, is given by the convolution of the expected flux from the source, $\phi_{\nu_\mu}(E) + \phi_{\bar{\nu}_\mu}(E)$, and the detector effective area, $A_{\rm eff}(E)$, through the relation

[5]http://www.ecap.nat.uni-erlangen.de/members/schmid/Doktorarbeit/JuliaSchmidDissertation.pdf.

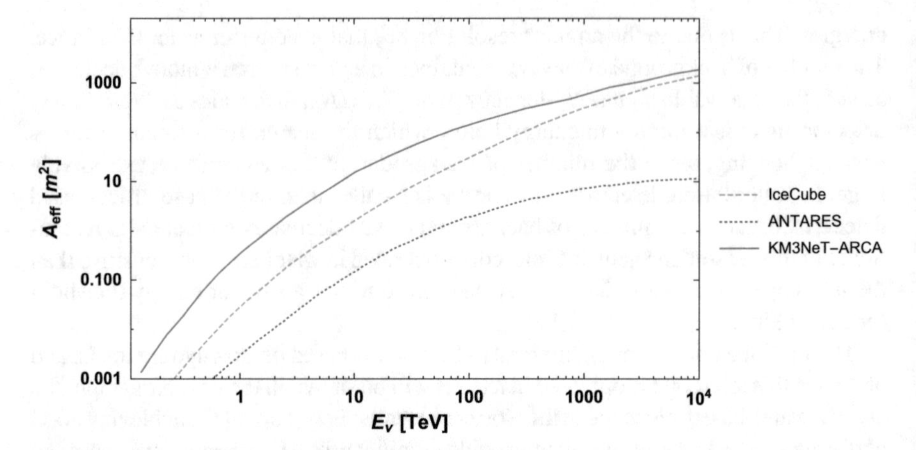

Fig. 4.7 Muon neutrino effective areas for the search of point-like sources. The one adopted for ANTARES is that in the declination band $-45° < \delta < 0°$, the one for IceCube is for $-30° < \delta < 0°$, the one for KM3NeT-ARCA is a declination-average value. See the text for references and for discussion. Figure reproduced from Celli et al. [80], under the CC BY LICENSE

$$N = T \int [\phi_{\nu_\mu}(E) + \phi_{\bar{\nu}_\mu}(E)]\, A_{\text{eff}}(E)\, dE \tag{4.19}$$

where T is the observation time. The integrand in this formula, namely the product among the neutrino flux and the effective area, is referred to as the parent distribution. Therefore, by using the neutrino fluxes of Fig. 4.6, it is possible to evaluate the neutrino energies that mostly contribute to the point-like source (PS) signal in IceCube, ANTARES and KM3NeT-ARCA. The results are reported in Fig. 4.8 and in Table 4.4, proving that the signal expected from Sgr A* is at relatively low energies. As discussed above, KM3NeT and ANTARES can look at the Galactic Center using upward-going muons, while IceCube has to use downward-going muons subject to the condition that their vertex is contained in the detector. Moreover, the first type of telescopes has a much better angular resolution, which allows them to reduce the background considered in the analysis, thus increasing the signal to noise ratio. Despite the fact that the dimensions of KM3NeT-ARCA and IceCube are comparable, the effective areas differ significantly *at low energies,* as can be ascertained from Fig. 4.7. Note, however, that the two effective areas become very similar around PeV energies, as expected from the similar physical sizes of these two neutrino telescopes. The different effective areas lead to a substantial difference in the number of events expected in IceCube and KM3NeT, which amounts to about one order of magnitude, as shown in Table 4.5.

Remarks One cautionary remark on the KM3NeT-ARCA effective area is in order. At the time of the publication, [210] was the only public source of an effective area for point-like source searches with the KM3NeT-ARCA neutrino telescope. Such an effective area was used here to evaluate the expected signal from the Galactic Center

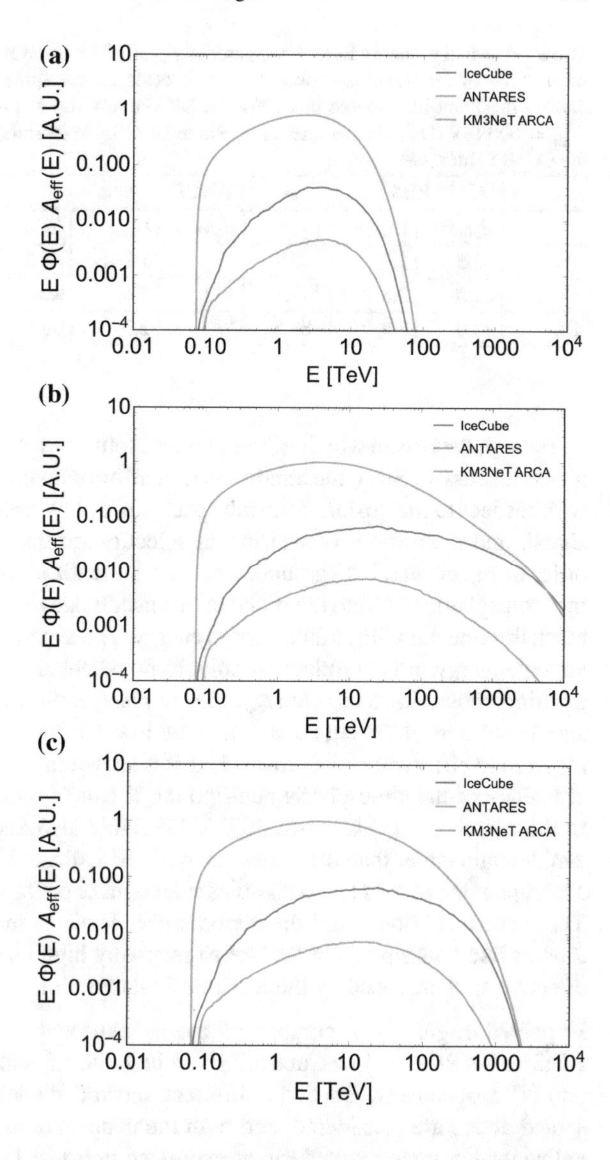

Fig. 4.8 Parental distributions of the neutrino signal, in arbitrary units. From top to bottom: the Point-like Source case (PS), the Diffuse case without any cut-off (D) and the Diffuse with a cut-off at $E^{\gamma}_{cut} = 600$ TeV (DC). See also Table 4.4. Figure reproduced from Celli et al. [80], under the CC BY LICENSE

region, being the best information available at present. However, the experimental cuts adopted in future releases by the Collaboration and consequently the effective area will likely change. The existing effective area is quite large, e.g., in comparison to the one of IceCube, but as will be demonstrated below, it corresponds to a signal of only a few events per year. Thus, it will be important to know whether the experimental cuts, that will be eventually implemented by the KM3NeT Collaboration for the search of the signal from the GC, will be compatible with similarly large effective area or will imply its revision.

Table 4.4 Median energy $E_{50\%}$ of the parental signal distribution and energy interval $[E_{16\%}-E_{84\%}]$ where 68% of the signal is expected to be detected for the same models considered in Fig. 4.8, namely the Point-like Source flux (PS), the Diffuse flux (D) and the Diffuse flux with Cut-off at $E^{\gamma}_{cut} = 600$ TeV (DC). All energies are given in TeV. Table reproduced from Celli et al. [80], under the CC BY LICENSE

	ANTARES			ARCA			IceCube		
	$E_{50\%}$	$E_{16\%}$	$E_{84\%}$	$E_{50\%}$	$E_{16\%}$	$E_{84\%}$	$E_{50\%}$	$E_{16\%}$	$E_{84\%}$
PS	3.3	1.0	9.2	3.4	1.1	9.2	1.9	0.5	6.3
D	21.9	2.8	179.1	15.0	2.6	89.2	4.8	0.8	34.4
DC	14.8	2.3	88.5	12.1	2.3	60.1	4.3	0.8	26.3

Finally, the reason why IceCube cannot usefully exploit the data set at lower energies is discussed. Since the atmospheric neutrino background has a steeper spectrum with respect to the cosmic neutrino spectrum, at low energies it dominates over the signal, and therefore a very stringent selection on data has to be implemented in order to reject such background. This can be realized, for example, through a tag: the atmospheric neutrino tag based on the accompanying muons works if the muons reach the detector with sufficient high energy. The accompanying muons should have enough energy in the production point to be revealed, which in turn means that the neutrino should have high energy, too. In the specific case under exam, however, a significant part of the signal is below the lower value of 10 TeV indicated in [211] required to effectively adopt this tag, as it is shown in Table 4.4. This implies that the efficiency of the atmospheric neutrino tag is smaller compared with that exploited in the search of HESE above 30 TeV [2]. Note also that Sgr A* is observed at a zenith angle larger than the lower value of 60° indicated in [211], and therefore, the accompanying muons lose a significant amount of energy before reaching IceCube.[6] These considerations limit the region where IceCube may conveniently search for a point-like emission from Sgr A* to relatively high energies, as quantified by the effective area produced by the IceCube Collaboration.

Expected Signal The number of events expected in one year in ANTARES, KM3NeT-ARCA and IceCube are given in Table 4.5 with the names $N^{ANTARES}$, N^{ARCA} and N^{IC} respectively. Here, the different spectral models of gamma-ray data presented above are considered, and both the contributions from muon neutrinos and antineutrinos are accounted for, as expressed in Eq. (4.17). Baikal-GVD will have a threshold of few TeV and a volume similar to KM3NeT-ARCA, so the results are expected to be similar, though a precise evaluation of the signal cannot be provided since its effective area is not available yet. As it can be seen from Table 4.5, the detectors located in the Northern Hemisphere are better suited for neutrino searches from Sgr A*. In fact, when the source is below the horizon, they can observe the Galactic Center region through upward-going track events, that are not polluted by

[6] IceCube is at a depth of 1.45 km $< h <$ 2.82 km; thus, muons traverse a relatively large amount of ice, $h/\cos\theta_Z \approx 2 \times h$.

Table 4.5 *First 4 columns*: Spectral parameters assumed for the gamma-ray fluxes, consistent with the H.E.S.S. observations as explained in the text: the search region (PS = Point-like Source or D = Diffuse), the spectral index Γ, the flux normalization ϕ_0 in units of 10^{-12} TeV^{-1} cm^{-2} s^{-1} and the energy cut-off E_{cut} in TeV (see Eq. (4.1)). For the PS and the D models, also the maximum and minimum expected values are shown. For the PS* model, a scenario with an increased non-standard gamma-ray absorption is assumed. *Last 4 columns*: Expected number of $\nu_\mu + \bar{\nu}_\mu$ events per year: downward-going tracks and HESE events in IceCube, upward-going tracks in ANTARES and ARCA. The significant increase of the event rate passing from the PS (1st row) to the PS* (4th one) model is linked to the non-standard gamma-ray absorption. Table reproduced from Celli et al. [80], under the CC BY LICENSE

	γ rays			$\nu_\mu + \bar{\nu}_\mu$			
	Γ	ϕ_0	E_{cut}	$N^{ANTARES}$	N^{ARCA}	N^{IC}	N^{IC}_{HESE}
PS	2.14	2.55	10.7	$6.2 \cdot 10^{-3}$	1.1	$5.2 \cdot 10^{-2}$	$1.4 \cdot 10^{-6}$
"	2.04	2.92	13.6	$9.5 \cdot 10^{-3}$	1.5	$8.2 \cdot 10^{-2}$	$6.1 \cdot 10^{-6}$
"	2.24	2.18	7.8	$3.9 \cdot 10^{-3}$	0.7	$3.2 \cdot 10^{-2}$	$1.9 \cdot 10^{-7}$
PS*	2.14	2.55	100	$1.7 \cdot 10^{-2}$	2.1	$1.5 \cdot 10^{-1}$	$5.0 \cdot 10^{-4}$
D	2.32	1.92	–	$1.2 \cdot 10^{-2}$	1.4	$1.3 \cdot 10^{-1}$	$2.2 \cdot 10^{-3}$
"	2.20	2.21	–	$2.1 \cdot 10^{-2}$	2.2	$2.6 \cdot 10^{-1}$	$5.5 \cdot 10^{-3}$
"	2.44	1.63	–	$7.5 \cdot 10^{-3}$	1.0	$7.4 \cdot 10^{-2}$	$8.8 \cdot 10^{-4}$
DC	2.32	1.92	400	$1.0 \cdot 10^{-2}$	1.3	$9.7 \cdot 10^{-2}$	$6.8 \cdot 10^{-4}$
DC	2.32	1.92	600	$1.1 \cdot 10^{-2}$	1.3	$1.0 \cdot 10^{-1}$	$8.8 \cdot 10^{-4}$
DC	2.32	1.92	2900	$1.2 \cdot 10^{-2}$	1.4	$1.2 \cdot 10^{-1}$	$1.6 \cdot 10^{-3}$

atmospheric muons. Such detectors are ANTARES, Baikal-GVD and KM3NeT. As visible from the Table, the expected yearly rates in ANTARES are just one order of magnitude smaller than those expected in IceCube with downward-going events: this result is well in agreement with the estimates presented in [221].

For completeness, the yearly rate of the expected HESE track events are also given in Table 4.5. It is indicated with the name of N^{IC}_{HESE} and it was obtained by using the effective areas reported in http://icecube.wisc.edu/science/data. The counting rates are, in the best case, few times 10^{-3} HESE per year. Therefore, this approach does not allow IceCube to search for neutrinos from the Galactic Center. As a comparison, the expected rate corresponding to Eq. (4.16) (namely assuming a E^{-2} distribution) in IceCube amounts to 3.8 per year, namely, more than one order of magnitude above the values of Table 4.5. The difference in the rate of neutrinos expected by assuming either the theoretical or the experimental upper bound illustrates the great importance of investigating the gamma-ray spectral energy distribution at higher energies than those currently observed, such that more stringent constraints are possibly derived in the VHE domain.

Discussion Among the gamma-ray models presented in this table, the most plausible ones are those described by a power law *with* a cut-off. In the diffuse case, even considering the less favorable case (the one with the lowest energy cut-off, which implies a cut-off in the primary spectrum of protons at about 4 PeV, where the *knee* of the Earth-observed CR spectrum is located), the predictions are such that the upcoming km^3-class detectors in the Northern Hemisphere as KM3NeT-ARCA could measure these neutrinos with a rate of few events per year. Several years of data-taking will be in any case needed in order to establish the presence of a proton Galactic accelerator up to PeV energies. In the case of non-detection, however, strong constraints will be derived concerning the proton acceleration efficiency of this poorly-understood source. A similar conclusion holds for the point-like source case. Within the assumption that the cut-off measured by H.E.S.S. is due to the absorption by a non-standard background radiation field, the muon neutrino signal increases. For instance, by comparing the first row (PS case) and the fourth one (PS*) of Table 4.5, an increase by 40–50% is realized: note that the exponential cut-off has been moved from $E_\gamma = 10.7\,\text{TeV}$ in the PS case to $E_\gamma = 100\,\text{TeV}$ in the PS* case. Remarkably, in the latter case, the protons accelerated in the source would reach energies up to the PeV scale. Unfortunately, these predictions cannot be tested yet with current neutrino telescopes. Anyway, since most of the signal is expected in the 1–100 TeV energy band, the Northern Hemisphere telescopes have to deal with a large background of atmospheric neutrino events.

Chapter 5
Sensitivity Studies for Gamma-Ray and Neutrino Telescopes

The progress made in ground-based gamma-ray astronomy over the last two decades has lead to the detection of more than 200 VHE ($E \geq 200$ GeV) sources reported by the H.E.S.S. [83], MAGIC [50], VERITAS [139] and HAWC [17] Collaborations. Recently, success has been reported also in neutrino astronomy by the detection of a diffuse flux of multi-TeV neutrinos of 'extra-terrestrial origin' by the IceCube Collaboration [2]. In the feasible future, the upgraded IceCube and the under construction KM3NeT neutrino telescopes will serve as the major tools of VHE neutrinos. Apparently, the identification of objects contributing to the reported diffuse neutrino flux, as well as the discovery of discrete sources of VHE neutrinos is the major objective of neutrino astronomy for the coming years. Despite the broad class of potential neutrino sources and the different possible scenarios of neutrino production in astrophysical environments, the production mechanisms of VHE neutrinos are connected to the hadronic interactions of ultra-relativistic protons with the ambient gas and radiation. Since these processes are accompanied by the production and decay of π^0-mesons, the VHE gamma rays and neutrinos are produced at comparable rates. Consequently, one would expect similar fluxes of gamma rays and neutrinos. On the other hand, the ground-based gamma-ray detectors, in particular the current arrays of IACTs, namely H.E.S.S., MAGIC and VERITAS, provide lower flux sensitivities for point-like sources around 1 TeV, compared to the sensitivities of the present IceCube and the forthcoming KM3NeT neutrino detectors. This circumstance reduces the chances of detection of discrete VHE neutrino sources, except for compact objects or sources located at cosmological distances. TeV gamma-ray fluxes from these objects are indeed expected to be suppressed because of both internal and intergalactic absorption, through photon-photon pair production interactions (see Sect. 4.3). In this regard, hidden sources constitute an interesting possibility for the explanation of the measured IceCube neutrino flux: these cosmic-ray

---

Part of this chapter has already been published in Ambrogi L., Celli S. & Aharonian F., 'On the potentials of the Cherenkov Telescope Arrays and KM3 Neutrino Telescopes for the detection of extended sources', Astroparticle Physics Journal, Vol. 100, Issue 69A, p. 69–79, and it is here reprinted with permission from Elsevier.

S. Celli, *Gamma-ray and Neutrino Signatures of Galactic Cosmic-ray Accelerators*, Springer Theses, https://doi.org/10.1007/978-3-030-33124-5_5

accelerators, being surrounded by very dense environments, cannot be probed by gamma rays, while transparent to neutrinos. Among them, choked GRBs and supermassive black-hole cores have widely been discussed in literature [63, 184, 215]. In these cases, neutrinos constitute crucial probes to shed light on the central engine activity. Otherwise, the VHE gamma-ray fluxes should be taken as a robust criterion regarding the expectations of discovery of discrete VHE neutrino sources. Given the difference in the TeV flux sensitivities of IACT arrays and km^3-scale neutrino detectors, the gamma-ray fluxes are especially constraining for point-like sources. In this chapter, a point-like source is called an astronomical object with angular extension less than the typical angular resolution of IACT arrays ($\sim$0.1°). For mildly-extended sources with an angular size $\sim$1°, which is one order of magnitude larger than the point spread function of IACTs but still comparable to the PSF of VHE neutrino detectors, the gamma-ray flux sensitivity degrades, while the flux sensitivity of neutrino detectors does not change significantly. Presently, this leaves room for the discovery of extended neutrino sources in our Galaxy, given also the fact that the Galactic Disc has not been homogeneously covered by the current IACT arrays. On the other hand, in-depth surveys of the Galactic Disc in coming years by CTA could significantly improve this situation. Alternatively, in a more optimistic scenario, CTA could reveal on the sky bright extended regions of multi-TeV gamma rays, and thus indicate the sites of potential detectable sources of VHE neutrinos. Here, this question is studied based on the comparative analysis of the sensitivities of CTA and KM3NeT towards extended sources: for this purpose, a common approach has been developed for calculations of sensitivities of the two instruments. The method is based on the analytical parametrization of the main quantities (as functions of energy) characterizing the process of detection of gamma rays and neutrinos: the effective detection area, the point spread function, the energy resolution and the background rates. The functions related to CTA have been provided in a previous work [53] using the results of the publicly available simulations performed by the CTA Consortium. Here, similar parametrizations are presented for the neutrino detector based on the simulation results published by the KM3NeT Collaboration [27]. Similar results are expected to hold for the IceCube-Gen2 detector [9], whose performances however are not yet publicly available. Hence, a focus on KM3NeT only is provided here.

The chapter is structured as follows: in Sects. 5.1 and 5.2 the performances of CTA and KM3NeT are discussed, in particular their angular resolution, effective area and expected background rates. Then, in Sect. 5.3 the procedure defined to compute the instrument sensitivity is described, considering different sizes of the sources and analyzing the different impact they have on the sensitivity of these instruments. As an application of this study, in Sect. 5.4 the case of two Galactic objects, for which the gamma-ray and neutrino connection has been widely discussed in literature [34, 148, 240, 241] is considered. The young SNR RX J1713-3946 is presented in Sect. 5.4.1, while the region of the Galactic Center Ridge is investigated in Sect. 1.1, being both realistic candidate neutrino sources [19, 28, 52, 87]. In addition to these scenarios, the second HAWC catalog of TeV sources is considered in Sect. 5.5, where potential sources for a neutrino detection are highlighted.

In the following, the sensitivities of CTA and KM3NeT are discussed. Both instruments are based on the Cherenkov technique, detecting the light induced by the passage of an ultra-relativistic charged particle in a given medium: the air in the case of IACTs and the water or the ice in the case of neutrino detectors. While the direction of the radiated photons is $\sim 42°$ away from the direction of the charged particle in water, the same angle is reduced to $\sim 1°$ when in the air. Although the same physical principle is applied, the reconstruction of the signal parameters and the background rejection are quite different. Both telescopes operate in the TeV domain, reaching the best performance between 1 and 10 TeV in the case of CTA and 10 to 100 TeV in the case of KM3NeT. It is worthwhile to mention here that, within this study, different possible improvements of the sensitivities of both detectors are not explored. They could however be achieved by applying dedicated tools for the background rejection and for the reconstruction of the gamma-ray and neutrino induced events from extended sources. Further details on these dedicated tools are given in Sect. 5.3. Therefore, some deviations are expected among these results and the upcoming, more detailed and sophisticated studies by the CTA and KM3NeT Consortia. In this regard, the results presented here can be considered as conservative estimates of sensitivities of both CTA and KM3NeT.

5.1 The Cherenkov Telescope Array

As introduced in Chap. 1, the CTA Southern array is aimed at studying the major fraction of the Galactic Plane, including the Galactic Center. One of the proposed layouts for the Southern observatory, the so-called *2-Q* layout, will consist of [134]:

(i) 4 large size telescopes (LSTs), 23 m class, field of view (FoV) of the order of 4.5 deg, optimized for detections below 100 GeV;
(ii) 24 medium size telescopes (MSTs), 12 m class, FoV of 7 deg, covering the core energy of CTA, i.e. 100 GeV to 10 TeV;
(iii) 72 small size telescopes (SSTs), 4 m class, FoV ranging from 9.1 to 9.6 deg, sensitive to energies above 10 TeV.

For this configuration, publicly available instrument response functions (IRFs) have been released by the CTA Consortium,[1] obtained through detailed Monte Carlo simulations of a point-like object placed at the center of the FoV and observed at a zenith angle of 20 deg (averaged between north/south-wise in azimuth). In [53] these IRFs were parametrized by simple analytical functions of energy. The results are presented in Table 5.1 and Fig. 5.1. It should be noted that the IRFs released by the CTA Collaboration, and here considered, are the derived best responses which maximize at each energy bin the CTA differential sensitivity to point-like sources.

[1] The publicly available CTA performance files can be accessed at https://www.cta-observatory.org/science/cta-performance/.

Table 5.1 Energy-dependent analytical parameterizations for the angular resolution (σ_{PSF}), the effective area (A_{eff}) and the background rate per solid angle after the rejection cuts (BgRate) of CTA, as reported in [53]. These formula have been obtained as best-fit to the publicly available IRFs for a possible layout of the CTA Southern array (https://www.cta-observatory.org/science/cta-performance/). The energy range of validity of the parametrization is from 50 GeV to 100 TeV. For the details of the IRFs as obtained by the CTA Collaboration, the reader is referred to [66, 134]. Reprinted from Ambrogi et al. [54] with permission from Elsevier

	CTA response parameterization, with $x = \log_{10}(E/1\text{ TeV})$
σ_{PSF} (deg)	$\sigma_{PSF}(x) = A \cdot \left[1 + \exp\left(-\frac{x}{B}\right)\right]$ $A = 2.71 \cdot 10^{-2}$ deg $B = 7.90 \cdot 10^{-1}$
A_{eff} (m^2)	$A_{eff}(x) = A \cdot \left[1 + B \cdot \exp\left(-\frac{x}{C}\right)\right]^{-1}$ $A = 4.36 \cdot 10^6$ m^2 $B = 6.05$ $C = 3.99 \cdot 10^{-1}$
BgRate (Hz/deg^2)	$\text{BgRate}(x) = A_1 \cdot \exp\left(-\frac{(x-\mu_1)^2}{2 \cdot \sigma_1^2}\right) + A_2 \cdot \exp\left(-\frac{(x-\mu_2)^2}{2\sigma_2^2}\right) + C$ $A_1 = 3.87 \cdot 10^{-1}$ Hz/deg^2 $\mu_1 = -1.25$ $\sigma_1 = 2.26 \cdot 10^{-1}$ $A_2 = 27.4$ Hz/deg^2 $\mu_2 = -3.90$ $\sigma_2 = 9.98 \cdot 10^{-1}$ $C = 3.78 \cdot 10^{-6}$ Hz/deg^2

Therefore, an improvement of the instrument performance is expected when using analysis cuts aimed at maximizing the telescope potential to extended objects, which are the main topic of this chapter.

5.2 The KM3-Neutrino Telescope

In order to identify neutrino cosmic sources, including those responsible for the extra-terrestrial flux of neutrinos detected by IceCube [2], a good angular resolution is a necessary requirement. Next-generation water-based telescopes, as KM3NeT, will be able to reach an angular resolution as low as 0.2° at 10 TeV in the track channel, as shown in Fig. 5.1 (top left). However, for very extended sources (more

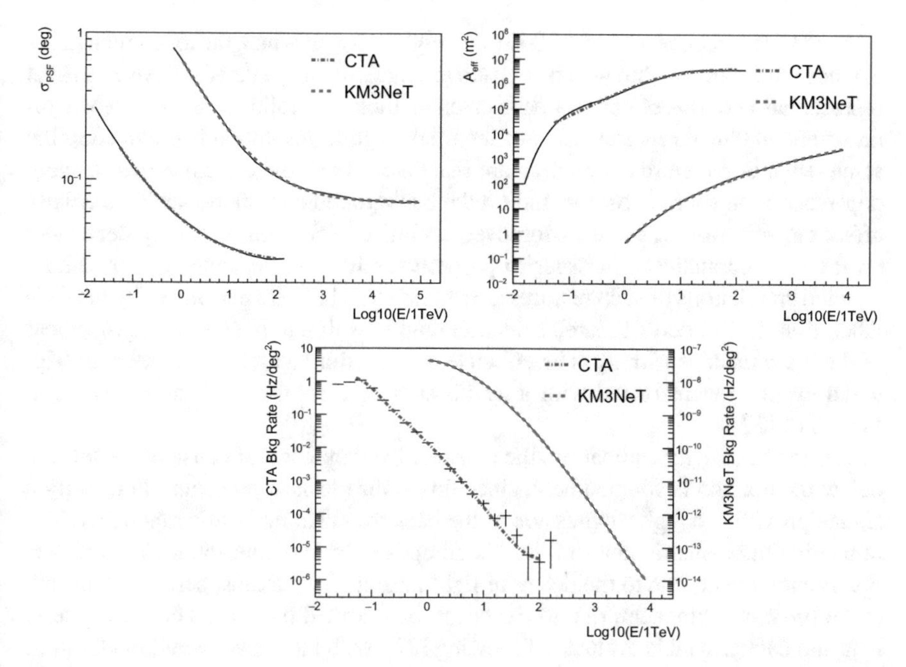

Fig. 5.1 Energy-dependent performance of the CTA (blue dot-dashed lines) and KM3NeT (red dashed lines) telescopes. *Top left*: Angular resolution. *Top right*: Effective area. *Bottom*: Background rates per unit of solid angle. For both the two telescopes, the black lines corresponds to the publicly available instrument responses, respectively for CTA Southern array (https://www.cta-observatory.org/science/cta-performance/) and KM3NeT [27]. Dotted curves are the best fit, valid in the energy range $E \in [0.05–100]$ TeV for CTA and above 1 TeV for KM3NeT. The corresponding analytical parameterizations are reported in Tables 5.1 and 5.2 for CTA and KM3NeT, respectively. The KM3NeT angular resolution is for ν_μ charged current events and the muon neutrino effective area (six blocks) corresponds to triggered events with a zenith angle greater than 80°, averaged over both ν_μ and $\bar{\nu}_\mu$. The same trigger conditions are exploited for the background rate computation. The atmospheric muon neutrino background here considered accounts for the conventional component from light meson decay [140] and for contribution from heavy hadrons [104]. For the details on the CTA IRFs and their parameterizations, the reader is referred to [53]. Reprinted from Ambrogi et al. [54] with permission from Elsevier

than 2° in radius), also cascade events could be used in principle for astronomy: the atmospheric background for shower events is significantly smaller than that for muon neutrinos, allowing for a clearer signal detection. Finally, the energy resolution is a very important goal as well: in neutrino telescopes, the energy of the muon is reconstructed through the energy deposited in the detector, therefore it is only a lower limit to the true neutrino energy. The energy resolution obtained for muon events fully contained in the detector is 0.27 units in $\log_{10}(E_\mu)$ for $10\,\text{TeV} \leq E_\mu \leq 10\,\text{PeV}$ [27]. The case of cascade events provides a better energy resolution, given that they develop entirely near the interaction point.

The effective area of the detector to up-going events is given in Fig. 5.1 (top right): it refers to the 6 building block configuration of the KM3NeT detector. This will correspond to an effective area of $\sim$1000 m^2 at high energies, where the long muon range extends the volume within which neutrino interactions can be detected. An analytical representation of this effective area is given in Table 5.2, valid for $E_\nu \geq 1$ TeV; similarly, an analytical representation of the track angular resolution is provided in the same table. Some words of caution are mandatory here: the effective area strongly depends on the source position and on the background conditions, which crucially affect the selection of events. Moreover, optimized selection is usually dependent on the specific analysis: in order to properly evaluate the detector performances, detailed simulations of such features are necessary, which are performed by the Collaboration itself. Such a tailored selection might result into a relevant improvement of the instrument sensitivity. The effective area used in the following refers to triggered events, reconstructed with a zenith angle greater than 80°, as presented in Fig. 32 of [27].

For the sensitivity estimation, the conservative approach of considering sources below the horizon is adopted here, since this is the cleanest procedure to identify a signal, providing a high suppression of the background atmospheric muon flux. The remaining atmospheric neutrino flux is composed of two contributions: a *conventional* component, due to the decay of light mesons from atmospheric air showers, and a *prompt* component, due to the decay of charmed hadrons. The atmospheric neutrino background is evaluated following [27], with the conventional model from [140] and the prompt model from [104], which becomes dominant over the conventional flux at $E_\nu > 1$ PeV: the expected up-going neutrino background rate is shown in Fig. 5.1 (bottom), while the corresponding analytical parameterization is provided in Table 5.2.

5.3 Sensitivity to Extended Sources

A common procedure for gamma-ray and neutrino telescopes is here introduced for the computation of the sensitivity curves: the same analytical approach is applied to the two detectors in order to calculate their sensitivities and facilitate their comparison. For calculations of the minimum detectable fluxes of gamma rays and neutrinos, the relevant functions provided in Tables 5.1 and 5.2 are used. Note that the curves shown in Figs. 5.2, 5.3, 5.4, 5.5, 5.6, and 5.7 correspond to the differential sensitivities: the per bin sensitivity allows not only the identification of a source but also its spectroscopic analysis. Since the sensitivity curves have to be compared point by point, the same energy binning is used for both the gamma-ray and the neutrino sensitivities. Three bins per logarithmic decade are set, so that the energy resolution of both instruments is covered in each bin.

The minimum detectable flux is defined as the flux that gives in each energy bin:

(1) a minimum number of signal events, $N_s^{\min}$;
(2) a minimum significance level of background rejection, $\sigma_{\min}$;
(3) a minimum signal excess over the background uncertainty level.

Table 5.2 Energy-dependent analytical parameterizations for the angular resolution (σ_{PSF}), the effective area (A_{eff}) and the background rate per solid angle (BgRate) to track-like events for the six building block configuration of KM3NeT. The effective area and background refer to the event selection after the cut on the zenith angle: only reconstructed upward-going muons are considered. See [27] for the details. The energy range of validity of the parametrization is $E_\nu \geq 1$ TeV. Reprinted from Ambrogi et al. [54] with permission from Elsevier

	KM3NeT response parameterization, with $x = \log_{10}(E/1\ \mathrm{TeV})$
σ_{PSF} (deg)	$\sigma_{PSF}(x) = A \cdot \exp\left(-\frac{x}{B}\right) + C$ $A = 5.88 \cdot 10^{-1}$ deg $B = 7.19 \cdot 10^{-1}$ $C = 6.95 \cdot 10^{-2}$ deg
A_{eff} (m^2)	$A_{eff}(x) = A \cdot (1+x)^B$ $A = 0.43$ m^2 $B = 5.51$
BgRate (Hz/deg^2)	$\mathrm{BgRate}(x) = A_1 \cdot \exp\left(-\frac{(x-\mu_1)^2}{2 \cdot \sigma_1^2}\right) + A_2 \cdot \exp\left(-\frac{(x-\mu_2)^2}{2\sigma_2^2}\right) + C$ $A_1 = 6.76 \cdot 10^{-10}$ Hz/deg^2 $\mu_1 = 2.89 \cdot 10^{-1}$ $\sigma_1 = 7.55 \cdot 10^{-1}$ $A_2 = 4.58 \cdot 10^{-8}$ Hz/deg^2 $\mu_2 = -2.37 \cdot 10^{-1}$ $\sigma_2 = 6.61 \cdot 10^{-1}$ $C = 3.53 \cdot 10^{-15}$ Hz/deg^2

Therefore, the instrument sensitivity is fixed by the one condition among the three listed above which dominates over the other two. The number of signal events, N_s, is obtained by folding an E^{-2} power-law spectrum with the instrument response. $N_s^{min} = 10$ is set for CTA and $N_s^{min} = 1$ for KM3NeT. The significance level of the detection is expressed by the standard deviation σ, defined as:

$$\sigma = \frac{N_s}{\sqrt{N_b}} \tag{5.1}$$

where N_b is the number of background events in the energy bin. The threshold on the minimum number of σ is set to $\sigma_{min} = 5$ for the gamma-ray telescope and to $\sigma_{min} = 3$ for the neutrino telescope. The values of N_s^{min} and σ_{min} are reduced in the case of neutrino telescopes in order to investigate the limits of the source detection capability, given that neutrino astronomy is not properly yet at its dawn. However, one has to keep in mind that each energy bin is satisfying all the above criteria: therefore, for instance, a differential 3σ requirement corresponds to an actual higher significance in the energy bins where this is not the dominant condition.

Concerning the condition on the background uncertainty, in the case of CTA a 1% systematic uncertainty is assumed on the modeling of the background and a signal of at least five times this background accuracy level is required, i.e. $N_s/N_b \geq 0.05$, following the approach adopted by CTA [66]. In the case of neutrinos, instead, the approach adopted by KM3NeT in [27] is followed and a 25% background systematic uncertainty is assumed, mainly related to uncertainty in the theoretical modeling of the atmospheric neutrino background. This uncertainty might be reduced in the future, adopting a data-based evaluation of the background. Therefore, in the neutrino case, a signal at least three times higher than the background accuracy is assumed: this converts into requiring a signal to background ratio of at least $N_s/N_b \geq 0.75$. Moreover, to account for the statistical fluctuations of the background, the number N_b is randomized according to a Poissonian distribution, with the results being averaged over 1000 realizations of the sensitivity estimation. An observation time of 50 h is assumed for the gamma-ray telescope, while 10 years are assumed for the neutrino telescope.

Note that the sensitivities are calculated without any optimization of the tools for the reconstruction of the primary particle characteristics and without exploring different dedicated background-rejection methods. A precise estimation of the CTA capabilities for the detection of extended sources would require a complete 3D analysis and the study of sub-structures on arcminute scales, as the arcminute PSF of CTA would permit to resolve the morphological details of many extended sources beyond the disc-like structure here considered. To our knowledge, such studies are currently being conducted by the CTA Consortium. However, morphological studies are not an easy task for neutrino astronomy, because of the fainter flux when only a reduced part of the source is considered. As from the neutrino side, tailored background rejection techniques targeted to the suppression of atmospheric muons and neutrinos might be included at the analysis level: for instance, vetoing few external layers of the detector has been demonstrated to be effective in rejecting background muons [2]. Furthermore, the same technique constitutes a powerful method to reject also downward-going atmospheric neutrinos, through the identification of the accompanying muon, as proposed in [211]. However, since a comparison between gamma-ray and neutrino telescopes is here proposed, estimations of the instrument potentials will be limited to the analysis technique previously described.

5.3.1 Source Angular Extension

The radial dimension of the extended source strongly affects the sensitivity for their detection. In the following, eight different source sizes are considered, i.e.

$$R_{\text{src}} = [0.1, 0.2, 0.5, 0.8, 1.0, 1.2, 1.5, 2.0] \text{ deg}.$$

The largest size has been fixed as a conservative threshold value for which the degradation of the CTA response with the off-axis angle does not play a significant effect and therefore the IRFs meant for sources located close to the center of the FoV are still valid (see e.g. [224] for a study on the expected CTA off-axis performance). Publicly available IRFs for objets placed off-axis have not been released by the CTA Consortium yet. However, an estimation of the worsening of the CTA sensitivity due to the off-axis pointing of point-like sources is presented in https://www.cta-observatory.org/science/cta-performance/. On the basis of this result, is it possible to evaluate a correction factor to the flux sensitivity for objects with an extension of up to 2 deg. As presented in Sect. 5.3.3, in those energy bins where the CTA results are available, the sensitivity worsening always results lower than a factor of two, even for sources as large as 2 deg.

In the sensitivity computation, the angular resolution σ_{PSF} affects the actual size of the observed region of interest (ROI). The radius of the ROI is defined as:

$$R_{\mathrm{ROI}} = \sqrt{\sigma_{\mathrm{PSF}}^2 + R_{\mathrm{src}}^2} \tag{5.2}$$

Spherical sources placed at the center of the FoV are considered, covering a solid angle $\Omega = \pi R_{\mathrm{ROI}}^2$ for the background computation. The resulting sensitivity curves are shown in Fig. 5.2a for the gamma-ray telescope and in Fig. 5.2b for the neutrino detector. Previous works on the potential of current neutrino telescopes towards the observation of extended sources [4, 26] show that the next-generation neutrino telescope will push sensitivity limits down by more than an order of magnitude with respect to current instrument sensitivities.

5.3.2 Discussion on Sensitivity Curves

Figures 5.2a and b demonstrate that the deterioration of the sensitivity with the source size shows for both instruments a dependence on energy. In principle, a simple re-scaling of the point-like source sensitivity according to the actual extension of the source (i.e. through an energy-dependent scaling-factor proportional to $R_{\mathrm{ROI}}/\sigma_{\mathrm{PSF}}$), would predict a stronger deterioration of the sensitivity for extended objects at energies at which the angular resolution is smaller. Thus, since the angular resolution improves with energy, one would expect a stronger effect at higher energies. Nevertheless, Fig. 5.2 does not show such a tendency for both telescopes. The reason being that at very high energies the detection of the signal proceeds at the very low background rate, thus the detection condition is determined by the signal statistics rather than by the background, i.e. by condition (1) listed in Sect. 5.3. Indeed, it is seen from Fig. 5.2 that the sensitivities become almost independent of the source extension at a few tens of TeV for both gamma rays and neutrinos.

According to the same arguments, one would expect a reduced dependence on the object size at low energies, where the angular resolution worsens.

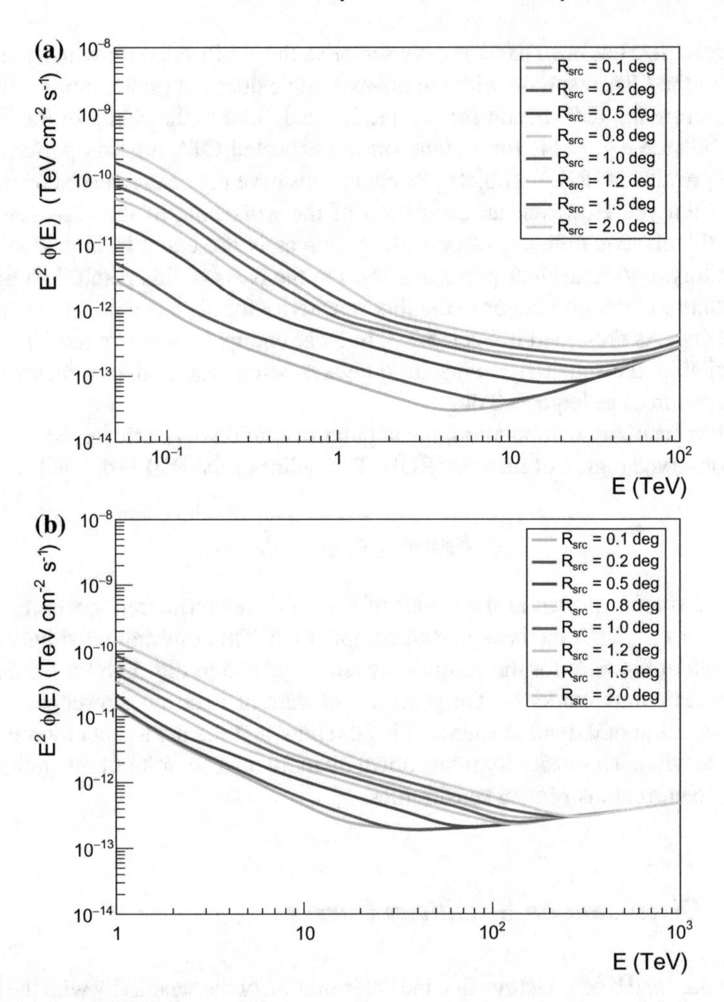

Fig. 5.2 Minimum detectable flux computed according to the procedure described in Sect. 5.3 in the case of extended sources for: **a** CTA (50 h observation time) and **b** KM3NeT (10 years observation time). Reprinted from Ambrogi et al. [54] with permission from Elsevier

Nevertheless, at these energies, the maximum deviation between different sizes is realized. This is because at lower energies, the sensitivity depends on the background as $N_s/\sqrt{N_b} \propto N_s/R_{\rm ROI}$ when the condition (2) holds (at intermediate energies), while it follows a $N_s/N_b \propto N_s/R^2_{\rm ROI}$ dependence when the background systematics are mainly affecting the signal identification, such that condition (3) dominates (at the lowest energies). Consequently, the increase of degradation with source size is maximized at the lowest energy, as the sensitivity deteriorates as $\propto R^2_{\rm ROI}$. This is especially evident in the case of CTA, while a less significant degradation with the source size takes play in the case of KM3NeT. The reason for this difference lies in the fact that, contrary to CTA, the ROI that defines the KM3NeT angular search window at the lowest energies is dominated by the instrument PSF. Consequently, the KM3NeT angular resolution affects the minimum detectable flux as long as $R_{\rm src} < \sigma_{\rm PSF}$, with $\sigma_{\rm PSF} \simeq 0.7$ deg around 1 TeV.

In summary, a deviation with increasing extension of the source size is observed, which is maximum at low energies, reduces towards higher energies, and eventually disappears at the highest energies.

5.3.3 *CTA Off-Axis Sensitivity*

The sensitivity study presented in the previous section assumes that the source is located in the center of the CTA FoV. However, in the case of extended sources, a degradation of the telescope performances should be accounted for, since part of the source would result displaced from the FoV center.

In this section, a correction factor to be applied to the CTA sensitivity curves shown in Sect. 5.3 is estimated, in order to take into account the degradation of the instrument response due to the offset of the source. The publicly available results of the CTA Southern array are considered for what concerns the point-like source detectability off-axis, as reported in https://www.cta-observatory.org/science/cta-performance/. Here, the instrument point-source off-axis sensitivity relative to the one at the center of the FoV is presented in four different energy bins (i.e. 50–80 GeV, 0.5–0.8 TeV, 5–8 TeV and 50–80 TeV). In the same energy intervals, the correction factor $\langle F \rangle$ is defined as the average value of the CTA expectations over the total extension of the objects. The objects with total radius R_{ROI} (see Eq. (5.2)) are treated as composed by concentric annuluses with size $2\sigma_{PSF}$, where σ_{PSF} is the mean value of the angular resolution in each of the four energy bins. Opting for a conservative approach, a set of N annuluses is considered, with N the smallest integer such that the radius of the source can be expressed as a finite multiple of the instrument angular resolution, so that $N \cdot (2\sigma_{PSF}) \geq R_{ROI}$. The contribution of each annulus to $\langle F \rangle$ is then weighted according to the area of the ring A_{ring} itself, yielding to the definition of $\langle F \rangle$ as:

$$\langle F \rangle = \frac{\sum_{i=1}^{N} f_i \cdot A_{ring,i}}{\sum_{i=1}^{N} A_{ring,i}} \tag{5.3}$$

where f_i is the value of relative worsening at a fixed distance from the camera center, as inferred from https://www.cta-observatory.org/science/cta-performance/. The estimated correction factor $\langle F \rangle$ is shown in Fig. 5.3 for sources with an extension $\geq$0.5 deg, as for smaller distances to the camera center the telescope sensitivity does not suffer any significant worsening (i.e. $\langle F \rangle = 1$). The worsening of sensitivity is less than a factor of two, even for the largest source extension considered in this work, i.e. 2 deg, in the lowest energy interval.

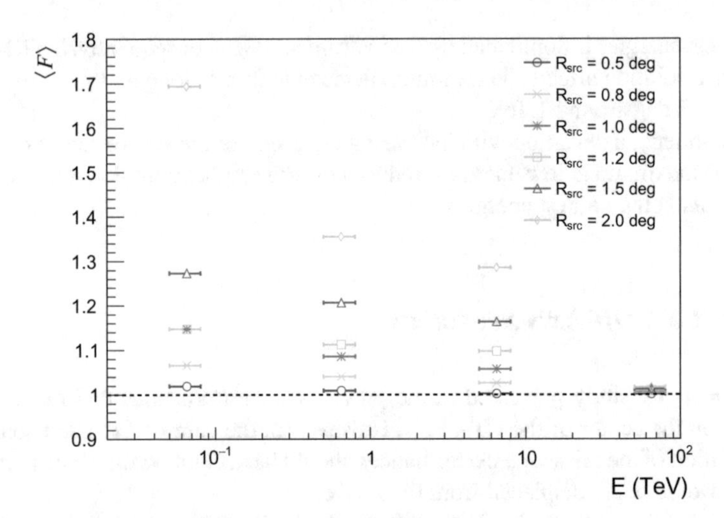

Fig. 5.3 CTA-South correction factor $\langle F \rangle$ to account for the off-axis degradation of the instrument sensitivity to extended sources. See the text for a detailed description of the factor $\langle F \rangle$, as defined in Eq. (5.3). The widths of the energy bins correspond to the intervals for which the CTA point-like source off-axis estimations are available. Reprinted from Ambrogi et al. [54] with permission from Elsevier

5.4 The Case of RX J1713.7-3946 and the Galactic Center Ridge

As the origin of the CR flux measured at Earth is still a matter of debate, it is mandatory to investigate sources which might be responsible for it. Galactic sources are believed to contribute up to energies of about 1 PeV, in correspondence of the *knee*. Among them, young SNRs represent promising candidates, given that the strong shocks produced during the supernova explosion might be able to accelerate particles. However, the lack of observational evidence for the presence of PeV protons in such objects does not permit yet to firmly establish the SNR paradigm for the origin of Galactic cosmic rays, and future gamma-ray and neutrino observations are needed to further constrain theoretical models. In this regard, the recent claim of a PeVatron in the center of our own Galaxy [19] opens a new possibility to explain the flux of CRs below the *knee* and it deserves a deeper investigation, both from the gamma-ray and the neutrino side. For these reasons, in this section, the case of two bright extended gamma-ray sources is discussed: the young SNR RX J1713.7-3946 and the Galactic Center Ridge. In the following, the muon neutrino fluxes expected from these sources are computed according to the model in [238], as presented in Chap. 4, assuming a 100% hadronic origin of the measured gamma-ray flux and no internal absorption of the photon flux. Both the measured gamma-ray and expected neutrino fluxes are shown together with the detector sensitivity curves. Fluxes are reported in a binned form, such that it is directly possible to compare the expected source flux in a given energy range with the detector sensitivity in the same band. These binned fluxes are defined as:

$$\phi(E) = \frac{1}{\Delta E} \int_{E_{\min}}^{E_{\max}} \frac{dN}{dE} dE$$

where $\Delta E = E_{\max} - E_{\min}$ represents the amplitude of the logarithmic bins used in the sensitivity computation. The fluxes are reported together with the associated error bands. In the case of gamma rays, the statistical errors of the estimated spectral parameters are considered and then, for each energy bin, the upper/lower band are computed as the curve defined by the combination of parameters ($\pm$ statistical errors) for which the flux is maximized/minimized. The same approach is adopted also for the neutrino flux error bands: a scanning of the neutrino fluxes resulting from all the different combinations of gamma-ray parameters is performed and the maximum/minimum neutrino flux are computed.

5.4.1 RX J1713.7-3946

The case of this source is of great interest for neutrino telescopes, given that it is the brightest SNR observed until now in the TeV sky. Moreover, its location in the sky makes it observable with up-going events at the latitude of KM3NeT for 70% of the time. The recent data from the H.E.S.S. Collaboration [12] suggest a spectrum in the form of an exponentially-suppressed power law with:

$$\frac{dN_\gamma}{dE}(E) = \phi_0 \left(\frac{E}{E_0}\right)^{-\alpha} \exp\left[-\frac{E}{E_{\text{cut}}}\right] \tag{5.4}$$

with $E_0 = 1$ TeV, $\phi_0 = 2.3 \times 10^{-11}$ TeV^{-1} cm^{-2} s^{-1}, $\alpha = 2.06$ and E_{cut} TeV for the best-fit model. Note that the best-fit flux of the source, published in the previous paper [40] of the H.E.S.S. Collaboration, with the parameters $\phi_0 = 2.13 \times 10^{-11}$ TeV^{-1} cm^{-2} s^{-1}, $\alpha = 2.04$ and $E_{\text{cut}} = 17.9$ TeV, predicts a noticeably higher flux of neutrinos at the most relevant energies for the detection of neutrinos, $E_\gamma \geq 10$ TeV. This is visible in Fig. 5.4a, where the expected neutrino fluxes from both H.E.S.S. measurements are shown together with the flux sensitivities of the two instruments for an extended source with a radius of 0.6 deg. High quality spectroscopic measurements of gamma rays at the highest energies are still missing, due to the limited sensitivity of current instruments. The uncertainty on the gamma-ray flux above ~10 TeV is expected to be substantially diminished through CTA observations on the source, in a rather short time. Remarkably, even for the lowest predicted neutrino flux, a statistically significant detection of the latter by KM3NeT seems realistic for timescales of 10 years. This is well in agreement with the lack of events from RX J1713.7-3946 by the current generation of neutrino telescopes, whose upper limits on its neutrino flux amounts to 6.7×10^{-12} TeV cm^{-1} s^{-1} in the case of ANTARES [47], and to 9.2×10^{-12} TeV cm^{-1} s^{-1} for the IceCube detector [7]. In the following, the possibility of a neutrino detection from the SNR with the

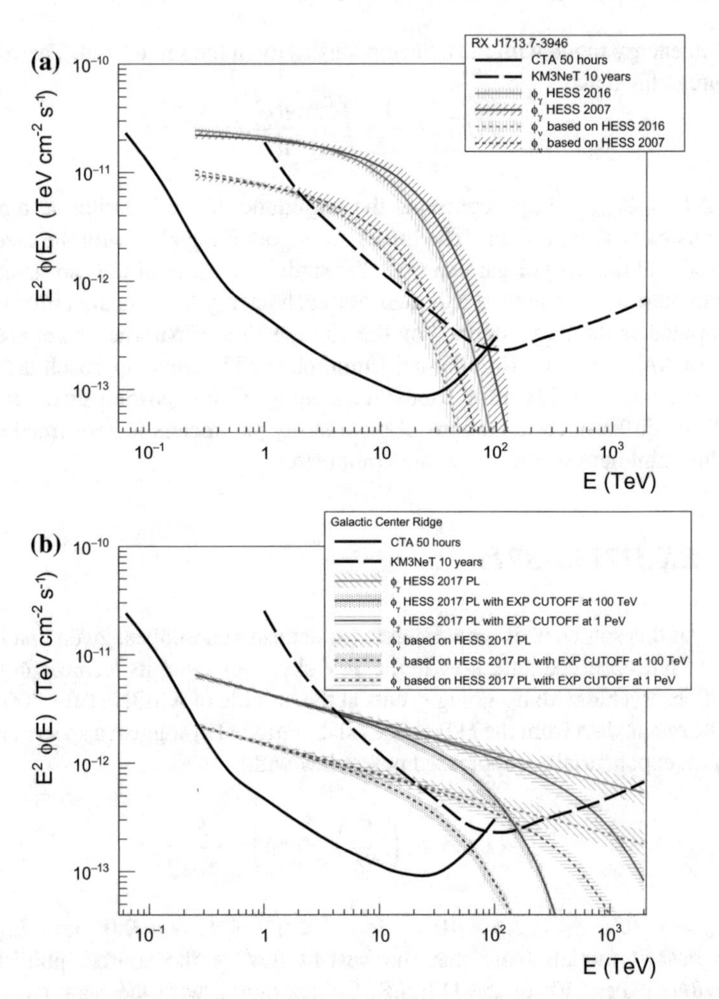

Fig. 5.4 Minimum detectable flux computed according to the procedure described in Sect. 5.3 for CTA and KM3NeT to: **a** the extended SNR RX J1713.7-3946 (spherical source with radius of 0.6 deg) and **b** the diffuse emission from the Galactic Center Ridge (rectangular box with longitudinal size of 2.0 deg and latitudinal size of 0.6 deg). The binned gamma-ray fluxes are shown as colored solid lines. The dashed curves are the binned muon neutrino fluxes computed according to the model in [238] in the hypothesis of 100% hadronic gamma rays. Reprinted from Ambrogi et al. [54] with permission from Elsevier

KM3NeT/ARCA detector is investigated, since it represents a timely study given that its completion is planned within 2020, as reported in [27].

The ARCA Sensitivity A less stringent requirement for the source detection is derived when demanding for the achievement of a given significance level by integrating over the whole energy range, rather than on individual bins. This approach is usually adopted when statistics is not large enough for spectroscopic studies. The low rate of events connected to neutrino interaction properties makes the case for integrated sensitivity analyses, even though one should carefully evaluate the instrument performances in those specific energy intervals where most of the signal is expected. Thus, in the following, the specific ARCA effective area released by the KM3NeT Collaboration towards the location of the source is considered, with the experimental cuts reported in Fig. 40 of [27] (either the one labeled as Λ or the Boosted Decision Tree (BDT)). These cuts refer to different stages of the analysis: Λ is an estimator for the quality of track reconstruction, while the BDT represents a more stringent event selection, being the result of a multivariate technique including the Λ parameter as well as other estimators (as the energy, the vertex position, etc.). The resulting effective area for ARCA is shown in Fig. 5.5a. Concerning the signal neutrino flux expected from the SNR, the best-fit gamma-ray modeling of the source emission is considered, as described in Sect. 2.6 with the parameter $B_c = 10\,\mu\text{G}$, and the muon neutrino flux is derived according to the model in [238]. The parent energy distribution, obtained by convolving the signal flux and the instrument effective area, is shown in Fig. 5.5b: as visible, the maximum of the function is expected within 10 and 20 TeV. The evaluation of the background follows the approach described in Sect. 5.2, with both a conventional and a prompt component for the atmospheric contribution. Figure 5.5c shows the parent energy distribution for the background flux, which is also shown to be maximum in the same energy region of the signal parent function, highlighting the importance of accessing this low-energy range with a clean event sample. The number of signal and background events expected in 5 years of observation is given in Table 5.3, where the source extension is assumed to be $R_{\text{src}} = 0.6°$: as visible, a 3σ detection is foreseen in about 5 years of operation of KM3NeT-ARCA. If no signal will be found during this time, the hadronic fraction of the gamma-ray flux can be constrained to be lower than 50% [45].

5.4.2 *The Galactic Center Ridge*

Another promising object from the point of view of multi-TeV neutrino detection is the Galactic Center Ridge. The observations with the H.E.S.S. telescope revealed the presence of a diffuse emission component from a 200 pc region around the Galactic Center [19, 39]. The hard energy spectrum of the emission extends well above 10 TeV without any indication of a spectral break or cut-off [19]. If such emission has a hadronic origin, neutrinos are expected as well and this source is another candidate to be detected by KM3NeT, as introduced in Chap. 4. The H.E.S.S. measurements of the Galactic Center Ridge spectrum, as recently reported in [11], point toward an unbroken power law of the form:

$$\frac{dN_\gamma}{dE}(E) = \phi_0 \left(\frac{E}{E_0}\right)^{-\alpha} \tag{5.5}$$

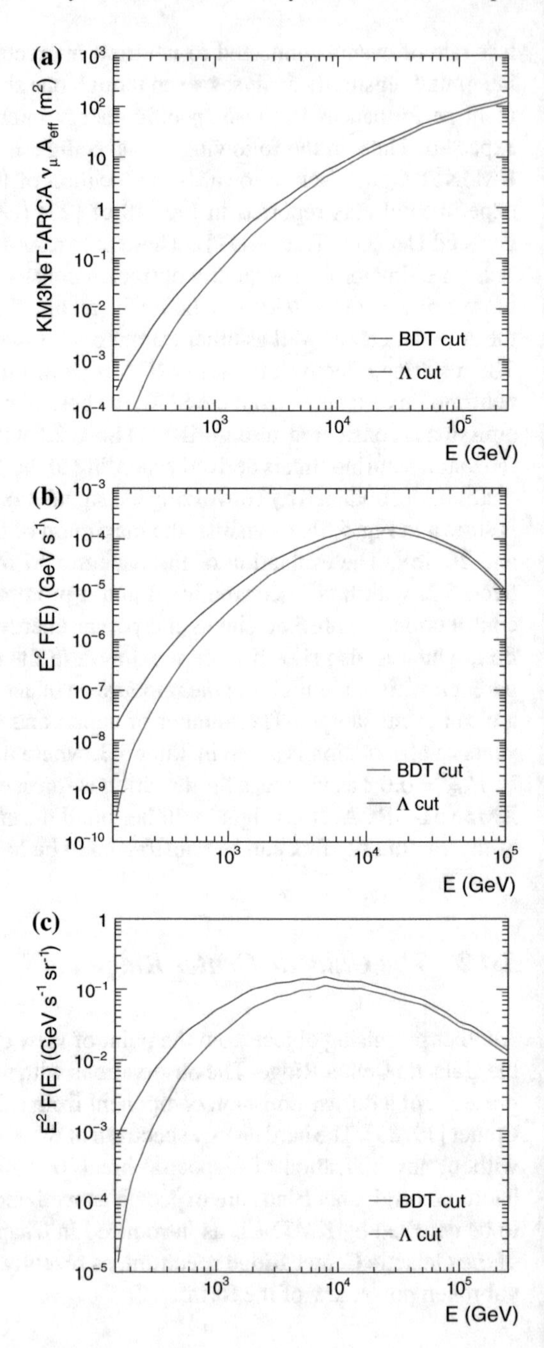

Fig. 5.5 Neutrino analysis tailored at the SNR RX J1713.7-3946. **a** KM3NeT-ARCA effective area, for the specific source declination and different analysis cuts (Λ and BDT, see text). **b** Parent energy distribution of the signal for muon tracks in KM3NeT-ARCA. **c** Parent energy distribution of the atmospheric background (both conventional and prompt neutrinos included)

Table 5.3 Number of muon neutrino (and antineutrino) events expected from the SNR RX J1713.7-3946 in the KM3NeT-ARCA detector during 5 years of operation. Different analyses cuts are considered (Λ and BDT, see text): N_s is the number of signal events, N_b is the number of background events and $\sigma = N_s/\sqrt{N_b}$ is the significance level

Cut	N_s	N_b	σ
Λ	9.2	8.3	3.2
BDT	5.9	3.9	3.0

with $E_0 = 1$ TeV, $\phi_0 = 1.2 \times 10^{-8}$ TeV^{-1} cm^{-2} sr^{-1} s^{-1}, $\alpha = 2.28$. This spectrum corresponds to the region $|l| \leq 1.0$ deg, $|b| \leq 0.3$ deg. The estimation of the sensitivities of CTA and KM3NeT is done for the same sky region, namely a rectangular box with longitudinal size of 2.0 deg and latitudinal size of 0.6 deg. The results are shown in Fig. 5.4b, where three neutrino spectra are calculated: the unattenuated power law that directly follows the gamma-ray measurements and two more spectra obtained by assuming an exponential cut-off in gamma rays respectively at 100 TeV and 1 PeV. One can see that only in the case for which the location of the gamma-ray cut-off energy is beyond 100 TeV, does the gamma-ray data guarantee a statistically significant detection of the counterpart neutrinos by KM3NeT. The gamma-ray flux above 100 TeV is too weak to be detected by the H.E.S.S. telescopes, even after a decade of continuous monitoring of this region. The exploration of this energy domain requires a more powerful gamma-ray instrument such as CTA. This can be seen in Fig. 5.4b. Note that, even in the case of effective gamma-ray production above 100 TeV, these photons could hardly escape the Ridge because of the absorption in the interactions with the enhanced far infrared radiation field in the central 200 pc region. Thus, it is likely that the neutrinos remain the only messengers of information about the CR protons with energies larger than 1 PeV. This opens a unique opportunity for KM3NeT to provide a major contribution to the exploration of the CR PeVatron in the Galactic Center.

5.5 The Second HAWC Catalog

Beyond the search for neutrino emission from individual sources, population studies are expected to provide a deep comprehension of the acceleration mechanism acting at the source. In this section, the recently published second High Altitude Water Cherenkov (HAWC) catalog of TeV sources, namely the 2HWC catalog reported in [18], is considered and the capability for neutrino telescopes to investigate such fluxes are discussed. A total of 39 VHE gamma-ray sources are reported in the catalog, both Galactic and extra-galactic ones: of these, two are associated with blazars, two with SNRs, seven with PWNe. Of the rest, 14 have possible associations with SNRs, PWNe and molecular clouds, while the remaining 14 are still unidentified. With respect to other TeV catalogs from IACT arrays, a HAWC extended source is actually a sky region where more sources might be overlapping, given the limited angular resolution with respect to IACTs. In the following, all the emission coming from such a region is considered as a single source, given that neutrino tele-

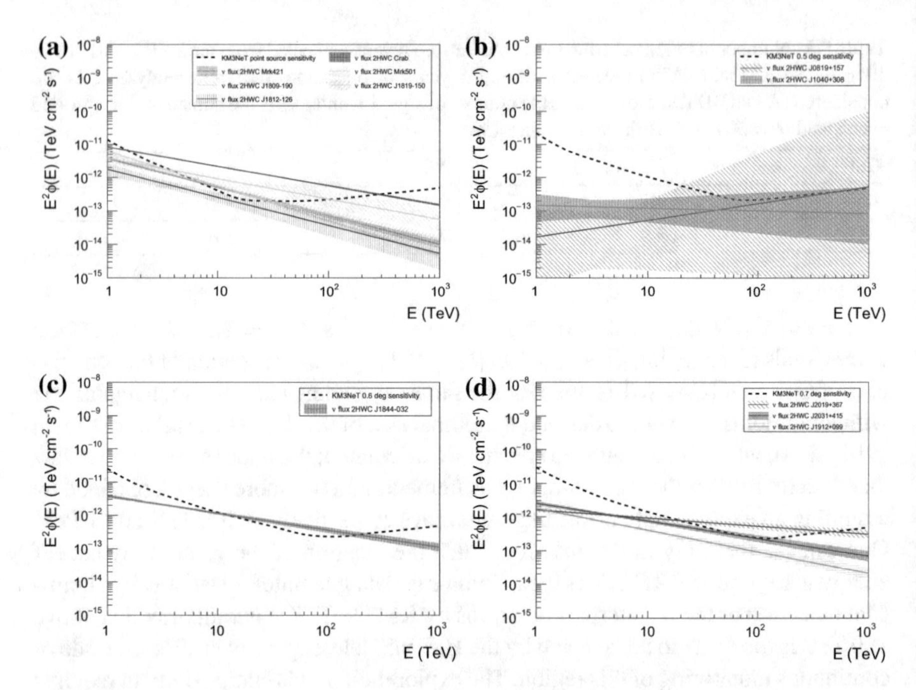

Fig. 5.6 Muon neutrino (and antineutrino) fluxes (solid lines) of the sources reported in the second HAWC catalog [18] and KM3NeT minimum detectable flux (dashed line), as computed according to the procedure described in Sect. 5.3, towards: **a** point-like sources, **b** 0.5 deg extended sources, **c** 0.6 deg extended sources and **d** 0.7 deg extended sources. The error bands on neutrino fluxes are computed as described in the text, accounting for the statistical error on the gamma-ray observation. Reprinted from Ambrogi et al. [54] with permission from Elsevier

scopes have a quite similar angular resolution to water-based Cherenkov gamma-ray instruments. Several among the sources reported in [18] are tested under different angular extension hypotheses, leading to different spectral fits: the spectral fit corresponding to the more extended source assumption is considered here. Furthermore, Geminga is flagged twice in the 2HWC catalog (both as a point-like and as a 2 deg extended object): therefore, the final list counts a total of 40 gamma-ray emitters. In the view of a neutrino detection, the sensitivity of the KM3NeT detector to upward-going track events from the 2HWC sources has been studied. Although this is not the best source sample to be investigated with KM3NeT through upgoing muons, given their sky position, nonetheless they are here considered since it is interesting to exploit sources not previously detected in the TeV energy region. The sensitivity of KM3NeT is reported in Figs. 5.6 and 5.7, together with the expected neutrino fluxes. The case of point-like sources with fluxes in the reach of KM3NeT is reported as well. In particular Fig. 5.6 shows the neutrino expectations for: Fig. 5.6a 2HWC point-like sources, Fig. 5.6b for source with an extension of 0.5 deg, Fig. 5.6c for 0.6 deg sources; Fig. 5.6d for 0.7 deg sources. Analogously Fig. 5.7a shows the results for 0.8 deg sources, Fig. 5.7b for 0.9 deg sources; Fig. 5.7c for 1.0 deg and

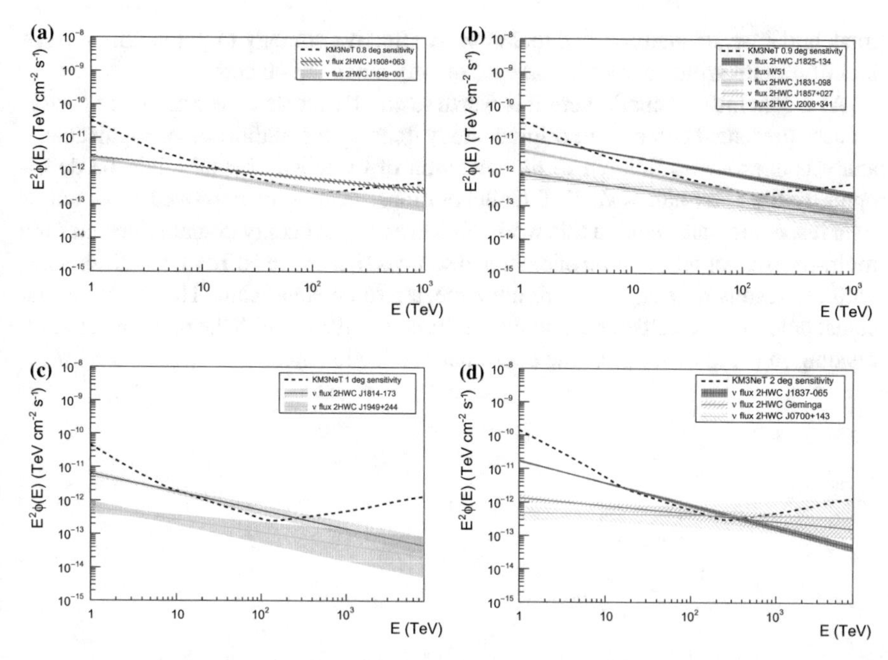

Fig. 5.7 Muon neutrino (and antineutrino) fluxes (solid lines) of the sources reported in the second HAWC catalog [18] and KM3NeT minimum detectable flux (dashed line), as computed according to the procedure described in Sect. 5.3, towards: **a** 0.8 deg extended sources, **b** 0.9 deg extended sources, **c** 1.0 deg extended sources and **d** 2.0 deg extended sources. The error bands on neutrino fluxes are computed as described in the text, accounting for the statistical error on the gamma-ray observation. Reprinted from Ambrogi et al. [54] with permission from Elsevier

Fig. 5.7d for 2.0 deg sources. The computation of neutrino spectra is realized without accounting for possible absorption of the gamma rays, which might be relevant for extra-galactic sources: in this case, neutrino fluxes would increase since neutrinos do not suffer of absorption. As visible in Figs. 5.6 and 5.7, promising sources for a neutrino detection are represented by 2HWC J1809-190, 2HWC J1819-150, Crab, Mrk 421, 2HWC J1844-032, 2HWC J2019+367, 2HWC J1908+063, 2HWC J1825-134, 2HWC J1814-173 and 2HWC J1837-065. Since only sources below the horizon are considered, the source visibility needs to be taken into account. The sources with more than 50% visibility at the KM3NeT latitude are constituted by 2HWC J1809-190, 2HWC J1819-150, 2HWC J1814-173, 2HWC J1825-134, 2HWC J1844-032 and 2HWC J1837-065. The inclusion of down-going events, with specific analysis features allowing the reduction of the atmospheric muon background, will permit to include in the analysis the whole data-taking period: in this case, however, the selection efficiency would further reduce the triggered sample by a factor of a few. Finally, in the case of two degree extended sources, the cascade channel might be added, given that the source dimensions are comparable to the shower angular resolution at high energies (for KM3NeT, about 2° above $E_\nu = 70$ TeV). A combined track and cascade analysis is thus the most effective strategy to pursue the goal of identifying neutrino sources in the case of very extended objects.

As a conclusive remark, here it is worth to emphasize that the instrument performances presented come as an original study done by the author: more sophisticated analysis approaches tailored to the detection of extended objects might be developed by the CTA and KM3NeT Collaborations, leading to improved sensitivities with respect to that found in this work. However, in the energy domain relevant for a multi-messenger observation of extended sources (i.e. above 10 TeV), the differences with the results here reported are not expected to be significant. This is due to the signal detection being determined by the limited statistics, with the main uncertainty residing in the knowledge of the instrument collection area.

Chapter 6
Summary and Conclusions

This thesis concerns the study of Galactic CRs through their gamma-ray and neutrino signatures: in fact, a solid connection exists between CR physics and VHE gamma-ray and neutrino astronomy. This association is a result of the fact that the sources of CRs are also expected to be bright VHE gamma-ray and neutrino emitters, due to the decay of neutral and charged pions produced in the interactions between the accelerated particles and the gas surrounding the acceleration site. SNRs are believed to be the main contributors to the Galactic CR flux, though so far their role in the production of PeV particles has not yet been observationally established. In fact, even the brightest TeV SNR detected so far does not seem to act as a PeVatron, namely a CR accelerator producing a hard spectrum of protons extending to PeV energies without any break or cut-off. Hence, a first issue arising in the SNR paradigm for the origin of Galactic CRs is connected with the maximum energy that these accelerators are able to achieve. A second concern refers to the shape of the VHE spectra observed in SNRs, as they systematically show a differential energy spectrum different from the E^{-2} power law expected from particle acceleration within the test-particle DSA theory. In particular, very young SNRs ($\sim$300 yr old) have generally steeper spectra than young SNRs ($\sim 10^3$ yr old). On the other hand, middle-aged SNRs ($\sim 10^4$ yr old) show even steeper spectra. Such a variety is not easily explained by the DSA theory, even when non-linear effects are taken into account. Furthermore, the origin of the VHE radiation observed in SNRs, either leptonic or hadronic, is practically under debate for all sources, as the broad-band spectrum is not easily accommodated by simplified one-zone models. Possibly, the solution to these inconsistencies might reside in the fact that the current acceleration theory does not account for realistic environments where SN explosions happen. Moreover, a complete theory, able to describe the full process of particle acceleration, propagation around the sources and escape from these is still nowadays missing.

A major limitation of the phenomenological description of particle acceleration and propagation at SNR shocks is connected with the assumption that the remnants are expanding into a uniform ISM. A more realistic situation is instead represented by a multi-phase medium, composed by dense structures surrounded by a more homogenous plasma. In Chap. 2, I investigated the effects of a shock propagating through

S. Celli, *Gamma-ray and Neutrino Signatures of Galactic Cosmic-ray Accelerators*, Springer Theses, https://doi.org/10.1007/978-3-030-33124-5_6

dense inhomogeneities, by numerically solving the MHD equations describing the evolution of the background plasma. On top of this, the numerical solution of the particle diffusive transport equation was obtained in order to describe the accelerated particle distribution. The presence of dense molecular clumps in shock environments strongly affects the plasma properties, in that the large scale magnetic field results amplified around them, due to both field compression and the shear motion. As a consequence, the propagation of particles accelerated at the shock proceeds in such a way that low-energy particles need more time to penetrate the clumps compared to high-energy ones. The resulting energy spectrum of particles contained inside the clump is significantly harder than the spectrum accelerated at the shock through DSA. Given that clumps contain most of the target mass, the gamma-ray spectrum produced in hadronic collisions of accelerated particles appears to be much harder than the parent spectrum. It is hence necessary to account for the inhomogeneous CSM to correctly predict the gamma-ray spectrum. Such a scenario is very common in the case of type II SNe, which due to the short life of their progenitor, typically expand in a region where the molecular clouds responsible for the progenitor formation still exist. I demonstrated that this effect might be relevant for the brightest SNR observed in TeV gamma rays, RX J1713.7-3946. The cumulative contribution of clumps embedded between the contact discontinuity and the current shock position is able to reproduce the observed GeV hardening. Remarkably, for the gas density assumed here inside the clumps ($n_c = 10^3\ \mathrm{cm}^{-3}$) the evaporation time is much longer than the SNR age. As a consequence, the clumps crossed by the forward shock do not produce significant thermal X-ray emission, in agreement with observations. Additionally, the proposed scenario can naturally account for the fast variability observed in the non-thermal X-ray emission of several hot-spots inside RX J1713.7-3946, as due to the electron synchrotron losses in the amplified magnetic field around the clumps. The detection of such small structures by means of molecular transition lines is within the reach of the ALMA telescope, and it appears extremely relevant to confirm the hadronic origin of the VHE radiation. An independent signature of the 'clumpy' origin of the gamma-ray emission could be revealed from a morphological study performed with the next-generation gamma-ray telescope, CTA. The superb sensitivity and the high angular resolution of CTA could allow to resolve regions small enough that they contain only few clumps, possibly revealing large spectral fluctuations in the gamma-ray spatial profile, unlike a scenario where the SNR is expanding into a uniform medium.

The deviation observed in the HE and VHE gamma-ray spectra of SNRs, particularly in middle-aged ones, with respect to the simple spectral shape predicted by the DSA theory, might possibly be connected to the particle escape process from the shock region. The escaping process of particles accelerated at SNR shocks remains one of the less understood piece of the shock acceleration theory. As a consequence, this aspect is often neglected, though it represents a fundamental part of the process, needed to explain the CR spectrum observed at Earth. In Chap. 3, I developed a phenomenological model for the description of particle escape from an SNR shock, aimed at evaluating the effects produced by the escape process on the gamma rays emitted by the source. In particular, when particles are not confined any more by the shock, they start to freely diffuse in the CSM, eventually escaping the accelerator. The streaming of CRs is able to excite the resonant instability, thus amplifying the turbulence necessary to confine the accelerated particles around the SNR for long time [93]. Since the escape process is not instantaneous, a relevant fraction of high-energy particles can still be located inside the SNR or close to it even once they are not confined anymore by the shock turbulence, producing diffuse TeV halos around the remnant. This process has at least two important consequences on gamma rays: (i) the spectrum from the SNR interior observed at a fixed time presents a steepening above the maximum energy of particles accelerated at that time, and (ii) the spectrum emitted from the halo around the remnant shows a low-energy cut-off in correspondence of the minimum energy of the escaping particles at the observation time. While the second aspect could be tested with future gamma-ray telescopes (for instance CTA), the former could have already been detected. I showed that accounting for both the contribution from confined particles, which are still undergoing the acceleration process, and non-confined ones, which are diffusing close to the source, it is possible to reproduce the spectra of some interesting middle-aged SNRs. A description of the observed VHE radiation has been achieved in the case of IC 443, W 51C and W 28N through the assumption of a diffusion coefficient reduced by a factor 10–100 with respect to the average Galactic value. Furthermore, all of these SNRs require an acceleration spectrum steeper than what is predicted by the test-particle DSA theory, in order to reproduce the HE gamma-ray data. In addition to the effects produced on the gamma-ray spectrum, I calculated the total CR spectrum injected in the Galaxy by an individual SNR, evolving in the ST phase. For an acceleration spectrum $\propto p^{-\alpha}$, under the assumption that a fixed fraction of the shock kinetic energy is converted into accelerated particles at every time, the spectrum injected into the Galaxy by a single SNR turns out to be: (i) at $p \gg m_p c$, $f_{\rm inj}(p) \propto p^{-4}$ if $\alpha < 4$ or $f_{\rm inj}(p) \propto p^{-\alpha}$ if $\alpha > 4$, and (ii) at $p \ll m_p c$, $f_{\rm inj}(p) \propto p^{-5}$ if $\alpha < 5$ and $f_{\rm inj}(p) \propto p^{-\alpha}$ if $\alpha > 5$. This result is independent of the temporal behavior of the maximum energy at the shock, but it strongly depends on the assumption that the particles are released entirely while the SNR is in its ST stage.

The lack of an observational proof of the SNR paradigm for the origin of GCRs provides a hint for considering among the responsible sources other objects than SNRs, for instance the black-hole at the center of the Galaxy or massive stellar clusters. To this extent, the recent claim by the H.E.S.S. Collaboration concerning the presence of a proton PeVatron in the center of our Galaxy [19] is critically considered in view of a neutrino counterpart, that would unambiguously identify the presence of hadronic emitters. The Galactic Center region contains indeed several possible high-energy emitters, including the closest super-massive black-hole, strong star-forming regions, massive and dense molecular clouds, numerous supernova remnants and pulsar wind nebulae, arc-like radio filaments, as well as the bottom part of what might be associated to large-scale Galactic outflows, i.e. the so-called Fermi bubbles. The detection of neutrinos from the Galactic Center region is crucial to confirm or discard the hadronic origin of gamma rays. In case of non-detection, however, neutrino telescope will be able to put severe constraints on the efficiency of hadronic acceleration in this region. In order to predict the expected neutrino flux, it is necessary to know the flux of gamma rays at the source, and for this reason it is crucial to evaluate correctly the effect of the absorption due to the interaction between the gamma rays and the background radiation fields. As I demonstrated in Chap. 4, the diffuse high-energy gamma-ray flux measured by H.E.S.S. is not affected by the absorption on standard known radiation fields, namely the CMB and the optical and IR radiation backgrounds. On the contrary, the gamma-ray flux from the central point-like source could be affected by an intense infrared radiation field, possibly located in the close vicinity of the Galactic Center. The absorption effect is compatible with an unbroken power-law distribution of the emitted gamma rays from the central point-like source, and it would increase the observable neutrino signal by 40–50% with respect to current estimates. This new hypothesis motivates further studies with IR telescopes and with 100 TeV gamma-ray instruments, as CTA. I obtained a precise upper limit on the expected neutrino flux from the regions close to the Galactic Center, by assuming that the gamma rays recently observed by H.E.S.S. are produced in CR collisions. The corresponding maximum signal is of few track (muon signal) events per year in the upcoming KM3NeT detector. In view of these results, I concluded that the KM3NeT detector has the best chances to observe neutrinos from Sgr A*, even if, in order to accumulate a large sample of signal events, several years of exposure will be necessary. Besides the analyses of the track-like events, also the analyses of shower-like events will contribute to advance the study of Sgr A*, thanks to the favorable location of the KM3NeT detector and to its good angular resolution. On the contrary, the expected signal in IceCube is smaller and unlikely to be observed in view of the larger background rate caused by the atmospheric muons. I examined the reasons of uncertainties in the expectations for the high-energy neutrinos. While a leptonic contribution in the observed gamma rays would decrease the observable signal, several other reasons could increase it, including: (i) the possibility of gamma-ray absorption, (ii) an extended angular region, with respect to that observed by H.E.S.S. around the Galactic Center, where the emission is sizeable, and (iii) a speculative E^{-2} behavior of the spectra at higher energies than presently measured with gamma rays, as often assumed in standard data analysis of neutrino observatories. These considerations emphasize the importance of extending the programs for the search and study of VHE gamma rays and neutrinos.

A very relevant aspect for the identification of CR sources in terms of gamma-ray and neutrino signatures concerns the potentials of future instruments. Both CTA and KM3NeT are expected to achieve significant performance improvements with respect to existing instruments, because of their larger effective areas and better angular and energy resolution. In particular, in Chap. 5, I explored the discovery potential of the future KM3NeT, in relation to the constraining power of CTA, for the detection of extended sources, as they are particularly relevant in the case of Galactic studies. In fact, unless multi-TeV photons are absorbed inside the sources or during their propagation through the interstellar or intergalactic radiation fields, gamma-ray observations can safely be considered a powerful tool to explore the potential for finding astronomical neutrino sources. The synergy of the two observatories is highlighted in that they will investigate the processes operating the non-thermal Universe through a complementary energy range. This comparative study is based on simple analytical representations of the basic telescope characteristics, as publicly available for both CTA and KM3NeT. I derived the following conclusions: (i) the sensitivity to extended sources shows a degradation with increasing source angular size such that it is maximum at low energies, it reduces at intermediate energies and it tends to disappear at very high energies; (ii) the most important energy region for the detection of neutrino sources is above 10 TeV, though the access to this low-energy threshold is crucial, in that most of the VHE Galactic sources appear to have a spectrum with a cut-off around an energy of 10–50 TeV. In this energy region, the CTA minimum detectable gamma-ray flux does not show a strong dependence with the source size, while a comparison of the performances of the two instruments shows that above this energy a joint exploration of the VHE sky in gamma rays and neutrinos will be possible. Nowadays, gamma-ray observations above 10 TeV are available, as the surveys of the Galactic Plane realized by the current ground-based telescopes, HAWC and H.E.S.S.: these observations are already constraining from the point of view of their counterpart neutrino detection. However, some room is still available for the presence of Galactic neutrino emitters. In the near future, CTA will explore the entire Galactic Plane in the 10–100 TeV energy domain. Therefore, CTA will significantly reduce the limits on the neutrino source expectations, setting conclusions on possible Galactic neutrino emitters through its extremely strong constraining power. My analysis showed that, by assuming a source emitting a $\propto E^{-2}$ differential energy spectrum in gamma rays through a fully hadronic mechanism, a minimum gamma-ray flux of $E^2\phi_\gamma(10\ \text{TeV}) > 1 \times 10^{-12}\,\text{TeV}\,\text{cm}^{-2}\ \text{s}^{-1}$ is necessary in order for its neutrino counterpart to be detectable with a 3σ significance on a timescale of 10 years with KM3NeT. This result assumes that the source has an angular size of $R_{\text{src}} = 0.1$ deg. In the extreme case of a source with a radial extent of $R_{\text{src}} = 2.0$ deg, only sources brighter than $E^2\phi_\gamma(10\ \text{TeV}) > 2 \times 10^{-11}\,\text{TeV}\,\text{cm}^{-2}\ \text{s}^{-1}$ will be within the reach of neutrino telescopes. These estimates are very weakly dependent on the source spectral index and are consistent with previous evaluations performed in [243] for the case of point-like sources. In particular, RX J1713.7-3946 and the Galactic Center Ridge remain potential sources for neutrinos. I found that a decade of observations is required for a 3σ (in each energy bin) neutrino detection from the SNR and from the most optimistic set of parameters considered for the Galactic Center Ridge (a cut-off in the gamma-ray spectrum at energies above 100 TeV).

Appendix A
MHD Shock-Clump Simulations with PLUTO

This appendix provides the reader with a brief review of the laws of physics governing the motion of fluids. In Sect. A.1 the key aspects of hydrodynamics are presented, with the aim of introducing the formation of shocks and the Rankine–Hugoniot conditions regulating it. Subsequently, the effects of the presence of electromagnetic fields are investigated, by moving towards the MHD equations for fluids: this is realized in Sect. A.2. Here, an insight into the numerical code adopted for the evolution of the system is discussed, namely the PLUTO code is presented. In particular, an individual shock-clump interaction is simulated with different setups. Three distinct magnetic field configurations have been considered: a magnetic field parallel to the shock normal in Sect. A.2.2, perpendicular in Sect. A.2.3 and oblique in Sect. A.2.4. Finally, several density contrasts among the clump and the CSM are explored in Sect. A.2.5.

A.1 The Hydrodynamics of Shocks

Hydrodynamics considers a fluid as a macroscopic object, i.e. as a continuous medium. Its local thermodynamical properties (namely pressure P, density ρ and temperature T) are required to describe the fluid in a state of rest: by assuming an equation of state, it is necessary to provide only two of the three fundamental thermodynamical quantities. Conversely, a generic fluid that is not in a state of rest will be also described by its instantaneous speed of motion [237]. In the following, all these quantities (P, ρ, T, $\mathbf{v}$) are assumed as continuous functions of space and time. Note that the velocity of the fluid refers to a fixed point in space and not to specific particles of the fluid, which in the course of time will move through space [161]. The same remark applies to pressure and density. The fundamental laws of ideal fluids, namely fluids not subject to dissipative phenomena (as friction and thermal conduction), express conservation of mass, momentum and energy of a fluid volume, which read respectively as

S. Celli, *Gamma-ray and Neutrino Signatures of Galactic Cosmic-ray Accelerators*, Springer Theses, https://doi.org/10.1007/978-3-030-33124-5

$$
\begin{cases}
\dfrac{d\rho}{dt} + \nabla \cdot (\rho \mathbf{v}) = 0 \\
\dfrac{\partial}{\partial t}(\rho v_i) + \nabla \cdot \hat{R} = 0 \\
\dfrac{\partial}{\partial t}\left(\rho \dfrac{v^2}{2} + \dfrac{P}{(\gamma - 1)}\right) + \nabla \cdot \left(\rho \dfrac{v^2}{2} + \dfrac{P\gamma}{(\gamma - 1)}\right) \mathbf{v} = 0
\end{cases}
\tag{A.1}
$$

where the fluid politropic index γ was introduced as well as the *Reynolds's stress tensor*, defined as

$$
R_{ik} \equiv P\delta_{ik} + \rho v_i v_k \tag{A.2}
$$

The first equation expresses mass conservation in differential form: mass can change within a volume only if it is either added or removed by a mass flow ($\rho\mathbf{v}$) crossing the surface of the given volume. The second equation is the *Euler's equation* applied to the i-th component of the momentum density vector $\rho\mathbf{v}$, where all external forces have been neglected. The last equation represents the conservation of total energy, kinetic plus internal, which applies to non-radiative cases. Note that if radiative processes take place on definite timescales, this equation can still be valid if all phenomena are taking place on shorter timescales, hence they are essentially adiabatic (i.e. without heat loss).

One of the peculiarities of hydrodynamics is that it allows discontinuous solutions, namely solutions such that all physical quantities are discontinuous on certain special surfaces, called surfaces of discontinuity. In the mathematical sense, these solutions are simply step functions, namely infinitely steep discontinuities. On the other hand, from the physical point of view, the discontinuity is thin compared with respect to all other physical dimensions, so that it is reasonable to assume the mathematical limit of an infinitely steep discontinuity. Two different kinds of discontinuities exist. The first one is called *tangential discontinuity*, that develops when two separate fluids lie one beside the other, and the surface between them is not crossed by a flux of matter. This kind of discontinuity is unstable and hence short-lived. The second kind of discontinuity, also called a *shock wave* or *shock*, is a surface of separation between two fluids, where there is a flux of mass, momentum and energy through the surface. Shock waves are naturally produced within a wide range of phenomena: they are practically unavoidable when the perturbations to which a hydrodynamic system is exposed are not infinitesimal. In other words, when the perturbations to which a system is exposed are small, sound waves are generated. If, on the other hand, perturbations are finite (i.e. not infinitesimal), shocks form.

In order to derive a solution for the system of Eq. (A.1), a reference frame moving with the surface of discontinuity is set in the following. Furthermore, plane symmetry (all quantities only depend on one coordinate, z, which is perpendicular to the discontinuity surface) and stationarity are assumed. Thus, the hydrodynamic equations presented in Eq. (A.1) can be written as

$$
\frac{dJ}{dz} = 0 \tag{A.3}
$$

where J is any kind of flux (mass, momentum or energy). As a consequence, these equations require continuous fluxes. In other words, physical quantities can be discontinuous, provided that fluxes are continuous: mass, momentum and energy cannot be created inside the surface of discontinuity. More quantitatively, for a one-dimensional (along z) stationary shock motion, the following set of equations holds:

$$\begin{cases} \frac{\partial}{\partial z}(\rho v) = 0 \\ \frac{\partial}{\partial z}(\rho v^2 + P) = 0 \\ \frac{\partial}{\partial z}\left(\frac{1}{2}\rho v^3 + \frac{\gamma}{\gamma - 1} P v\right) = 0 \end{cases} \tag{A.4}$$

Integrating these equations across the discontinuity surface, two solutions can be derived, describing respectively tangential discontinuities and shock waves. By adopting the standard notation where the region 1 is the shock upstream, while the region 2 is the downstream (as in Fig. 1.2), a solution of Eq. (A.4) is represented by $\rho_1 v_1 = \rho_2 v_2 = 0$: since $\rho_1 \neq 0$ and $\rho_2 \neq 0$, then $v_1 = v_2 = 0$ is obtained. Consequently, $P_1 = P_2$. Thus, the trivial solution reads as $v_{1z} = v_{2z} = 0$ and $P_1 = P_2$, while all the other thermodynamic properties can be discontinuous. Hence, this kind of discontinuity is the tangential one: a contact discontinuity is a subclass of the tangential discontinuities, where velocities are continuous but density is not [161]. In the second kind of discontinuity, namely shock waves, the mass flux does not vanish, therefore v_{1z} and v_{2z} cannot be vanishing. Now the tangential velocity is continuous across the shock, and consequently through all space. It holds that

$$\begin{cases} \rho_1 v_1 = \rho_2 v_2 \\ \rho_1 v_1^2 + P_1 = \rho_2 v_2^2 + P_2 \\ \frac{1}{2}\rho_1 v_1^3 + \frac{\gamma}{\gamma - 1} P_1 v_1 = \frac{1}{2}\rho_2 v_2^3 + \frac{\gamma}{\gamma - 1} P_2 v_2 \end{cases} \tag{A.5}$$

These conditions of continuity constitute what hydrodynamics imposes on shock waves: they are called Rankine–Hugoniot (RH) conditions. The RH conditions allow us to determine the thermodynamics properties of the fluid behind the shock, once the conditions ahead of the shock are known. Defining the sonic Mach number of a shock as $M = v/c_s$, where c_s is the sound speed, then Eq. (A.5) can be rewritten as

$$\begin{cases} \frac{\rho_2}{\rho_1} = \frac{v_1}{v_2} = \frac{(\gamma + 1) M_1^2}{(\gamma - 1) M_1^2 + 2} \\ \frac{P_2}{P_1} = \frac{2\gamma M_1^2}{\gamma + 1} - \frac{\gamma - 1}{\gamma + 1} \\ \frac{T_2}{T_1} = \frac{[2\gamma M_1^2 - (\gamma - 1)][(\gamma - 1) M_1^2 + 2]}{(\gamma + 1)^2 M_1^2} \end{cases} \tag{A.6}$$

where $M_1 = v_1/c_{s,1}$ is the Mach number in the upstream. On the other hand, the Mach number downstream reads as

$$M_2^2 = \frac{2 + (\gamma - 1)M_1^2}{2\gamma M_1^2 - \gamma + 1} \tag{A.7}$$

In the limit $M_1 \gg 1$, the *strong shock solution* is obtained, which reads as

$$\begin{cases} \dfrac{\rho_2}{\rho_1} = \dfrac{v_1}{v_2} = \dfrac{\gamma + 1}{\gamma - 1} \\ \dfrac{P_2}{P_1} = \dfrac{2\gamma M_1^2}{\gamma + 1} \\ \dfrac{T_2}{T_1} = \dfrac{2\gamma(\gamma - 1)}{(\gamma + 1)^2} M_1^2 \\ M_2^2 = \dfrac{\gamma - 1}{2\gamma} \end{cases} \tag{A.8}$$

Strong shocks compress moderately the unperturbed gas: for a monoatomic gas with $\gamma = 5/3$, a maximum factor of $\rho_2/\rho_1 = 4$ is achieved. However, they can heat the gas to a high temperature, since $T_2/T_1 \propto M_1^2$. Obviously, the source of energy for this heating must be the bulk kinetic energy of the incoming fluid. In fact, the speed of the fluid after the shock is smaller than the one before the shock, by the same factor by which density increases. In summary, a shock appears as a discontinuity in the thermal properties of a plasma. Two main effects of the presence of a shock are influencing the plasma, namely:

- The plasma behind the shock is compressed and slowed down;
- The plasma behind the shock is heated ($T > 10^8$ K), which typically leads to an intense X-ray emission.

Therefore, the shock transforms bulk kinetic energy into internal (thermal) energy of the outgoing fluid, with creation of entropy. The generation of entropy is due to collisions between atoms or molecules of the fluid in question. As an order of magnitude estimate, the thickness of the shock is given by the mean free path λ of a particle, since the speed of atoms (or molecules) is changed by 90° on a scale length of that order of magnitude. In astrophysical systems, matter is mostly ionized, and electrons and atomic nuclei are subject to accelerations and deflections due to electric and magnetic fields. Thus, a mixture of disordered and partly transient electric and magnetic fields is responsible for making the nuclei's momenta isotropic. The shock thickness must therefore be comparable with the Larmor radius of a proton, since the proton, which carries most of the energy and momentum of incoming matter, is deflected by the typical magnetic field over a distance of this order of magnitude. Therefore one gets

$$\lambda \simeq r_{\rm L,th} = 10^{10}\ {\rm cm}\left(\frac{v}{10^4\ {\rm km/s}}\right)\left(\frac{\mu{\rm G}}{B}\right) \tag{A.9}$$

where standard values are used for the speed of a proton emitted by a supernova, and for a typical Galactic magnetic field. This thickness is such that the idealization of an infinitely thin discontinuity results acceptable. If the agents responsible for the isotropization of the bulk kinetic energy are electromagnetic fields, shocks are called non-collisional. The best known example is the shock between the solar wind and the Earth magnetosphere, at about 10^5 km from the Earth.

A.2 The MHD of Shocks and the PLUTO Code

In the previous section, the equations of hydrodynamics have been introduced, neglecting all electromagnetic phenomena related to the fluids in question. In fact, in high-energy astrophysics, the fluid temperatures are very high, and most atoms are completely ionized: in this situation, electric fields are irrelevant, thanks to matter charge neutrality and to the abundance of free charges. However, if the fluid is immersed in a magnetic field, its motion relative to B generates an electric field, and this in turn generates currents. These currents are affected by magnetic fields and generate new magnetic fields, thus creating a complex and interesting physical situation.

MHD describes the behavior of fluids, which are at least partly ionized, in the presence of electromagnetic fields. This description holds in the limit of mean free paths short with respect to all macroscopic lengths of the problem. By mean free path it is meant not only that between collisions among particles of various species, but also the Larmor radius for particles in the magnetic field. With respect to the hydrodynamics equations, the equation expressing mass conservation continues to hold without any change. On the other hand, the equation expressing the conservation of momentum needs a correction: indeed a current $\mathbf{J}$ immersed in a magnetic field is subject to a force per unit volume (which is actually the Lorentz force per unit volume, in the absence of a net charge density) $\mathbf{J} \times \mathbf{B}/c$. The equation of energy conservation must be modified too, since the presence of currents implies a resistance, and resistance implies dissipation and therefore heating. Furthermore, Maxwell equations $\nabla \cdot \mathbf{B} = 0$ and $\nabla \times \mathbf{B} = 4\pi \mathbf{J}/c$ should be included to close the system.

The MHD equations discussed so far have been already introduced in Eq. (2.6). They are implemented in the PLUTO code [174], including some additional terms, as external forces (the gravitational one). Thus, the time evolution of the thermodynamics variables defining the status of the background plasma are given, in case of a magnetized plasma, as solution of the of the following equations

$$\frac{\partial}{\partial t}\begin{pmatrix}\rho \\ \mathbf{m} \\ E+\rho\phi \\ \mathbf{B}\end{pmatrix} + \nabla\cdot\begin{pmatrix}\rho\mathbf{v} \\ \mathbf{m}\mathbf{v}-\mathbf{B}\mathbf{B}+I P_t \\ (E+P_t+\rho\phi)\mathbf{v}-\mathbf{B}(\mathbf{v}\cdot\mathbf{B}) \\ \mathbf{v}\mathbf{B}-\mathbf{B}\mathbf{v}\end{pmatrix}^T = \begin{pmatrix}0 \\ -\rho\nabla\phi+\rho\mathbf{g} \\ \mathbf{m}\cdot\mathbf{g} \\ \mathbf{0}\end{pmatrix}$$

where ρ is the mass density, $\mathbf{m} = \rho\mathbf{v}$ is the momentum density, $\mathbf{v}$ is the velocity, $P_t = P + \mathbf{B}^2/2$ is the total pressure (thermal and magnetic), $\mathbf{B}$ is the magnetic field and E is the total energy density

$$E = \rho e + \frac{\mathbf{m}^2}{2\rho} + \frac{\mathbf{B}^2}{2}$$

where an additional equation of state $\rho e = \rho e(P, \rho)$ provides the closure. The source term on the right hand side is written in terms of the time-independent gravitational potential ϕ and the acceleration vector $\mathbf{g}$.

A particular package of PLUTO implements the interaction among a shock and a ionized clump, providing the user with the time evolution of all the thermodynamics variables of interest. The interaction between the shock and a dense clump is divided into two phases: (1) the collapse stage, where the front of the clump is strongly compressed and two fast shocks are generated, and (2) the re-expansion phase, which begins when the transmitted fast shock overtakes the back of the clump. In the following simulations $n_c = 10^3\,\text{cm}^{-3}$ is assumed for the clump density (unless a different value is specified, as in Sect. A.2.5).

A.2.1 Simulation Setup

The MHD simulations here performed are mainly aimed at answering two main questions, whose answers determine the propagation of CRs through a clumpy medium: (1) whether the clump evaporates on a timescale longer than the time it takes to be completely engulfed by the contact discontinuity; (2) under which conditions efficient magnetic field amplification around the clump is realized. To this purpose, several MHD simulations of a shock interacting with a clump have been performed, in 3D cartesian coordinates, aiming at exploring different magnetic field configurations. The simulations are performed within a uniform grid with x, y and z all ranging from 0 to 2 pc. The system evolution is followed in the clump reference system, with the clump located in $x_0 = y_0 = z_0 = 1$ pc. A third order Runge–Kutta method is selected for the temporal integration of Eq. (2.6), with a Courant condition of $C_a = 0.4$. The shock moves along the z-direction with a velocity $v_s = 4.4 \times 10^8\,\text{cm}\,\text{s}^{-1}$, the temperature upstream is fixed to $T = 10^6$ K, so that a sonic Mach number of $M_1 \simeq 37$ characterizes the motion and conditions for strong shocks apply (see Eq. (A.8)). Thus, fixing a low-density medium in the upstream with $n_{\text{up}} = 10^{-2}\,\text{cm}^{-3}$, the downstream numerical density results $n_{\text{down}} = 4 \times 10^{-2}\,\text{cm}^{-3}$. Boundary conditions are such that an outflow is present at all boundaries, except in the downstream of the

Table A.1 Parallel shock configuration, initial conditions for the simulation ($T_{up} = 10^6$ K, $v_s = 4.4 \times 10^8\,\mathrm{cm\,s^{-1}}$)

	Upstream	Downstream
$\rho\,(m_p/g)(\mathrm{g/cm^3})$	0.01	0.04
P	ρ_{up}/T_{up}	$2\rho_{up}v_s^2/(\gamma+1)$
v_x	0	0
v_y	0	0
v_z	0	$3v_s/4$
B_x (μG)	0	0
B_y (μG)	0	0
B_z (μG)	5	5

Table A.2 Perpendicular shock configuration, initial conditions for the simulation ($T_{up} = 10^6$ K, $v_s = 4.4 \times 10^8\,\mathrm{cm\,s^{-1}}$)

	Upstream	Downstream
$\rho\,(m_p/g)(\mathrm{g/cm^3})$	0.01	0.04
P	ρ_{up}/T_{up}	$2\rho_{up}v_s^2/(\gamma+1)$
v_x	0	0
v_y	0	0
v_z	0	$3v_s/4$
B_x (μG)	$5\sqrt{2}/2$	$10\sqrt{2}$
B_y (μG)	$5\sqrt{2}/2$	$10\sqrt{2}$
B_z (μG)	0	0

grid, where an incoming flux is set. Within this setup, three different situations for the magnetic field configuration are shown in the following: (i) the parallel case ($\mathbf{B} \parallel \mathbf{v}_s$), where $\mathbf{B} = (0, 0, B)$; (ii) the perpendicular configuration ($\mathbf{B} \perp \mathbf{v}_s$), where $\mathbf{B} = (B_x, B_y, 0)$; (iii) the oblique configuration, to represent the most generic case, where $\mathbf{B} = (B_x, 0, B_z)$. The initial conditions for all the configurations are reported in Tables A.1, A.2 and A.3 respectively: they reflect the jump conditions reported in Eq. (A.8), when the magnetic field is dynamically negligible. Such conditions derive from the RH equations, as presented in Sect. A.1.

A.2.2 Parallel Shock

In the parallel configuration, the magnetic field is fully directed along the direction of motion of the shock: no field compression is therefore realized at the shock surface. The temporal evolution of the mass density is reported in Fig. A.1a–f: the clump maintains its density, even when the shock passes all around it, without any visible

Table A.3 Oblique shock configuration, initial conditions for the simulation ($T_{up} = 10^6$ K, $v_s = 4.4 \times 10^8$ cm s^{-1})

	Upstream	Downstream
ρ (m_p/g)(g/cm^3)	0.01	0.04
P	ρ_{up}/T_{up}	$2\rho_{up} v_s^2/(\gamma+1)$
v_x	0	0
v_y	0	0
v_z	0	$3v_s/4$
B_x (μG)	$5\sqrt{2}/2$	$10\sqrt{2}$
B_y (μG)	0	0
B_z (μG)	$5\sqrt{2}/2$	$5\sqrt{2}/2$

mass loss, at least during the timescale that is relevant for the simulation ($\tau_{age} \leq$ 300 yr, as discussed in Chap. 2). The time behavior of the modulus of the velocity is shown in Fig. A.2a–f: here it is quite visible that the shock is not crossing the clump at initial times, due to the very high density contrast. Moreover, a long halo is visible as soon as the shock has passed the clump, as a result of the Kelvin-Helmhotz instability. The magnetic energy density, defined as $U_B = B^2/(8\pi)$ is represented in Fig. A.3a–f: magnetic field amplification by a factor of few is realized only along the direction of motion of the shock, right in front of the clump. Finally, pressure conditions are shown in Fig. A.4a–f: here, the reflected bow shock generated at the time of the shock-clump contact is visible, as well as another transient shock in front of the clump.

A.2.3 Perpendicular Shock

In this case, the magnetic field is set in both the directions orthogonal to the shock speed: hence, compression is realized in both the components, with a compression factor $r = 4$ in each perpendicular direction. The time evolution of mass density is reported in Fig. A.5a–f. The main difference with respect to the parallel case is that a much stronger amplification in the magnetic field is realized around the clump: a sort of layer appears, whose typical dimensions are half of the clump size, as shown in Fig. A.6a–f and further discussed in Sect. 2.3. The same figures show that, in the clump vicinity, the magnetic field lines appear mostly directed in the direction tangential to the clump surface.

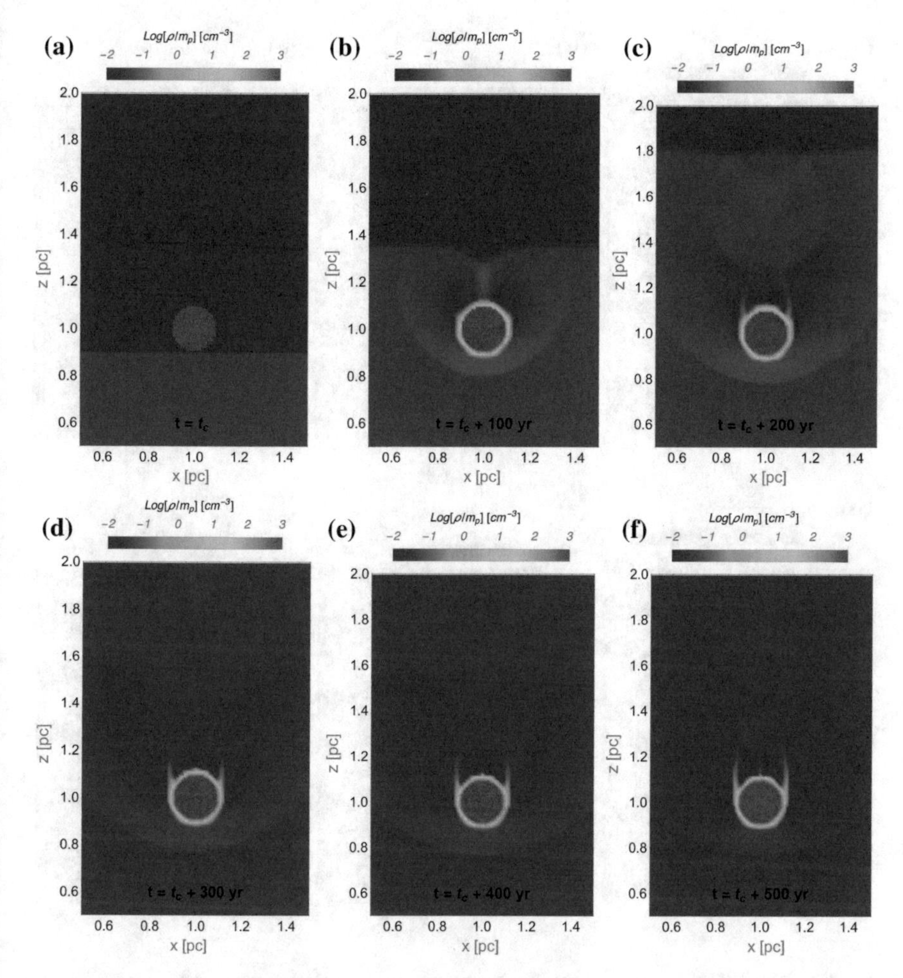

Fig. A.1 Parallel shock case, with density contrast $\chi = 10^5$: mass density in a 2D section along $y = 1$ pc, passing through the centre of the clump, at different times with respect to the first shock-clump interaction, occurring at $t = t_c$

A.2.4 Oblique Shock

Finally, in the case of an oblique shock, the magnetic field is set with one component along the direction of the shock motion and the other in the orthogonal direction: however, only the orthogonal component results to be compressed at the shock. The main difference with respect to the previous cases is that a much stronger amplification in the magnetic field is realized around the clump, as shown in Fig. A.9a–f. However, no large differences arise in the mass density, which is shown from Fig. A.7a–f, neither in the modulus of the velocity, reported in Fig. A.8a–f, or in the pressure.

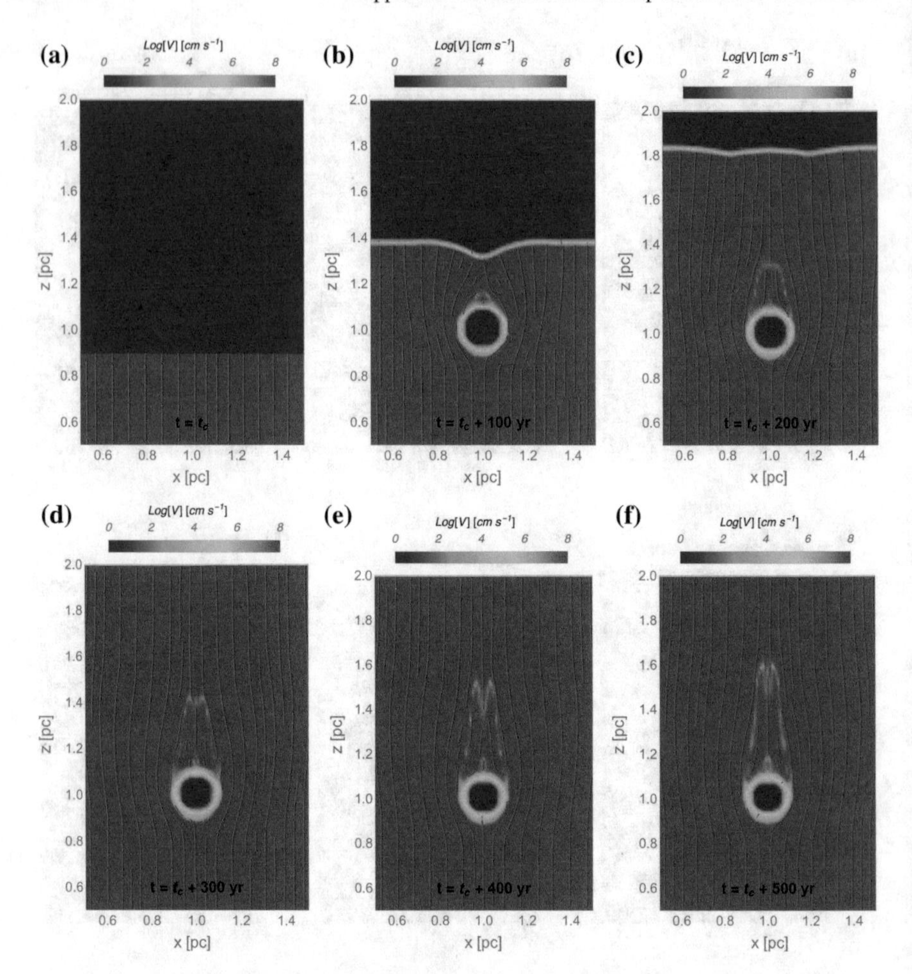

Fig. A.2 Parallel shock case, with density contrast $\chi = 10^5$: modulus of the velocity (color scale) in a 2D section along $y = 1\,\mathrm{pc}$, passing through the centre of the clump, at different times with respect to the first shock-clump interaction, occurring at $t = t_c$. Stream lines show the direction of the velocity field in the corresponding plane

It is interesting at this point to study how the magnetic field amplification changes as a function of the clump density, investigating in particular cases of lower density clouds. This is the case for instance of diffuse clouds, whose typical densities are around $100\,\mathrm{cm}^{-3}$. Thus, in the following, the oblique shock configuration is investigated with $\chi = 10^4$, $\chi = 10^3$ and $\chi = 10^2$.

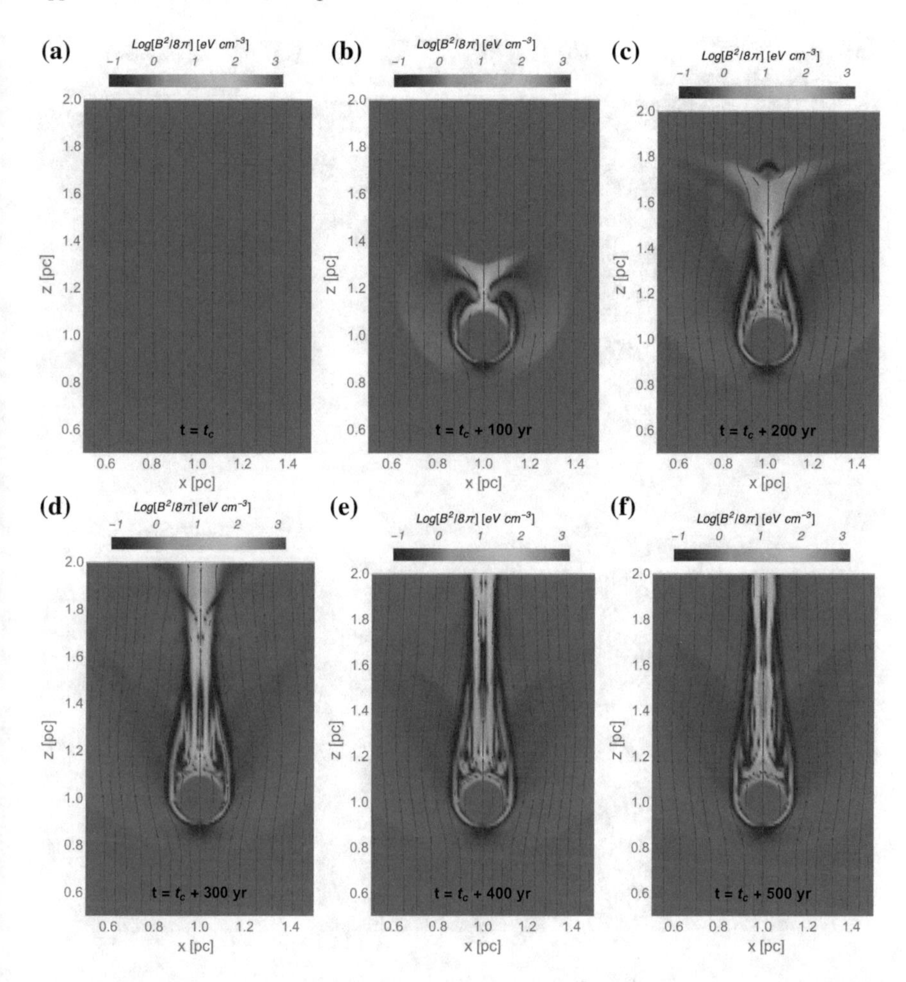

Fig. A.3 Parallel shock case, with density contrast $\chi = 10^5$: energy density of magnetic field (color scale) in a 2D section along $y = 1$ pc, passing through the centre of the clump, at different times with respect to the first shock-clump interaction, occurring at $t = t_c$. Stream lines show the direction of the regular magnetic field in the corresponding plane

A.2.5 *Effect of Density Contrast* χ *on Oblique Shocks*

In this section, different density contrasts among the clump and the CSM are simulated. To this purpose, the clump density is modified in order to reduce the density contrast, aiming at quantifying the magnetic field amplification and the evaporation timescale in different situations. The density maps are reported respectively in Fig. A.10a and d for the case $\chi = 10^4$, in Fig. A.10e and h for the case $\chi = 10^3$ and finally in Fig. A.10i and l for the case $\chi = 10^2$. The energy density of magnetic field for the case $\chi = 10^4$ is shown in Fig. A.11a–d, while for the case $\chi = 10^3$ it is shown

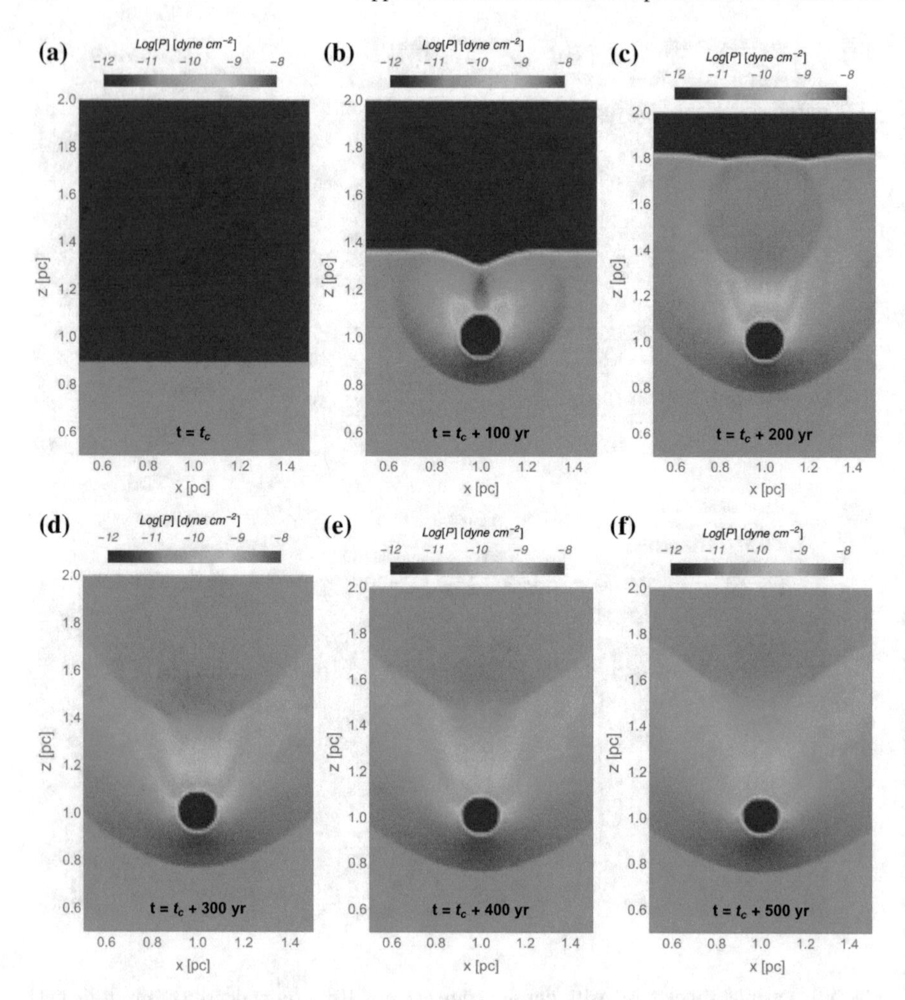

Fig. A.4 Parallel shock case, with density contrast $\chi = 10^5$: pressure in a 2D section along $y = 1$ pc, passing through the centre of the clump, at different times with respect to the first shock-clump interaction, occurring at $t = t_c$

in Fig. A.11e–h and finally for the case $\chi = 10^2$ in Fig. A.11i–l. It is quite visible that a factor 10 in magnetic field strength amplification can be easily obtained. However, for less dense clumps, the evaporation time is very much reduced: already with a density contrast of $\chi = 10^3$, the shock can easily penetrate the clump, whose spherical shape gets quickly destroyed. This implies that magnetic field lines penetrates the clump, where particle acceleration can be achieved, regardless of their energy. In such a situation, gamma rays would be uniformly produced within the clouds and no break would appear in their spectrum.

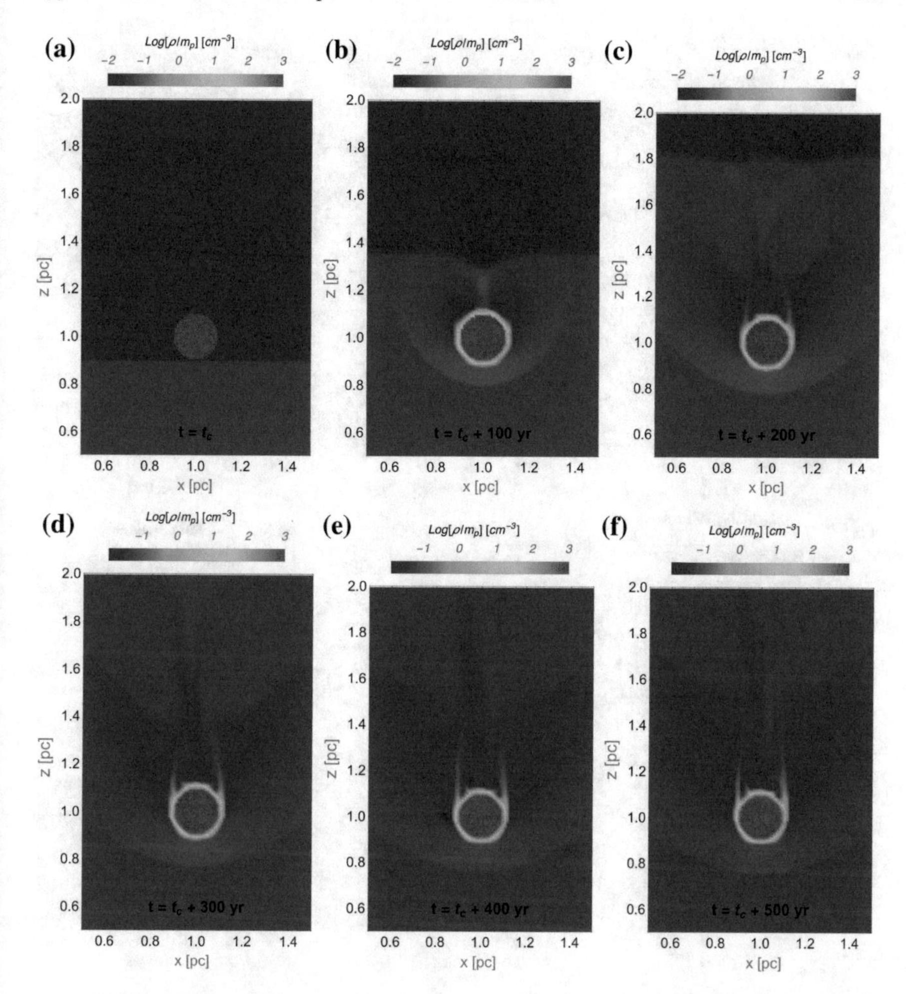

Fig. A.5 Perpendicular shock case, with density contrast $\chi = 10^5$: mass density in a 2D section along $y = 1$ pc, passing through the centre of the clump, at different times with respect to the first shock-clump interaction, occurring at $t = t_c$

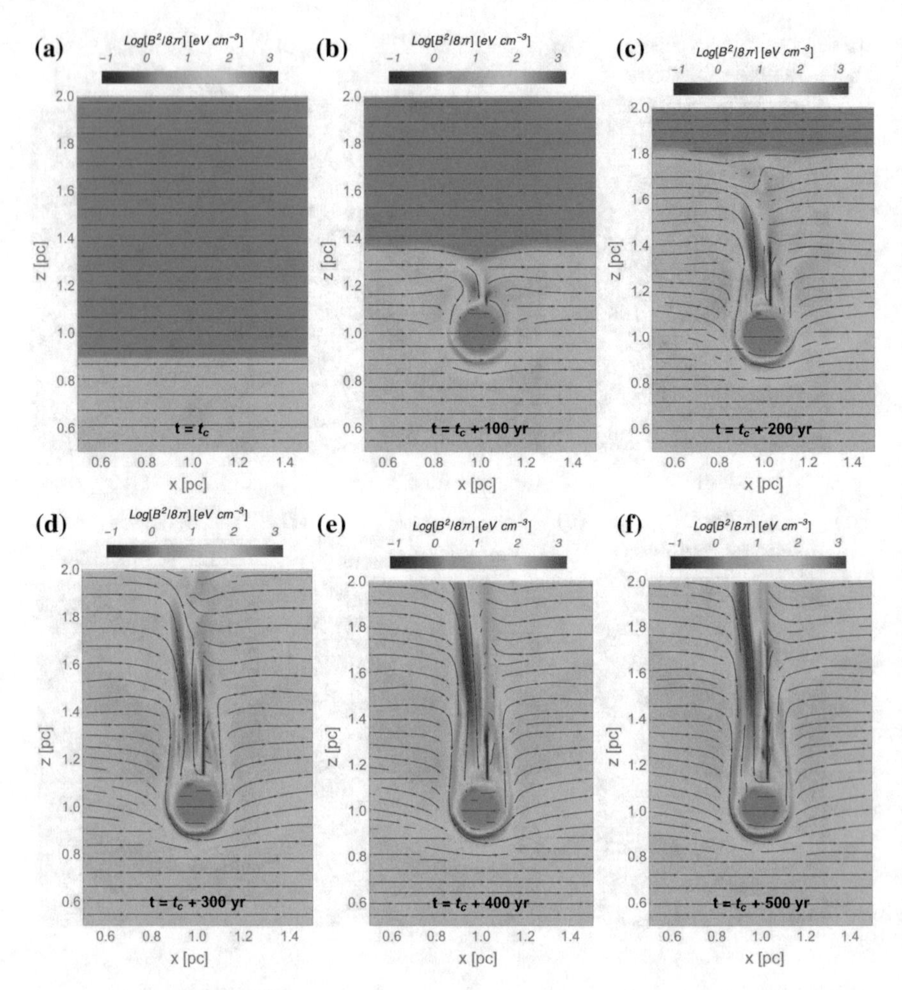

Fig. A.6 Perpendicular shock case, with density contrast $\chi = 10^5$: energy density of magnetic field (color scale) in a 2D section along $y = 1$ pc, passing through the centre of the clump, at different times with respect to the first shock-clump interaction, occurring at $t = t_{\rm c}$. Stream lines show the direction of the regular magnetic field in the corresponding plane

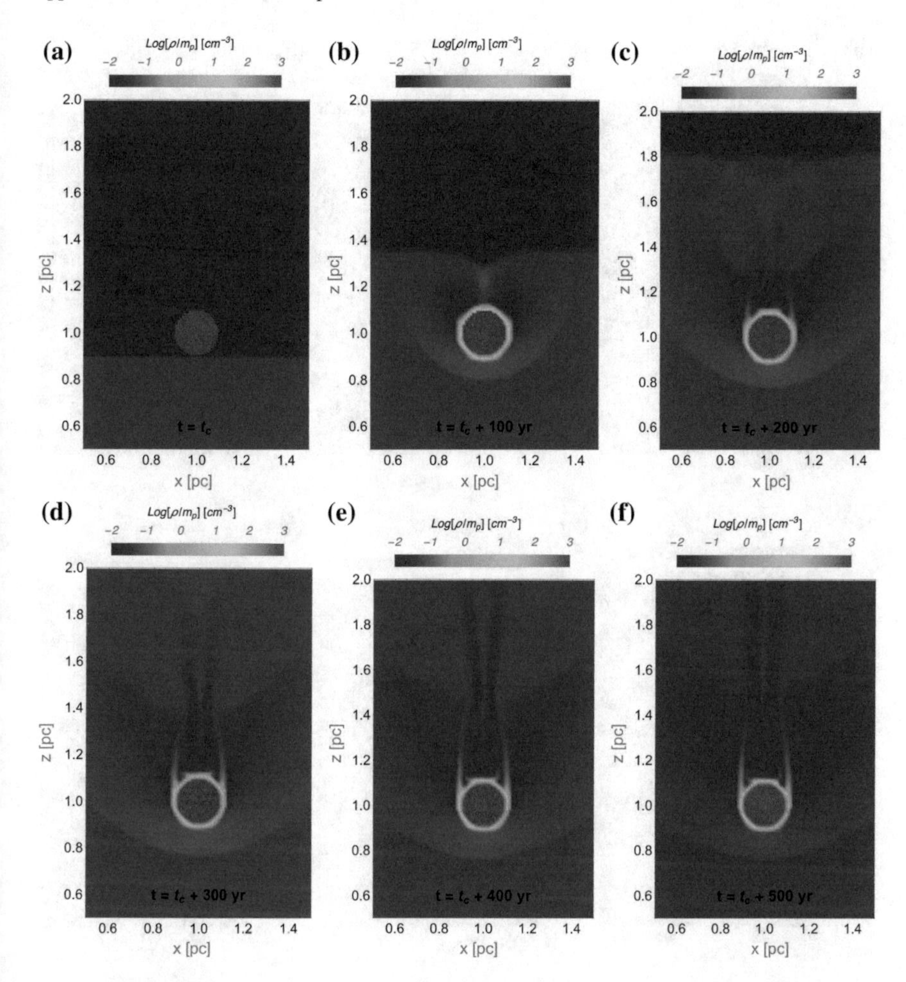

Fig. A.7 Oblique shock case, with density contrast $\chi = 10^5$: mass density in a 2D section along $y = 1$ pc, passing through the centre of the clump, at different times with respect to the first shock-clump interaction, occurring at $t = t_{\rm c}$

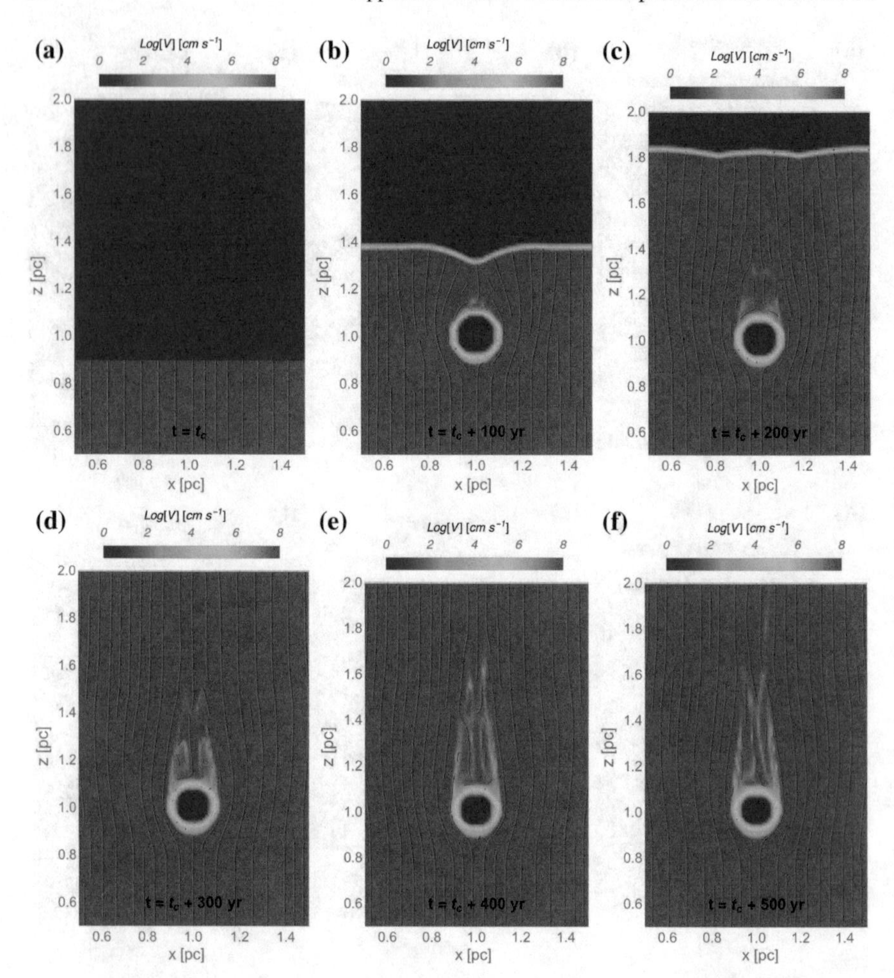

Fig. A.8 Oblique shock case, with density contrast $\chi = 10^5$: modulus of the velocity (color scale) in a 2D section along $y = 1$ pc, passing through the centre of the clump, at different times with respect to the first shock-clump interaction, occurring at $t = t_c$. Stream lines show the direction of the velocity field in the corresponding plane

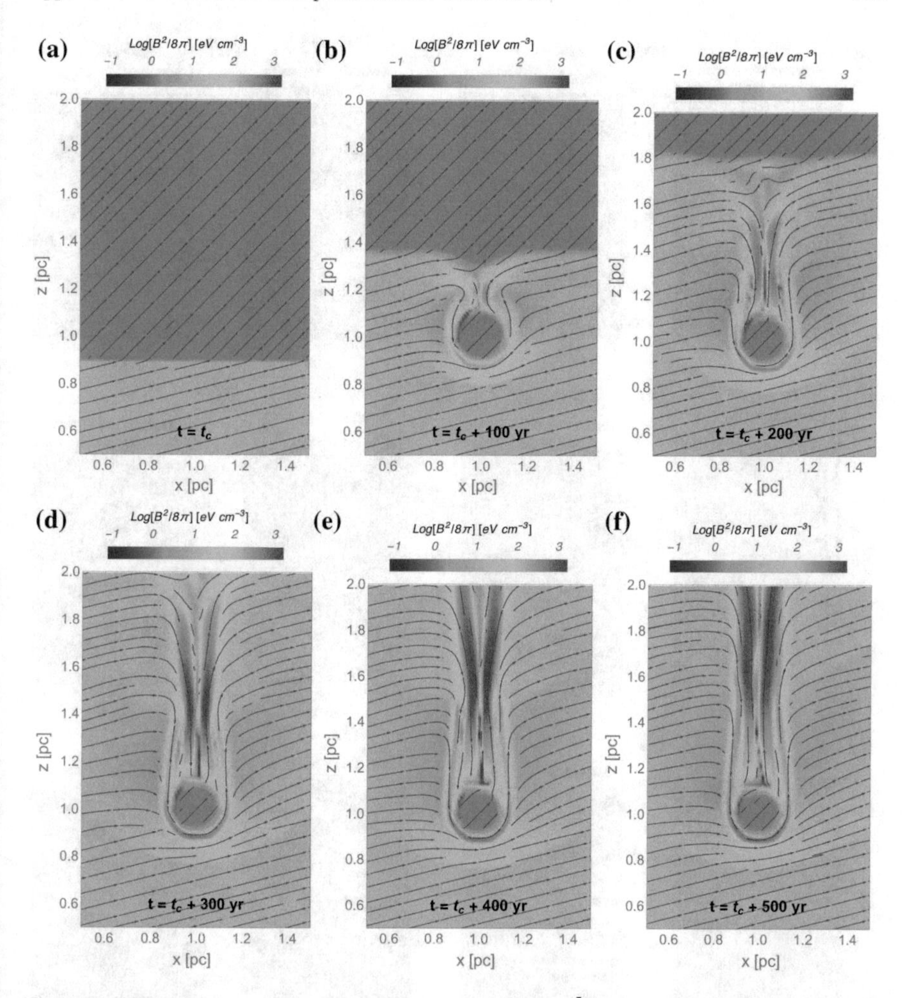

Fig. A.9 Oblique shock case, with density contrast $\chi = 10^5$: energy density of magnetic field (color scale) in a 2D section along $y = 1$ pc, passing through the centre of the clump, at different times with respect to the first shock-clump interaction, occurring at $t = t_c$

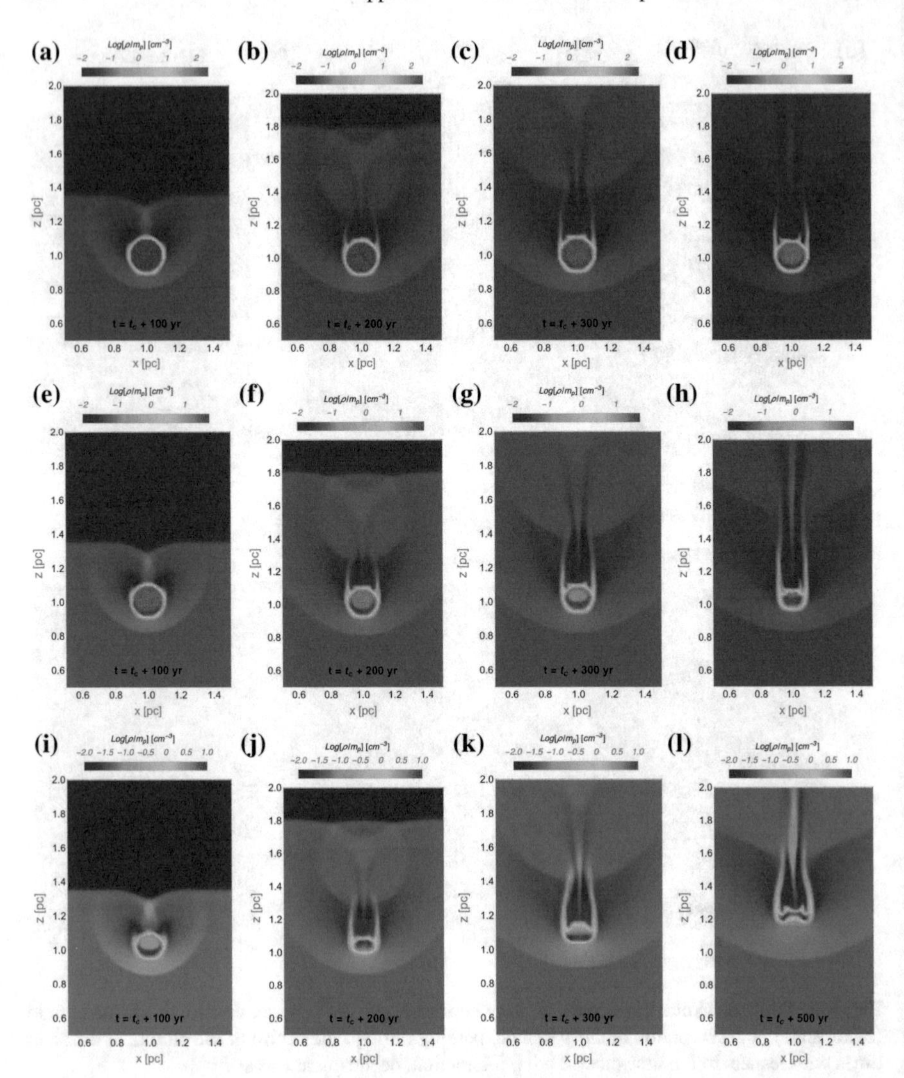

Fig. A.10 **Oblique shock case**: mass density in a 2D section along $y = 1$ pc, passing through the centre of the clump, at different times with respect to the first shock-clump interaction, occurring at $t = t_\mathrm{c}$. Panels refer to different density contrasts. *Top*: $\chi = 10^4$. *Middle*: $\chi = 10^3$. *Bottom*: $\chi = 10^2$

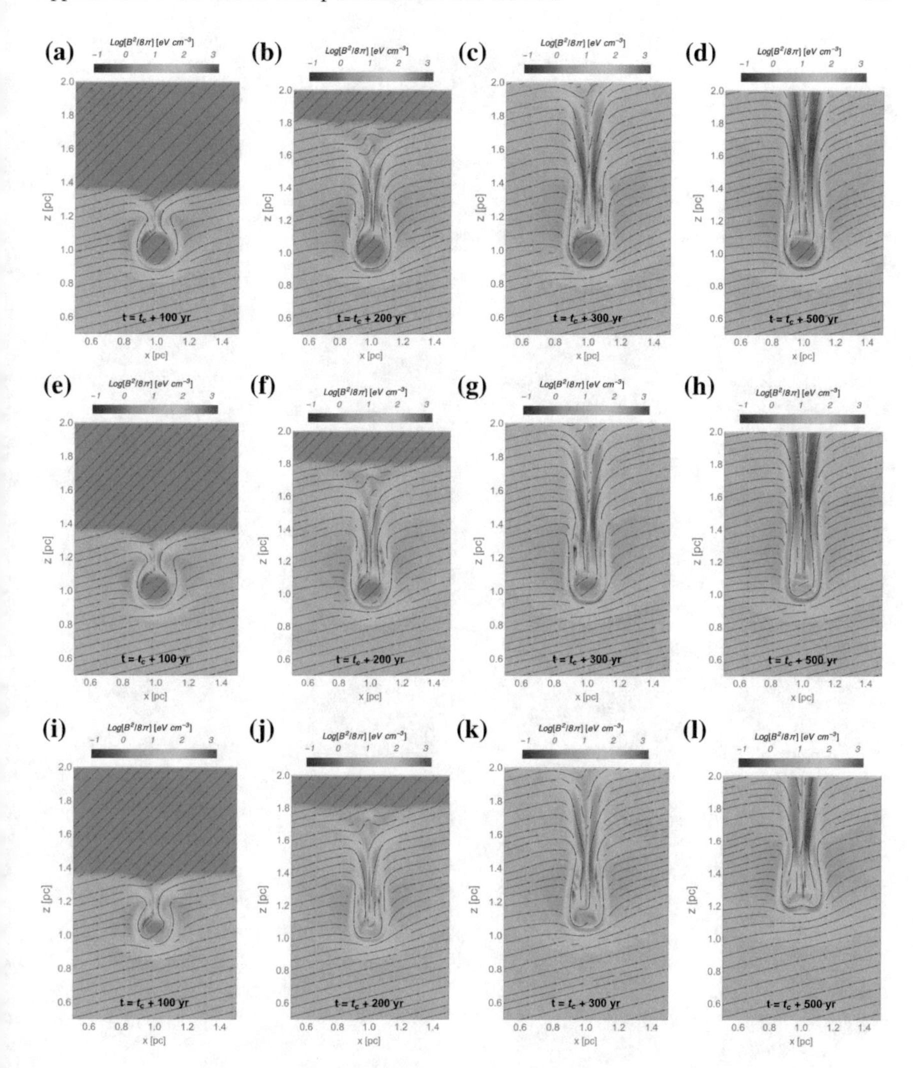

Fig. A.11 **Oblique shock case**: energy density of magnetic field (color scale) in a 2D section along $y = 1\,\mathrm{pc}$, passing through the centre of the clump, at different times with respect to the first shock-clump interaction, occurring at $t = t_\mathrm{c}$. Stream lines show the direction of the regular magnetic field in the corresponding plane. Panels refer to different density contrasts. *Top*: $\chi = 10^4$. *Middle*: $\chi = 10^3$. *Bottom*: $\chi = 10^2$

Appendix B
Numerical Algorithm for the Solution of the CR Proton Transport Equation

The work contained in this appendix is an original work, entirely developed by the author and here reported in order to facilitate the reproducibility of the results presented in Chap. 2. It concerns the numerical technique implemented in order to solve the transport equation for CR protons in the presence of clumpy inhomogeneities of the CSM, close to an astrophysical accelerator responsible for a shock wave.

Since the problem of a one-dimensional shock moving through a spherical clump presents cylindrical symmetry, a coordinate system involving the variables r and z (symmetric with respect to ϕ) is adopted for the description of the particle transport. Assuming f to be the isotropic component of the particle density function in the phase space, the transport equation reads as

$$\frac{\partial f}{\partial t} + \mathbf{v} \cdot \nabla f = \nabla \cdot [D \nabla f] + \frac{1}{3} p \frac{\partial f}{\partial p} \nabla \cdot \mathbf{v} \tag{B.1}$$

including spatial diffusion, advection and adiabatic compression. In particular, the last term describes the acceleration of CRs due to the fluid compression, as it might be relevant, especially at low energies. Note that the source term has been omitted here, as it only affects the normalization of f and hence it will be implemented separately in Sect. B.4. The momentum p here is in the frame of the Alfvén waves, which move within the plasma. However, since the Alfvén speed v_A is much lower than the shock speed v, p also corresponds to momentum in the fluid frame.

In the following, since the partial differential Eq. (B.1) is a multi-dimensional one, I will adopt the operator splitting technique for each term of the previous equation, in the form of the Alternating Direction Implicit (ADI) implementation [203]. It consists into dividing each time step Δt into two steps of size $\Delta t/2$: if n defines the index for the temporal evolution, the so-called 'half time step' allows to take the evolution from n to $n + 1/2$, while the so-called 'integer time step' moves the system from $n + 1/2$ to $n + 1$. In each sub-step, a different dimension is treated implicitly, with the advantage that each sub-step requires only the solution of a simple tridiagonal

S. Celli, *Gamma-ray and Neutrino Signatures of Galactic Cosmic-ray Accelerators*, Springer Theses, https://doi.org/10.1007/978-3-030-33124-5

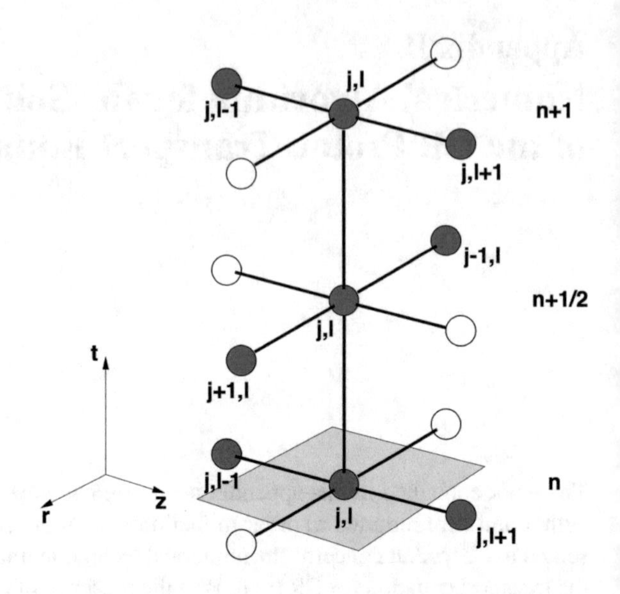

Fig. B.1 Schematic behavior of the ADI method: the index n refers to the temporal dimension, while j and l are respectively for the radial and the axial coordinates. Red dots mark the implicit coordinate at every half time step. Figure from WIKIPEDIA under the CC BY-SA 3.0 LICENSE

system. Through half time steps the radial dimension r (index j) is treated implicitly while the axial dimension z (index l) is left explicit, while the reverse holds in correspondence of integer time steps. The situation is schematically illustrated in Fig. B.1. Moreover, since the plasma velocity field is set with null divergence (as explained in Chap. 2), the adiabatic compression term vanishes. This appendix is organized as follows: in Sect. B.1 the numerical method for the discretization of the diffusive term is illustrated, then in Sect. B.2 the technique concerning the advective term is presented, and finally in Sect. B.3 the final algorithm is reported. The injection at the shock surface is discussed in Sect. B.4, while initial and boundary conditions are exploited respectively in Sects. B.5 and B.6. Finally, Sects. B.7 and B.8 concern the accuracy and stability of the algorithm.

B.1 Flux Conservative Diffusion Equation

The diffusion term is written in a flux conservative way, applying the Chang-Cooper method, such that the total number of particle remains constant on the grid. The vectorial equation reads as

$$\frac{\partial f}{\partial t} = \nabla \cdot [D \, \nabla f] \equiv \nabla \cdot \mathbf{F} \tag{B.2}$$

where the diffusive flux vector $\mathbf{F}$ was introduced for simplicity. In cylindrical coordinates, the equation reads as

$$\frac{\partial f}{\partial t} = \frac{1}{r}\frac{\partial}{\partial r}(rF_r) + \frac{\partial}{\partial z}F_z \tag{B.3}$$

where, in the case of isotropic diffusion, the component of **F** are

$$F_r = D\frac{\partial f}{\partial r} \qquad F_z = D\frac{\partial f}{\partial z} \tag{B.4}$$

B.1.1 *Half Time Step*

By discretizing Eq. (B.3) in the half-time step, one gets

$$\frac{f_{j,l}^{n+1/2} - f_{j,l}^{n}}{\Delta t/2} = \frac{1}{r_{j,l}^{n+1/2}}\left[\frac{(rF_r)_{j+1/2,l}^{n+1/2} - (rF_r)_{j-1/2,l}^{n+1/2}}{\Delta r_{j,l}^{n+1/2}}\right] + \left[\frac{(F_z)_{j,l+1/2}^{n} - (F_z)_{j,l-1/2}^{n}}{\Delta z_{j,l}^{n}}\right] \tag{B.5}$$

where

$$(rF_r)_{j+1/2,l}^{n+1/2} = r_{j+1/2,l}^{n+1/2} D_{j+1/2,l}^{n+1/2}\frac{(f_{j+1,l}^{n+1/2} - f_{j,l}^{n+1/2})}{\Delta r_{j+1/2,l}^{n+1/2}} =$$

$$= \frac{(r_{j+1,l}^{n+1/2} + r_{j,l}^{n+1/2})}{2}\frac{(D_{j+1,l}^{n+1/2} + D_{j,l}^{n+1/2})}{2}\frac{(f_{j+1,l}^{n+1/2} - f_{j,l}^{n+1/2})}{\Delta r_{j+1/2,l}^{n+1/2}}$$

$$(rF_r)_{j-1/2,l}^{n+1/2} = r_{j-1/2,l}^{n+1/2} D_{j-1/2,l}^{n+1/2}\frac{(f_{j,l}^{n+1/2} - f_{j-1,l}^{n+1/2})}{\Delta r_{j-1/2,l}^{n+1/2}} =$$

$$= \frac{(r_{j,l}^{n+1/2} + r_{j-1,l}^{n+1/2})}{2}\frac{(D_{j,l}^{n+1/2} + D_{j-1,l}^{n+1/2})}{2}\frac{(f_{j,l}^{n+1/2} - f_{j-1,l}^{n+1/2})}{\Delta r_{j-1/2,l}^{n+1/2}}$$

$$(F_z)_{j,l+1/2}^{n} = D_{j,l+1/2}^{n}\frac{(f_{j,l+1}^{n} - f_{j,l}^{n})}{\Delta z_{j,l+1/2}^{n}} = \frac{(D_{j,l+1}^{n} + D_{j,l}^{n})}{2}\frac{(f_{j,l+1}^{n} - f_{j,l}^{n})}{\Delta z_{j,l+1/2}^{n}}$$

$$(F_z)_{j,l-1/2}^{n} = D_{j,l-1/2}^{n}\frac{(f_{j,l}^{n} - f_{j,l-1}^{n})}{\Delta z_{j,l-1/2}^{n}} = \frac{(D_{j,l}^{n} + D_{j,l-1}^{n})}{2}\frac{(f_{j,l}^{n} - f_{j,l-1}^{n})}{\Delta z_{j,l-1/2}^{n}}$$

and

$$\Delta r_{j+1/2,l} = r_{j+1,l} - r_{j,l}$$

$$\Delta r_{j-1/2,l} = r_{j,l} - r_{j-1,l}$$

$$\Delta r_{j,l} = \frac{(r_{j+1,l} - r_{j-1,l})}{2}$$

Introducing these relations in Eq. (B.5), one derives

$$
\begin{aligned}
\frac{f_{j,l}^{n+1/2} - f_{j,l}^{n}}{\Delta t/2} = & \frac{1}{2r_{j,l}^{n+1/2}} \frac{(r_{j+1,l}^{n+1/2} + r_{j,l}^{n+1/2})}{(r_{j+1,l}^{n+1/2} - r_{j,l}^{n+1/2})} \frac{(D_{j+1,l}^{n+1/2} + D_{j,l}^{n+1/2})(f_{j+1,l}^{n+1/2} - f_{j,l}^{n+1/2})}{(r_{j+1,l}^{n+1/2} - r_{j-1,l}^{n+1/2})} + \\
& - \frac{1}{2r_{j,l}^{n+1/2}} \frac{(r_{j,l}^{n+1/2} + r_{j-1,l}^{n+1/2})}{(r_{j,l}^{n+1/2} - r_{j-1,l}^{n+1/2})} \frac{(D_{j,l}^{n+1/2} + D_{j-1,l}^{n+1/2})(f_{j,l}^{n+1/2} - f_{j-1,l}^{n+1/2})}{(r_{j+1,l}^{n+1/2} - r_{j-1,l}^{n+1/2})} + \\
& + \frac{1}{(z_{j,l+1}^{n} - z_{j,l}^{n})} \frac{(D_{j,l+1}^{n} + D_{j,l}^{n})(f_{j,l+1}^{n} - f_{j,l}^{n})}{(z_{j,l+1}^{n} - z_{j,l-1}^{n})} + \\
& - \frac{1}{(z_{j,l}^{n} - z_{j,l-1}^{n})} \frac{(D_{j,l}^{n} + D_{j,l-1}^{n})(f_{j,l}^{n} - f_{j,l-1}^{n})}{(z_{j,l+1}^{n} - z_{j,l-1}^{n})}
\end{aligned}
\tag{B.6}
$$

B.1.2 Integer Time Step

Similarly, the integer time step reads as

$$
\begin{aligned}
\frac{f_{j,l}^{n+1} - f_{j,l}^{n+1/2}}{\Delta t/2} = & \frac{1}{2r_{j,l}} \frac{(r_{j+1,l}^{n+1/2} + r_{j,l}^{n+1/2})}{(r_{j+1,l}^{n+1/2} - r_{j,l}^{n+1/2})} \frac{(D_{j+1,l}^{n+1/2} + D_{j,l}^{n+1/2})(f_{j+1,l}^{n+1/2} - f_{j,l}^{n+1/2})}{(r_{j+1,l}^{n+1/2} - r_{j-1,l}^{n+1/2})} + \\
& - \frac{1}{2r_{j,l}^{n+1/2}} \frac{(r_{j,l}^{n+1/2} + r_{j-1,l}^{n+1/2})}{(r_{j,l}^{n+1/2} - r_{j-1,l}^{n+1/2})} \frac{(D_{j,l}^{n+1/2} + D_{j-1,l}^{n+1/2})(f_{j,l}^{n+1/2} - f_{j-1,l}^{n+1/2})}{(r_{j+1,l}^{n+1/2} - r_{j-1,l}^{n+1/2})} + \\
& + \frac{1}{(z_{j,l+1}^{n+1} - z_{j,l}^{n+1})} \frac{(D_{j,l+1}^{n+1} + D_{j,l}^{n+1})(f_{j,l+1}^{n+1} - f_{j,l}^{n+1})}{(z_{j,l+1}^{n+1} - z_{j,l-1}^{n+1})} + \\
& - \frac{1}{(z_{j,l}^{n+1} - z_{j,l-1}^{n+1})} \frac{(D_{j,l}^{n+1} + D_{j,l-1}^{n+1})(f_{j,l}^{n+1} - f_{j,l-1}^{n+1})}{(z_{j,l+1}^{n+1} - z_{j,l-1}^{n+1})}
\end{aligned}
\tag{B.7}
$$

B.2 Upwind Advection Equation

The vectorial advection equation is

$$
\frac{\partial f}{\partial t} = -\mathbf{v} \cdot \nabla f \tag{B.8}
$$

which in cylindrical coordinates simply reads as

$$\frac{\partial f}{\partial t} = -v_r \frac{\partial f}{\partial r} - v_z \frac{\partial f}{\partial z} \tag{B.9}$$

The discretization here is implemented with an upwind method along the z direction, since $v_z = v_s > 0$. Along the r direction, instead, the method is set upwind (downwind) if $v_r > 0$ ($v_r < 0$). If the clump were absent, then $v_r = 0$ through all the space. However, in the presence of a clump (whose central position along the shock direction is $z = z_c$), the velocity filed will have $v_r > 0$ when $z < z_c$ and $v_r < 0$ when $z > z_c$.

B.2.1 *Half Time Step*

The upwind method in both r and z reads, when $v_r \geq 0$ and $v_z \geq 0$, as

$$\frac{f_{j,l}^{n+1/2} - f_{j,l}^{n}}{\Delta t/2} = -(v_r)_{j,l}^{n+1/2} \frac{(f_{j,l}^{n+1/2} - f_{j-1,l}^{n+1/2})}{(r_{j,l}^{n+1/2} - r_{j-1,l}^{n+1/2})} - (v_z)_{j,l}^{n} \frac{(f_{j,l}^{n} - f_{j,l-1}^{n})}{(z_{j,l}^{n} - z_{j,l-1}^{n})} \tag{B.10}$$

while, when $v_r < 0$ and still $v_z \geq 0$, it reads as

$$\frac{f_{j,l}^{n+1/2} - f_{j,l}^{n}}{\Delta t/2} = -(v_r)_{j,l}^{n+1/2} \frac{(f_{j+1,l}^{n+1/2} - f_{j,l}^{n+1/2})}{(r_{j+1,l}^{n+1/2} - r_{j,l}^{n+1/2})} - (v_z)_{j,l}^{n} \frac{(f_{j,l}^{n} - f_{j,l-1}^{n})}{(z_{j,l}^{n} - z_{j,l-1}^{n})} \tag{B.11}$$

B.2.2 *Integer Time Step*

The upwind method in both r and z reads as

$$\frac{f_{j,l}^{n+1} - f_{j,l}^{n+1/2}}{\Delta t/2} = -(v_r)_{j,l}^{n+1/2} \frac{(f_{j,l}^{n+1/2} - f_{j-1,l}^{n+1/2})}{(r_{j,l}^{n+1/2} - r_{j-1,l}^{n+1/2})} - (v_z)_{j,l}^{n+1} \frac{(f_{j,l}^{n+1} - f_{j,l-1}^{n+1})}{(z_{j,l}^{n+1} - z_{j,l-1}^{n+1})} \tag{B.12}$$

while the upwind method in z and the downwind method r is

$$\frac{f_{j,l}^{n+1} - f_{j,l}^{n+1/2}}{\Delta t/2} = -(v_r)_{j,l}^{n+1/2} \frac{(f_{j+1,l}^{n+1/2} - f_{j,l}^{n+1/2})}{(r_{j+1,l}^{n+1/2} - r_{j,l}^{n+1/2})} - (v_z)_{j,l}^{n+1} \frac{(f_{j,l}^{n+1} - f_{j,l-1}^{n+1})}{(z_{j,l}^{n+1} - z_{j,l-1}^{n+1})} \tag{B.13}$$

B.3 The Advection-Diffusion Algorithm

By merging Eq. (B.6) with Eq. (B.10) (or with Eq. (B.11) if $v_r < 0$), one obtains the final algorithm for the half time step. Analogously, by merging Eq. (B.7) with Eq. (B.12) (or with Eq. (B.13) if $v_r < 0$), the final algorithm for the integer time step is derived. It provides for every half time step a system of equations, which can be inverted in order to obtain the distribution function at every time and position on the grid. In particular, as the system can be described by a tridiagonal matrix acting on the vector f, which contains the unknown distribution function at the subsequent time step (either $n + 1/2$ or $n + 1$), it is possible to invert the matrix by means of the Thomas algorithm [228]. In the following, the terms of the tridiagonal matrix are indicated as: $a_{j,l}$ for the sub-diagonal terms, $b_{j,l}$ for for the diagonal ones and $c_{j,l}$ for the super-diagonal ones. The constant terms of the system, instead, are indicated with $d_{j,l}$. Following the ADI method implemented so far, the index l is left explicit when inverting the system of equations at the half time step, while the index j is fixed during the integer time step system inversion. The two complete algorithms are here provided to the interested reader.

B.3.1 Half Time Step

If $v_r > 0$ the algorithm is

$$
\begin{aligned}
& f_{j-1,l}^{n+1/2}\Bigg[-\frac{\Delta t}{4r_{j,l}^{n+1/2}}\frac{(r_{j,l}^{n+1/2}+r_{j-1,l}^{n+1/2})(D_{j,l}^{n+1/2}+D_{j-1,l}^{n+1/2})}{(r_{j,l}^{n+1/2}-r_{j-1,l}^{n+1/2})(r_{j+1,l}^{n+1/2}-r_{j-1,l}^{n+1/2})}-\frac{(v_r)_{j,l}^{n+1/2}\Delta t}{2(r_{j,l}^{n+1/2}-r_{j-1,l}^{n+1/2})}\Bigg]+\\
&+f_{j,l}^{n+1/2}\Bigg[1+\frac{\Delta t}{4r_{j,l}^{n+1/2}}\frac{(r_{j+1,l}^{n+1/2}+r_{j,l}^{n+1/2})(D_{j+1,l}^{n+1/2}+D_{j,l}^{n+1/2})}{(r_{j+1,l}^{n+1/2}-r_{j,l}^{n+1/2})(r_{j+1,l}^{n+1/2}-r_{j-1,l}^{n+1/2})}+\frac{(v_r)_{j,l}^{n+1/2}\Delta t}{2(r_{j,l}^{n+1/2}-r_{j-1,l}^{n+1/2})}+\\
&+\frac{\Delta t}{4r_{j,l}^{n+1/2}}\frac{(r_{j,l}^{n+1/2}+r_{j-1,l}^{n+1/2})(D_{j,l}^{n+1/2}+D_{j-1,l}^{n+1/2})}{(r_{j,l}^{n+1/2}-r_{j-1,l}^{n+1/2})(r_{j+1,l}^{n+1/2}-r_{j-1,l}^{n+1/2})}\Bigg]+\\
&+f_{j+1,l}^{n+1/2}\Bigg[-\frac{\Delta t}{4r_{j,l}}\frac{(r_{j+1,l}^{n+1/2}+r_{j,l}^{n+1/2})(D_{j+1,l}^{n+1/2}+D_{j,l}^{n+1/2})}{(r_{j+1,l}^{n+1/2}-r_{j,l}^{n+1/2})(r_{j+1,l}^{n+1/2}-r_{j-1,l}^{n+1/2})}\Bigg]=\\
&=f_{j,l-1}^{n}\Bigg[\frac{\Delta t}{2}\frac{(D_{j,l}^{n}+D_{j,l-1}^{n})}{(z_{j,l}^{n}-z_{j,l-1}^{n})(z_{j,l+1}^{n}-z_{j,l-1}^{n})}+\frac{(v_z)_{j,l}^{n}\Delta t}{2(z_{j,l}^{n}-z_{j,l-1}^{n})}\Bigg]+\\
&+f_{j,l}^{n}\Bigg[1-\frac{\Delta t}{2}\frac{(D_{j,l+1}^{n}+D_{j,l}^{n})}{(z_{j,l+1}^{n}-z_{j,l}^{n})(z_{j,l+1}^{n}-z_{j,l-1}^{n})}-\frac{\Delta t}{2}\frac{(D_{j,l}^{n}+D_{j,l-1}^{n})}{(z_{j,l}^{n}-z_{j,l-1}^{n})(z_{j,l+1}^{n}-z_{j,l-1}^{n})}+\\
&-\frac{(v_z)_{j,l}^{n}\Delta t}{2(z_{j,l}^{n}-z_{j,l-1}^{n})}\Bigg]+f_{j,l+1}^{n}\Bigg[\frac{\Delta t}{2}\frac{(D_{j,l+1}^{n}+D_{j,l}^{n})}{(z_{j,l+1}^{n}-z_{j,l}^{n})(z_{j,l+1}^{n}-z_{j,l-1}^{n})}\Bigg]
\end{aligned}
\tag{B.14}
$$

On the other hand, if $v_r < 0$ the algorithm is

$$
\begin{aligned}
& f_{j-1,l}^{n+1/2}\left[-\frac{\Delta t}{4r_{j,l}^{n+1/2}}\frac{(r_{j,l}^{n+1/2}+r_{j-1,l}^{n+1/2})(D_{j,l}^{n+1/2}+D_{j-1,l}^{n+1/2})}{(r_{j,l}^{n+1/2}-r_{j-1,l}^{n+1/2})(r_{j+1,l}^{n+1/2}-r_{j-1,l}^{n+1/2})}\right]+\\
& +f_{j,l}^{n+1/2}\left[1+\frac{\Delta t}{4r_{j,l}^{n+1/2}}\frac{(r_{j+1,l}^{n+1/2}+r_{j,l}^{n+1/2})(D_{j+1,l}^{n+1/2}+D_{j,l}^{n+1/2})}{(r_{j+1,l}^{n+1/2}-r_{j,l}^{n+1/2})(r_{j+1,l}^{n+1/2}-r_{j-1,l}^{n+1/2})}-\frac{(v_r)_{j,l}^{n+1/2}\Delta t}{2(r_{j+1,l}^{n+1/2}-r_{j,l}^{n+1/2})}\right.\\
& \left.+\frac{\Delta t}{4r_{j,l}^{n+1/2}}\frac{(r_{j,l}^{n+1/2}+r_{j-1,l}^{n+1/2})(D_{j,l}^{n+1/2}+D_{j-1,l}^{n+1/2})}{(r_{j,l}^{n+1/2}-r_{j-1,l}^{n+1/2})(r_{j+1,l}^{n+1/2}-r_{j-1,l}^{n+1/2})}\right]+\\
& +f_{j+1,l}^{n+1/2}\left[-\frac{\Delta t}{4r_{j,l}}\frac{(r_{j+1,l}^{n+1/2}+r_{j,l}^{n+1/2})(D_{j+1,l}^{n+1/2}+D_{j,l}^{n+1/2})}{(r_{j+1,l}-r_{j,l})(r_{j+1,l}^{n+1/2}-r_{j-1,l}^{n+1/2})}+\frac{(v_r)_{j,l}^{n+1/2}\Delta t}{2(r_{j+1,l}^{n+1/2}-r_{j,l}^{n+1/2})}\right]=\\
& =f_{j,l-1}^{n}\left[\frac{\Delta t}{2}\frac{(D_{j,l}^{n}+D_{j,l-1}^{n})}{(z_{j,l}^{n}-z_{j,l-1}^{n})(z_{j,l+1}^{n}-z_{j,l-1}^{n})}+\frac{(v_z)_{j,l}^{n}\Delta t}{2(z_{j,l}^{n}-z_{j,l-1}^{n})}\right]+\\
& +f_{j,l}^{n}\left[1-\frac{\Delta t}{2}\frac{(D_{j,l+1}^{n}+D_{j,l}^{n})}{(z_{j,l+1}^{n}-z_{j,l}^{n})(z_{j,l+1}^{n}-z_{j,l-1}^{n})}-\frac{\Delta t}{2}\frac{(D_{j,l}^{n}+D_{j,l-1}^{n})}{(z_{j,l}^{n}-z_{j,l-1}^{n})(z_{j,l+1}^{n}-z_{j,l-1}^{n})}+\right.\\
& \left.-\frac{(v_z)_{j,l}^{n}\Delta t}{2(z_{j,l}^{n}-z_{j,l-1}^{n})}\right]+f_{j,l+1}^{n}\left[\frac{\Delta t}{2}\frac{(D_{j,l+1}^{n}+D_{j,l}^{n})}{(z_{j,l+1}^{n}-z_{j,l}^{n})(z_{j,l+1}^{n}-z_{j,l-1}^{n})}\right]
\end{aligned}
\tag{B.15}
$$

B.3.2 Integer Time Step

If $v_r > 0$ the algorithm is

$$
\begin{aligned}
& f_{j,l-1}^{n+1}\left[-\frac{\Delta t}{2}\frac{(D_{j,l}^{n+1}+D_{j,l-1}^{n+1})}{(z_{j,l}^{n+1}-z_{j,l-1}^{n+1})(z_{j,l+1}^{n+1}-z_{j,l-1}^{n+1})}-\frac{(v_z)_{j,l}^{n+1}\Delta t}{2(z_{j,l}^{n+1}-z_{j,l-1}^{n+1})}\right]+\\
& f_{j,l}^{n+1}\left[1+\frac{\Delta t}{2}\frac{(D_{j,l+1}^{n+1}+D_{j,l}^{n+1})}{(z_{j,l+1}^{n+1}-z_{j,l}^{n+1})(z_{j,l+1}^{n+1}-z_{j,l-1}^{n+1})}+\frac{\Delta t}{2}\frac{(D_{j,l}^{n+1}+D_{j,l-1}^{n+1})}{(z_{j,l}^{n+1}-z_{j,l-1}^{n+1})(z_{j,l+1}^{n+1}-z_{j,l-1}^{n+1})}+\right.\\
& \left.+\frac{(v_z)_{j,l}^{n+1}\Delta t}{2(z_{j,l}^{n+1}-z_{j,l-1}^{n+1})}\right]+f_{j,l+1}^{n+1}\left[-\frac{\Delta t}{2}\frac{(D_{j,l+1}^{n+1}+D_{j,l}^{n+1})}{(z_{j,l+1}^{n+1}-z_{j,l}^{n+1})(z_{j,l+1}^{n+1}-z_{j,l-1}^{n+1})}\right]=\\
& =f_{j-1,l}^{n+1/2}\left[\frac{\Delta t}{4r_{j,l}}\frac{(r_{j,l}^{n+1/2}+r_{j-1,l}^{n+1/2})(D_{j,l}^{n+1/2}+D_{j-1,l}^{n+1/2})}{(r_{j,l}^{n+1/2}-r_{j-1,l}^{n+1/2})(r_{j+1,l}^{n+1/2}-r_{j-1,l}^{n+1/2})}+\frac{(v_r)_{j,l}^{n+1/2}\Delta t}{2(r_{j,l}^{n+1/2}-r_{j-1,l}^{n+1/2})}\right]+\\
& +f_{j,l}^{n+1/2}\left[1-\frac{\Delta t}{4r_{j,l}^{n+1/2}}\frac{(r_{j+1,l}^{n+1/2}+r_{j,l}^{n+1/2})(D_{j+1,l}^{n+1/2}+D_{j,l}^{n+1/2})}{(r_{j+1,l}^{n+1/2}-r_{j,l}^{n+1/2})(r_{j+1,l}^{n+1/2}-r_{j-1,l}^{n+1/2})}-\frac{(v_r)_{j,l}^{n+1/2}\Delta t}{2(r_{j,l}^{n+1/2}-r_{j-1,l}^{n+1/2})}+\right.\\
& \left.-\frac{\Delta t}{4r_{j,l}^{n+1/2}}\frac{(r_{j,l}^{n+1/2}+r_{j-1,l}^{n+1/2})(D_{j,l}^{n+1/2}+D_{j-1,l}^{n+1/2})}{(r_{j,l}^{n+1/2}-r_{j-1,l}^{n+1/2})(r_{j+1,l}^{n+1/2}-r_{j-1,l}^{n+1/2})}\right]+\\
& +f_{j+1,l}^{n+1/2}\left[\frac{\Delta t}{4r_{j,l}^{n+1/2}}\frac{(r_{j+1,l}^{n+1/2}+r_{j,l}^{n+1/2})(D_{j+1,l}^{n+1/2}+D_{j,l}^{n+1/2})}{(r_{j+1,l}^{n+1/2}-r_{j,l}^{n+1/2})(r_{j+1,l}^{n+1/2}-r_{j-1,l}^{n+1/2})}\right]
\end{aligned}
\tag{B.16}
$$

On the other hand, if $v_r < 0$ the algorithm is

$$
\begin{aligned}
& f_{j,l-1}^{n+1}\left[-\frac{\Delta t}{2}\frac{(D_{j,l}^{n+1}+D_{j,l-1}^{n+1})}{(z_{j,l}^{n+1}-z_{j,l-1}^{n+1})(z_{j,l+1}^{n+1}-z_{j,l-1}^{n+1})}-\frac{(v_z)_{j,l}^{n+1}\Delta t}{2(z_{j,l}^{n+1}-z_{j,l-1}^{n+1})}\right]+ \\
& +f_{j,l}^{n+1}\left[1+\frac{\Delta t}{2}\frac{(D_{j,l+1}^{n+1}+D_{j,l}^{n+1})}{(z_{j,l+1}^{n+1}-z_{j,l}^{n+1})(z_{j,l+1}^{n+1}-z_{j,l-1}^{n+1})}+\frac{\Delta t}{2}\frac{(D_{j,l}^{n+1}+D_{j,l-1}^{n+1})}{(z_{j,l}^{n+1}-z_{j,l-1}^{n+1})(z_{j,l+1}^{n+1}-z_{j,l-1}^{n+1})}+\right. \\
& \left.+\frac{(v_z)_{j,l}^{n+1}\Delta t}{2(z_{j,l}^{n+1}-z_{j,l-1}^{n+1})}\right]+f_{j,l+1}^{n+1}\left[-\frac{\Delta t}{2}\frac{(D_{j,l+1}^{n+1}+D_{j,l}^{n+1})}{(z_{j,l+1}^{n+1}-z_{j,l}^{n+1})(z_{j,l+1}^{n+1}-z_{j,l-1}^{n+1})}\right]= \\
& =f_{j-1,l}^{n+1/2}\left[\frac{\Delta t}{4r_{j,l}^{n+1/2}}\frac{(r_{j,l}^{n+1/2}+r_{j-1,l}^{n+1/2})(D_{j,l}^{n+1/2}+D_{j-1,l}^{n+1/2})}{(r_{j,l}^{n+1/2}-r_{j-1,l}^{n+1/2})(r_{j+1,l}^{n+1/2}-r_{j-1,l}^{n+1/2})}\right]+ \\
& +f_{j,l}^{n+1/2}\left[1-\frac{\Delta t}{4r_{j,l}^{n+1/2}}\frac{(r_{j+1,l}^{n+1/2}+r_{j,l}^{n+1/2})(D_{j+1,l}^{n+1/2}+D_{j,l}^{n+1/2})}{(r_{j+1,l}^{n+1/2}-r_{j,l}^{n+1/2})(r_{j+1,l}^{n+1/2}-r_{j-1,l}^{n+1/2})}+\frac{(v_r)_{j,l}^{n+1/2}\Delta t}{2(r_{j+1,l}^{n+1/2}-r_{j,l}^{n+1/2})}+\right. \\
& \left.-\frac{\Delta t}{4r_{j,l}^{n+1/2}}\frac{(r_{j,l}^{n+1/2}+r_{j-1,l}^{n+1/2})(D_{j,l}^{n+1/2}+D_{j-1,l}^{n+1/2})}{(r_{j,l}^{n+1/2}-r_{j-1,l}^{n+1/2})(r_{j+1,l}^{n+1/2}-r_{j-1,l}^{n+1/2})}\right]+ \\
& +f_{j+1,l}^{n+1/2}\left[\frac{\Delta t}{4r_{j,l}^{n+1/2}}\frac{(r_{j+1,l}^{n+1/2}+r_{j,l}^{n+1/2})(D_{j+1,l}^{n+1/2}+D_{j,l}^{n+1/2})}{(r_{j+1,l}^{n+1/2}-r_{j,l}^{n+1/2})(r_{j+1,l}^{n+1/2}-r_{j-1,l}^{n+1/2})}-\frac{(v_r)_{j,l}^{n+1/2}\Delta t}{2(r_{j+1,l}^{n+1/2}-r_{j,l}^{n+1/2})}\right]
\end{aligned}
\tag{B.17}
$$

B.4 Injection at the Shock

The shock itself acts as a source injecting continuously particles towards the downstream. In order to derive the source term, the stationary transport equation is considered along the z direction

$$
v_s\frac{\partial f}{\partial z}=D(p)\frac{\partial^2 f}{\partial z^2}+Q_{\mathrm{CR}} \tag{B.18}
$$

where the source term Q_{CR} is injected at the shock location. Thus, integrating such equation around the shock, and taking into account that the distribution function is constant across the shock, the left side vanishes, while on the right side one gets (1 standing for the upstream, 2 standing for the downstream)

$$
Q_{\mathrm{CR}}=D(p)\left.\frac{\partial f}{\partial z}\right|_2-D(p)\left.\frac{\partial f}{\partial z}\right|_1=-D(p)\left.\frac{\partial f}{\partial z}\right|_1 \tag{B.19}
$$

because the downstream distribution function is homogeneous. Now, the equilibrium solution in the upstream is found through the integration of

$$v_s \frac{\partial f}{\partial z} = D(p) \frac{\partial^2 f}{\partial z^2} \Longrightarrow v_s f = D(p) \frac{\partial f}{\partial z} \Longrightarrow \frac{\partial f}{f} = \frac{v_s}{D(p)} \partial z$$

between z_s and z, yielding

$$\Longrightarrow f(z) = f_0(p) \exp\left[-\frac{(z - z_s)v_s}{D(p)} \right] \tag{B.20}$$

Then, its derivative looks like

$$\frac{\partial f}{\partial z} = -\frac{v_s}{D(p)} f_0(p) \exp\left[-\frac{(z - z_s)v_s}{D(p)} \right] \tag{B.21}$$

that computed at the shock is

$$\frac{\partial f}{\partial z}(z = z_s) = -\frac{v_s}{D(p)} f_0(p) \tag{B.22}$$

Finally, the source term in the shock frame amounts to

$$Q_{\mathrm{CR}} = v_s f_0(p) \delta(z - z_s) \tag{B.23}$$

This term should be injected at the shock position. However, the presence of a sharp injection due to the δ–Dirac term might produce numerical instabilities: such kind of difficulties are usually solved by injecting a tight Gaussian distribution. The approach adopted here, on the other hand, is to simply set the condition $f_0 = f_{\mathrm{inj}}$ for every momentum p, so that the shock remains sharp through the whole evolution. This is equivalent to impose, for every j in correspondence of $l = l_s$, the solution of the system of equations previously described (without solving for it).

B.5 CR Precursor

The presence of a shock precursor upstream of the shock is introduced by setting an initial particle distribution function equal to the expected precursor shape along the z dimension. In fact, as derived in Eq. (B.20), the spatial distribution of particles in the shock reference frame is provided by an exponential distribution along the shock direction, centered on the shock position. Such a distribution represents the initial condition of the problem, since the microphysics of the acceleration will not be exploited here. However, within the clump, no particles are set as initial conditions.

B.6 Boundary Conditions

A non-square grid is set with $0 \leq l < N$ and $0 \leq j < T$. The meshes are uniformly spaced along l ($\Delta z =$const), but logarithmically spaced along j (with a spacing β that depends on the particle momentum, as discussed later because of the accuracy condition). The logarithmic step satisfies the condition

$$r_{j+1,l} = r_{j,l} + \Delta r_{j,l} \implies \log(r_{j+1,l}) = \log(r_{j,l} + \Delta r_{j,l}) = \beta \log(r_{j,l})$$

Note that, since a logarithmic scale is set in r, it is not be possible to start from $r = 0$, but from some $r_0 > 0$. Thus, $j = -1$ corresponds to the preceding step in the log-scale with respect to $r_{0,l}$, namely

$$\log(r_{-1,l}) = \beta^{-1} \log(r_{0,l}) \tag{B.24}$$

while $j = T$ corresponds to the step following $r_{T-1,l}$.

$$\log(r_{T,l}) = \beta \log(r_{T-1,l}) \tag{B.25}$$

The boundary conditions should respect the reflective property of the grid, namely the fact that the particles, while diffusing, enter and exit the grid at the same rate. On the other hand, there will be a net flux of particles due to advection, which comes because of the injection rate. Such a condition is implemented in the form of a null diffusive flux: it corresponds to the presence of a generally non-vanishing net flux, due to advection. On the other hand, the axial upstream boundary should consider the presence of a precursor located at $z > z_s$, where the distribution function follows Eq. (B.20). Lastly, the system in cylindrical coordinates will appear symmetric with respect to the z-axis. In summary, the boundary conditions will read as:

(i) null diffusive flux in $l = -1/2$;
(ii) precursor shape in $l = N$;
(iii) symmetry in $j = 0$;
(iv) null diffusive flux in $j = T - 1/2$.

In the rest of the section, these conditions are discussed in details. Note that the lower boundary on z is defined by every j at $l = -1/2$ (the fractional element coming from the flux conservative algorithm) and it will be called $z1$, while the upper boundary is defined by every j at $l = N$ and it will be called $z2$. Analogously, the lower boundary on r is defined by every l with $j = 0$ and it will be called $r1$, while the upper boundary is defined by every l at $j = T - 1/2$ and it will be called $r2$.

B.6.1 Null Diffusive Flux at $z = z1$

The diffusive flux on $l = -1/2$ is set to zero by imposing that, at every j, it holds

$$f_{j,-1} = f_{j,0} \tag{B.26}$$

This condition will affect the constant term of the half mesh system as

$$d^n_{j,0} = f^n_{j,0}\left[1 - \frac{\Delta t}{2}\frac{(D^n_{j,1} + D^n_{j,0})}{(z^n_{j,1} - z^n_{j,0})(z^n_{j,1} - z^n_{j,-1})}\right] + f^n_{j,1}\left[\frac{\Delta t}{2}\frac{(D^n_{j,1} + D^n_{j,0})}{(z^n_{j,1} - z^n_{j,0})(z^n_{j,1} - z^n_{j,-1})}\right] \tag{B.27}$$

On the other hand, for every j, the matrix terms with $l = 0$ of the integer mesh system will be characterized by

$$\begin{cases} a_{j,0} &= 0 \\ b_{j,0} &= 1 + \frac{\Delta t}{2}\frac{(D^{n+1}_{j,1} + D^{n+1}_{j,0})}{(z^{n+1}_{j,1} - z^{n+1}_{j,0})(z^{n+1}_{j,1} - z^{n+1}_{j,-1})} \\ c_{j,0} &= -\frac{\Delta t}{2}\frac{(D^{n+1}_{j,1} + D^{n+1}_{j,0})}{(z^{n+1}_{j,1} - z^{n+1}_{j,0})(z^{n+1}_{j,1} - z^{n+1}_{j,-1})} \end{cases} \tag{B.28}$$

B.6.2 Precursor Shape at $z = z2$

This condition simply reads as

$$f_{j,N} = f(z_s)\exp\left[-\frac{(z(l = N) - z_s)v_s}{D}\right] = f(z_s)\exp\left[-\frac{(z_N - z_s)v_s}{D}\right] \tag{B.29}$$

As a consequence, at every j, the constant term at $l = N - 1$ of the half time step is affected as

$$\begin{aligned} d^n_{j,N-1} =& f^n_{j,N-2}\left[\frac{\Delta t}{2}\frac{(D^n_{j,N-1} + D^n_{j,N-2})}{(z^n_{j,N-1} - z^n_{j,N-2})(z^n_{j,N} - z^n_{j,N-2})} + \frac{(v_z)^n_{j,N-1}\Delta t}{2(z^n_{j,N-1} - z^n_{j,N-2})}\right] + \\ &+ f^n_{j,N-1}\left[1 - \frac{\Delta t}{2}\frac{(D^n_{j,N} + D^n_{j,N-1})}{(z^n_{j,N} - z^n_{j,N-1})(z^n_{j,N} - z^n_{j,N-2})} - \frac{(v_z)^n_{j,N-1}\Delta t}{2(z^n_{j,N-1} - z^n_{j,N-2})} + \right. \\ &\left. - \frac{\Delta t}{2}\frac{(D^n_{j,N-1} + D^n_{j,N-2})}{(z^n_{j,N-1} - z^n_{j,N-2})(z^n_{j,N} - z^n_{j,N-2})}\right] + \\ &+ f^n(z^n_s)\exp\left[-\frac{(z^n_N - z^n_s)v_s}{D^n_{j,N}}\right]\left[\frac{\Delta t}{2}\frac{(D^n_{j,N} + D^n_{j,N-1})}{(z^n_{j,N} - z^n_{j,N-1})(z^n_{j,N} - z^n_{j,N-2})}\right] \end{aligned} \tag{B.30}$$

as well as the constant term of the integer step, that becomes

$$d^{n+1/2}_{j,N-1} += f^{n+1}(z_s)\exp\left[-\frac{(z^{n+1}_N - z^{n+1}_s)v_s}{D^{n+1}_{j,N}}\right]\left[\frac{\Delta t}{2}\frac{(D^{n+1}_{j,N} + D^{n+1}_{j,N-1})}{(z^{n+1}_{j,N} - z^{n+1}_{j,N-1})(z^{n+1}_{j,N} - z^{n+1}_{j,N-2})}\right] \tag{B.31}$$

B.6.3 Symmetry at r = r1

Once $r_{-1,l}$ is defined as in Eq. (B.24), the symmetry with respect to the radial axis holds if

$$f_{-1,l} = f_{0,l} \tag{B.32}$$

This is equivalent to require that, at every fixed l, the distribution function is constant across the r-axis (including the spatial region inside the clump). This condition affects the matrix terms with $j = 0$ of the half time step as they become, if $v_r \geq 0$,

$$\begin{cases} a_{0,l} &= 0 \\ b_{0,l} &= 1 + \dfrac{\Delta t}{4r_{0,l}^{n+1/2}} \dfrac{(r_{1,l}^{n+1/2} + r_{0,l}^{n+1/2})(D_{1,l}^{n+1/2} + D_{0,l}^{n+1/2})}{(r_{1,l}^{n+1/2} - r_{0,l}^{n+1/2})(r_{1,l}^{n+1/2} - r_{-1,l}^{n+1/2})} \\ c_{0,l} &= -\dfrac{\Delta t}{4r_{0,l}^{n+1/2}} \dfrac{(r_{1,l}^{n+1/2} + r_{0,l}^{n+1/2})(D_{1,l}^{n+1/2} + D_{0,l}^{n+1/2})}{(r_{1,l}^{n+1/2} - r_{0,l}^{n+1/2})(r_{1,l}^{n+1/2} - r_{-1,l}^{n+1/2})} \end{cases} \tag{B.33}$$

as well as the constant term of the integer time step, that is, if $v_r \geq 0$,

$$\begin{aligned} d_{0,l}^{n+1/2} =& f_{0,l}^{n+1/2} \left[1 - \frac{\Delta t}{4r_{0,l}^{n+1/2}} \frac{(r_{1,l}^{n+1/2} + r_{0,l}^{n+1/2})(D_{1,l}^{n+1/2} + D_{0,l}^{n+1/2})}{(r_{1,l}^{n+1/2} - r_{0,l}^{n+1/2})(r_{1,l}^{n+1/2} - r_{-1,l}^{n+1/2})} \right] + \\ &+ f_{1,l}^{n+1/2} \left[\frac{\Delta t}{4r_{0,l}^{n+1/2}} \frac{(r_{1,l}^{n+1/2} + r_{0,l}^{n+1/2})(D_{1,l}^{n+1/2} + D_{0,l}^{n+1/2})}{(r_{1,l}^{n+1/2} - r_{0,l}^{n+1/2})(r_{1,l}^{n+1/2} - r_{-1,l}^{n+1/2})} \right] \end{aligned} \tag{B.34}$$

On the other hand, if $v_r < 0$, then the boundary on r1 affects the matrix terms of the half time step as

$$\begin{cases} a_{0,l} &= 0 \\ b_{0,l} &= 1 + \dfrac{\Delta t}{4r_{0,l}^{n+1/2}} \dfrac{(r_{1,l}^{n+1/2} + r_{0,l}^{n+1/2})(D_{1,l}^{n+1/2} + D_{0,l}^{n+1/2})}{(r_{1,l}^{n+1/2} - r_{0,l}^{n+1/2})(r_{1,l}^{n+1/2} - r_{-1,l}^{n+1/2})} - \dfrac{(v_r)_{0,l}^{n+1/2} \Delta t}{2(r_{1,l}^{n+1/2} - r_{0,l}^{n+1/2})} \\ c_{0,l} &= -\dfrac{\Delta t}{4r_{0,l}^{n+1/2}} \dfrac{(r_{1,l}^{n+1/2} + r_{0,l}^{n+1/2})(D_{1,l}^{n+1/2} + D_{0,l}^{n+1/2})}{(r_{1,l}^{n+1/2} - r_{0,l}^{n+1/2})(r_{1,l}^{n+1/2} - r_{-1,l}^{n+1/2})} + \dfrac{(v_r)_{0,l}^{n+1/2} \Delta t}{2(r_{1,l}^{n+1/2} - r_{0,l}^{n+1/2})} \end{cases} \tag{B.35}$$

and the constant term of the integer time step as

$$\begin{aligned} d_{0,l}^{n+1/2} =& f_{0,l}^{n+1/2} \left[1 - \frac{\Delta t}{4r_{0,l}^{n+1/2}} \frac{(r_{1,l}^{n+1/2} + r_{0,l}^{n+1/2})(D_{1,l}^{n+1/2} + D_{0,l}^{n+1/2})}{(r_{1,l}^{n+1/2} - r_{0,l}^{n+1/2})(r_{1,l}^{n+1/2} - r_{-1,l}^{n+1/2})} + \frac{(v_r)_{0,l}^{n+1/2} \Delta t}{2(r_{1,l}^{n+1/2} - r_{0,l}^{n+1/2})} \right] + \\ &+ f_{1,l}^{n+1/2} \left[\frac{\Delta t}{4r_{0,l}^{n+1/2}} \frac{(r_{1,l}^{n+1/2} + r_{0,l}^{n+1/2})(D_{1,l}^{n+1/2} + D_{0,l}^{n+1/2})}{(r_{1,l}^{n+1/2} - r_{0,l}^{n+1/2})(r_{1,l}^{n+1/2} - r_{-1,l}^{n+1/2})} - \frac{(v_r)_{0,l}^{n+1/2} \Delta t}{2(r_{1,l}^{n+1/2} - r_{0,l}^{n+1/2})} \right] \end{aligned} \tag{B.36}$$

B.6.4 Null Flux at r = r2

The diffusive flux $F_{T-1/2,l}$ vanishes when imposing that, at every fixed l, the following equality holds

$$f_{T-1,l} = f_{T,l} \tag{B.37}$$

It means that the surface at $r = \infty$ makes only the advective flux going out, while the diffusive one is reflected back. The previous condition acts on the matrix terms at $j = T - 1$ of the half time step as these become, if $v_r \geq 0$,

$$\begin{cases} a_{T-1,l} &= \left[-\frac{\Delta t}{4r^{n+1/2}_{T-1,l}} \frac{(r^{n+1/2}_{T-1,l} + r^{n+1/2}_{T-2,l})(D^{n+1/2}_{T-1,l} + D^{n+1/2}_{T-2,l})}{(r^{n+1/2}_{T-1,l} - r^{n+1/2}_{T-2,l})(r^{n+1/2}_{T,l} - r^{n+1/2}_{T-2,l})} - \frac{(v_r)^{n+1/2}_{T-1,l}\Delta t}{2(r^{n+1/2}_{T-1,l} - r^{n+1/2}_{T-2,l})} \right] \\ b_{T-1,l} &= \left[1 + \frac{\Delta t}{4r^{n+1/2}_{T-1,l}} \frac{(r^{n+1/2}_{T-1,l} + r^{n+1/2}_{T-2,l})(D^{n+1/2}_{T-1,l} + D^{n+1/2}_{T-2,l})}{(r^{n+1/2}_{T-1,l} - r^{n+1/2}_{T-2,l})(r^{n+1/2}_{T,l} - r^{n+1/2}_{T-2,l})} + \frac{(v_r)^{n+1/2}_{T-1,l}\Delta t}{2(r^{n+1/2}_{T-1,l} - r^{n+1/2}_{T-2,l})} \right] \\ c_{T-1,l} &= 0 \end{cases} \tag{B.38}$$

as well as on the constant term of the integer step that, if $v_r \geq 0$, is

$$\begin{aligned} d^{n+1/2}_{T-1,l} = f^{n+1/2}_{T-2,l} &\left[\frac{\Delta t}{4r^{n+1/2}_{T-1,l}} \frac{(r^{n+1/2}_{T-1,l} + r^{n+1/2}_{T-2,l})(D^{n+1/2}_{T-1,l} + D^{n+1/2}_{T-2,l})}{(r^{n+1/2}_{T-1,l} - r^{n+1/2}_{T-2,l})(r^{n+1/2}_{T,l} - r^{n+1/2}_{T-2,l})} + \frac{(v_r)^{n+1/2}_{T-1,l}\Delta t}{2(r^{n+1/2}_{T-1,l} - r^{n+1/2}_{T-2,l})} \right] + \\ + f^{n+1/2}_{T-1,l} &\left[1 - \frac{\Delta t}{4r^{n+1/2}_{T-1,l}} \frac{(r^{n+1/2}_{T-1,l} + r^{n+1/2}_{T-2,l})(D^{n+1/2}_{T-1,l} + D^{n+1/2}_{T-2,l})}{(r^{n+1/2}_{T-1,l} - r^{n+1/2}_{T-2,l})(r^{n+1/2}_{T,l} - r^{n+1/2}_{T-2,l})} - \frac{(v_r)^{n+1/2}_{T-1,l}\Delta t}{2(r^{n+1/2}_{T-1,l} - r^{n+1/2}_{T-2,l})} \right] \end{aligned} \tag{B.39}$$

On the other hand, if $v_r < 0$, then the boundary on r2 affects the matrix terms of the half time step as

$$\begin{cases} a_{T-1,l} &= \left[-\frac{\Delta t}{4r^{n+1/2}_{T-1,l}} \frac{(r^{n+1/2}_{T-1,l} + r^{n+1/2}_{T-2,l})(D^{n+1/2}_{T-1,l} + D^{n+1/2}_{T-2,l})}{(r^{n+1/2}_{T-1,l} - r^{n+1/2}_{T-2,l})(r^{n+1/2}_{T,l} - r^{n+1/2}_{T-2,l})} \right] \\ b_{T-1,l} &= \left[1 + \frac{\Delta t}{4r^{n+1/2}_{T-1,l}} \frac{(r^{n+1/2}_{T-1,l} + r^{n+1/2}_{T-2,l})(D^{n+1/2}_{T-1,l} + D^{n+1/2}_{T-2,l})}{(r^{n+1/2}_{T-1,l} - r^{n+1/2}_{T-2,l})(r^{n+1/2}_{T,l} - r^{n+1/2}_{T-2,l})} \right] \\ c_{T-1,l} &= 0 \end{cases} \tag{B.40}$$

and the constant term of the integer step as

$$\begin{aligned} d^{n+1/2}_{T-1,l} = f^{n+1/2}_{T-2,l} &\left[\frac{\Delta t}{4r^{n+1/2}_{T-1,l}} \frac{(r^{n+1/2}_{T-1,l} + r^{n+1/2}_{T-2,l})(D^{n+1/2}_{T-1,l} + D^{n+1/2}_{T-2,l})}{(r^{n+1/2}_{T-1,l} - r^{n+1/2}_{T-2,l})(r^{n+1/2}_{T,l} - r^{n+1/2}_{T-2,l})} \right] + \\ + f^{n+1/2}_{T-1,l} &\left[1 - \frac{\Delta t}{4r^{n+1/2}_{T-1,l}} \frac{(r^{n+1/2}_{T-1,l} + r^{n+1/2}_{T-2,l})(D^{n+1/2}_{T-1,l} + D^{n+1/2}_{T-2,l})}{(r^{n+1/2}_{T-1,l} - r^{n+1/2}_{T-2,l})(r^{n+1/2}_{T,l} - r^{n+1/2}_{T-2,l})} \right] \end{aligned} \tag{B.41}$$

B.7 Accuracy of the Algorithm

As previously introduced, a logarithmic grid is set along r and a linear one along z. In order to have a detailed description of the clump magnetic skin, the radial coordinate starts from $r_0 = 5 \times 10^{16}$ cm and goes up to $r_T = 6 \times 10^{18}$ cm, where the number of mesh points depends on the particle momentum. Then, in order to limit the boundary effects on z, a large value of points along it is set ($N = 600$ in the following, yielding $\Delta z = 1 \times 10^{16}$ cm and thus covering the region from $z = 0$ to $z = 6 \times 10^{18}$ cm$\sim$ 2 pc). The clump center is located in $z_c = 2 \times 10^{18}$ cm.

The logarithmic grid in r is set with a different spatial resolution (i.e. with a different number of points T) according to the simulated particle momentum. This is due to the fact that different length scales regulate the time-evolution of the particle distribution function at the different energies: in fact, the evolution of the low-energy particles is mainly regulated by advection, while that of high-energy particles is regulated by diffusion. In order to achieve a high precision along the whole simulated energy range, particularly on the resulting proton spectrum, different lengths have to be resolved in order to follow both the high and the low energies. Therefore, the radial step for an individual particle momentum is found by requiring convergence on the spectrum of protons contained inside the clump, namely that the true solution does not differ significantly from the numerical one. Here, the true solution is defined as the solution that, by changing the resolution of the grid, does not change by more than the given accuracy level. Convergence is achieved by performing, for each particle momentum, different simulations with decreasing step in the radial coordinate, and by subsequently computing the correspondent proton spectrum: the final step Δr is chosen as the one that provides at least a 5% accuracy on the proton spectrum, with respect to the previous simulation (which is equivalent to say that the same particle spectrum was obtained through the simulation with larger spatial step, at a fixed proton momentum). This is shown in Fig. B.2, where the final resolution adopted for each particle momentum is visible. The final setup is provided by:

- $1.0 \leq \log_{10}\left(p/\mathrm{GeVc}^{-1}\right) \leq 1.6 \longrightarrow \Delta r_{\min} = 1.2 \times 10^{14}$ cm ($T = 2000$);
- $1.7 \leq \log_{10}\left(p/\mathrm{GeVc}^{-1}\right) \leq 2.2 \longrightarrow \Delta r_{\min} = 1.6 \times 10^{14}$ cm ($T = 1500$);
- $2.3 \leq \log_{10}\left(p/\mathrm{GeVc}^{-1}\right) \leq 3.1 \longrightarrow \Delta r_{\min} = 2.4 \times 10^{14}$ cm ($T = 1000$);
- $3.2 \leq \log_{10}\left(p/\mathrm{GeVc}^{-1}\right) \leq 3.5 \longrightarrow \Delta r_{\min} = 4.8 \times 10^{14}$ cm ($T = 500$);
- $3.6 \leq \log_{10}\left(p/\mathrm{GeVc}^{-1}\right) \leq 7.7 \longrightarrow \Delta r_{\min} = 9.7 \times 10^{14}$ cm ($T = 250$);
- $4.8 \leq \log_{10}\left(p/\mathrm{GeVc}^{-1}\right) \leq 6.0 \longrightarrow \Delta r_{\min} = 5.0 \times 10^{15}$ cm ($T = 50$).

Note that the spatial step here reported refers to the smallest r-step along the grid (the so-called $r_{\min}$), since the logarithmic scale adopted extends the width of the radial bins when moving from j to j+1.

Particles ranging from $p = 10$ GeV/c to $p = 1$ PeV/c are simulated. The spectrum is then extended down to low-energies (down to 1 GeV) through a linear fitting procedure.

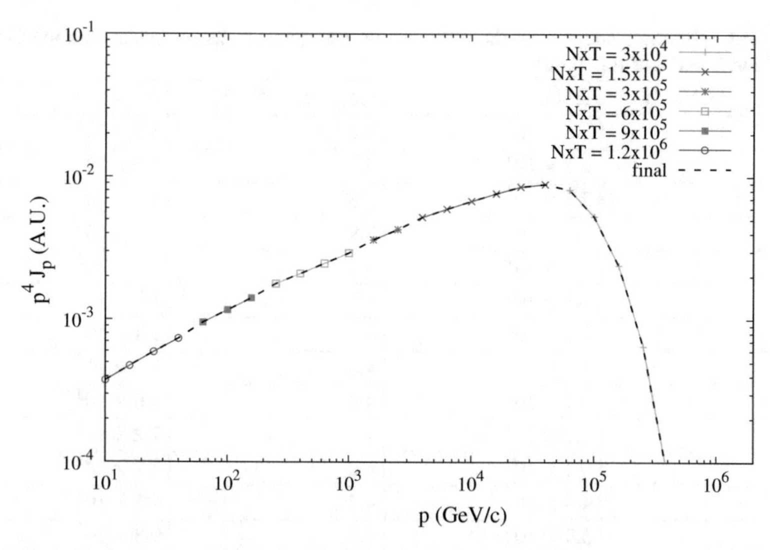

Fig. B.2 Proton spectrum obtained from the simulations with different radial step sizes. Each proton momentum is simulated with several Δr, which is then chosen as the largest value among the simulated ones reaching convergence on the proton spectrum within 5% accuracy. The colors here represent the various regime exploited in the spatial resolution, namely the lowest energy particles have been simulated with all the steps indicated in the legend. The total number of points included in the grid of the final simulation is shown in the legend (for both dimensions), while the dashed line defines the final spectrum

B.8 Stability of the Algorithm

The ADI method is a fully implicit method, and hence it results stable for whatever choice of the spatial and temporal step. However, stability conditions have to be satisfied in order to achieve convergence towards the correct solution. Thus, Courant conditions apply. In particular, both the diffusive energy-dependent Courant condition and the advective one have to be satisfied at each half time step Δt, which is therefore defined by the condition

$$\Delta t = \alpha \min(\tau_{\text{diff}}, \tau_{\text{adv}}) \tag{B.42}$$

where the diffusion timescale is defined as

$$\tau_{\text{diff}} = \frac{\Delta L^2}{D(p)} \tag{B.43}$$

while the advection timescale is

Table B.1 Time step adopted for the different particle momenta, as compliant with the Courant condition $\alpha = 0.8$ of Eq. (B.42)

$\log_{10}(p/\mathrm{GeVc}^{-1})$	Δt (s)	$\log_{10}(p/\mathrm{GeVc}^{-1})$	Δt (s)
1.0	3.5×10^4	3.6	5.7×10^3
1.1	2.7×10^4	3.7	4.5×10^3
1.2	2.2×10^4	3.8	3.6×10^3
1.3	1.7×10^4	3.9	2.8×10^3
1.4	1.4×10^4	4.0	2.3×10^3
1.5	1.1×10^4	4.1	1.8×10^3
1.6	8.7×10^3	4.2	1.4×10^3
1.7	1.2×10^4	4.3	1.1×10^3
1.8	9.7×10^3	4.4	9.0×10^2
1.9	7.7×10^3	4.5	7.2×10^2
2.0	6.2×10^3	4.6	5.7×10^2
2.1	4.9×10^3	4.7	4.5×10^2
2.2	3.9×10^3	4.8	9.5×10^3
2.3	6.9×10^3	4.9	7.6×10^3
2.4	5.5×10^3	5.0	6.0×10^3
2.5	4.4×10^3	5.1	4.8×10^3
2.6	3.5×10^3	5.2	3.8×10^3
2.7	2.8×10^3	5.3	3.0×10^3
2.8	2.2×10^3	5.4	2.4×10^3
2.9	1.7×10^3	5.5	1.9×10^3
3.0	1.4×10^3	5.6	1.5×10^3
3.1	1.1×10^3	5.7	1.2×10^3
3.2	3.5×10^3	5.8	9.5×10^2
3.3	2.8×10^3	5.9	7.6×10^2
3.4	2.2×10^3	6.0	6.0×10^2
3.5	1.8×10^3		

$$\tau_{\mathrm{adv}} = \frac{\Delta L}{v_s} \tag{B.44}$$

being ΔL the generic coordinate step (either Δr or Δz). Considering all the above, the time step is set by fixing $\alpha = 0.8$. Therefore, different time steps are then assigned to different energies, as reported in Table B.1.

Appendix C
The Escaping Particle Distribution Function

The aim of this appendix is to solve analytically the diffusion equation for non-confined particles, as presented in Eq. (3.58) of Chap. 3. Its content is an original work developed by the author, intended to guide the reader through the mathematical steps which lead to Eqs. (3.59), (3.62) and (3.69). If fact, the diffusive solution depends on the initial condition, which is provided by the density of confined particles at the escape time, which in turn depends on the acceleration spectrum. Thus the cases of acceleration spectra $\propto p^{-4}$ and $\propto p^{-(4+1/3)}$ are discussed respectively in Sects. C.1 and C.2. Moreover, the contribution of escaping particles from the precursor region is computed in Sect. C.3. It is worth to recall that the conclusions here derived strongly depend on the assumption of a remnant evolving in the ST phase, hence the formulas presented in this appendix only apply to middle-aged SNRs.

As non-confined particles are released by the shock after the escape time, the following discussion will refer to particles with momentum p in the evolutionary stage with $t > t_{\rm esc}(p)$. However, in order to facilitate the reading, in the following the term $\theta(t - t_{\rm esc}(p))$ will be dropped. By adopting the auxiliary variable $u(r,t) = rf(r,t)$, Eq. (3.58) can be solved introducing the Laplace operator

$$\bar{u}(r,s) = \int_0^\infty u(r,t)e^{-st}dt \tag{C.1}$$

in order to get rid of the time variable. Then, the integral over time is solved by parts as

$$\int_0^\infty \frac{\partial u}{\partial t} e^{-st}dt = -u_0(r) + s\int_0^\infty ue^{-st}dt = -u_0(r) + s\bar{u} \tag{C.2}$$

where $u_0(r) = rf_{\rm conf}(r, t_{\rm esc})$ represents the initial condition for diffusive escape, matching the confinement solution given in Eq. (3.48). The right hand side of Eq. (3.58) reads now as

S. Celli, *Gamma-ray and Neutrino Signatures of Galactic Cosmic-ray Accelerators*, Springer Theses, https://doi.org/10.1007/978-3-030-33124-5

$$\int_0^\infty \frac{\partial^2 u}{\partial r^2} e^{-st} dt = \frac{\partial^2}{\partial r^2} \int_0^\infty u e^{-st} dt = \frac{\partial^2 \bar{u}}{\partial r^2} \tag{C.3}$$

Thus, eventually, one obtains

$$D \frac{\partial^2 \bar{u}}{\partial r^2} = -u_0(r) + s\bar{u} \tag{C.4}$$

which can be rewritten as

$$\frac{\partial^2}{\partial r^2} \bar{u}(r, s) - \frac{s}{D} \bar{u}(r, s) = -\frac{u_0(r)}{D} \tag{C.5}$$

This is a non homogeneous second order partial differential equation: in order to solve it, one has to first find the solution $\bar{u}_h(r, s)$ of the associated homogeneous equation and then obtain the particular solution $\bar{u}_s(r, s)$, since

$$\bar{u}(r, s) = \bar{u}_\mathrm{h}(r, s) + \bar{u}_\mathrm{s}(r, s) \tag{C.6}$$

The general solution of the characteristic equation associated to the homogeneous partial differential equation is

$$\bar{u}_\mathrm{h}(r, s) = C_1 e^{\sqrt{\frac{s}{D}} r} + C_2 e^{-\sqrt{\frac{s}{D}} r} \tag{C.7}$$

where the constants C_1 and C_2 are found by imposing the boundary conditions: (i) since no particle can get to $r = \infty$, then $C_1 = 0$; (ii) when $r = 0$, $u(r) = 0$ so that $C_2 = -C_1 = 0$, providing with a null characteristic solution $\bar{u}_h(r, s) = 0$. Therefore, moving to the particular solution of Eq. (C.5), the second order differential equation is split into a system of two first order differential equations, which applies to $\bar{u}_\mathrm{s}(r, s)$ (though in the following the subscript s will be dropped), and it reads as

$$\begin{cases} \dfrac{d\bar{u}}{dr} &= -\omega\bar{u} + \gamma \\ \dfrac{d\gamma}{dr} &= \omega\gamma - \dfrac{u_0(r)}{D} \end{cases} \tag{C.8}$$

where $\omega = \sqrt{s/D}$ was defined. The solution of this system is given by

$$\begin{cases} \bar{u}(r) &= e^{-\omega r} \displaystyle\int_0^r \gamma(r') e^{\omega r'} dr' \\ \gamma(r) &= e^{\omega r} \displaystyle\int_r^\infty \frac{u_0(r')}{D} e^{-\omega r'} dr' \end{cases} \tag{C.9}$$

which can be finally written as

$$\bar{u}(r) = e^{-\omega r} \int_0^r dr' e^{2\omega r'} \int_{r'}^{\infty} dr'' \frac{u_0(r'')}{D} e^{-\omega r''} \tag{C.10}$$

Therefore, according to the radial dependence of the function u_0, a different solution for the escaping particle density function is found. The cases of $u_0(r) = \text{const}$ and $u_0(r) \propto r$ respectively refer to acceleration spectra $\propto p^{-4}$ and $p^{-(4+1/3)}$, as discussed in the next sections.

C.1 Particle Acceleration Spectrum $\propto p^{-4}$

For an acceleration spectrum as p^{-4}, the confined solution inside of the shocked region $f_{\text{conf}}(t, r, p)$ was found to be independent of r (see Eq. (3.44)). Therefore

$$f(t = 0, r) = \theta(R_s(t_{\text{esc}}) - r) f_{\text{conf}} = \theta(R_{\text{esc}} - r) f_{\text{conf}} \tag{C.11}$$

and therefore

$$u_0(r) = \theta(R_{\text{esc}} - r) r f_{\text{conf}} \tag{C.12}$$

Thus, Eq. (C.10) becomes

$$\begin{aligned} \bar{u}(r) &= \frac{1}{D} e^{-\omega r} \int_0^r dr' e^{2\omega r'} \int_{r'}^{\infty} dr'' f_{\text{conf}} r'' e^{-\omega r''} \theta(R_{\text{esc}} - r'') \\ &= \frac{1}{D} e^{-\omega r} \int_0^M dr' e^{2\omega r'} \int_{r'}^{R_{\text{esc}}} dr'' f_{\text{conf}} r'' e^{-\omega r''} \end{aligned} \tag{C.13}$$

where $M = \min(r, R_{\text{esc}})$. The physical meaning of this formula is straightforward: while the inner integral represents a source term, made by particles confined within $R_{\text{esc}}(p)$, the outer integral accounts for their diffusive propagation, smoothed by a decreasing exponential term. This integral can be solved by parts as

$$\begin{aligned} \frac{\bar{u}(r)}{f_{\text{conf}}} &= \frac{e^{-\omega r}}{\omega^2 D} \int_0^M dr' e^{2\omega r'} \left[e^{-\omega r'} \left(1 + \omega r'\right) - e^{-\omega R_{\text{esc}}} (1 + \omega R_{\text{esc}}) \right] = \\ &= \frac{e^{-\omega r}}{\omega^2 D} \left[\int_0^M dr' e^{\omega r'} \left(1 + \omega r'\right) - e^{-\omega R_{\text{esc}}} (1 + \omega R_{\text{esc}}) \int_0^M dr' e^{2\omega r'} \right] = \\ &= \frac{e^{-\omega r}}{\omega^2 D} \left[M e^{\omega M} - e^{-\omega R_{\text{esc}}} (1 + \omega R_{\text{esc}}) \frac{1}{2\omega} (e^{2\omega M} - 1) \right] = \\ &= \frac{1}{\omega^2 D} \left[M e^{\omega (M - r)} - \frac{(1 + \omega R_{\text{esc}})}{2\omega} \left(e^{\omega (2M - R_{\text{esc}} - r)} - e^{-\omega (R_{\text{esc}} + r)} \right) \right] \end{aligned} \tag{C.14}$$

and thus, by writing the complete dependency on the Laplace variable s, one gets

$$\frac{\bar{u}(r,s)}{f_{\rm conf}} = \frac{1}{s}\left[Me^{\sqrt{s/D}(M-r)} - \frac{(1+R_{\rm esc}\sqrt{s/D})}{2\sqrt{s/D}}\left(e^{\sqrt{s/D}(2M-R_{\rm esc}-r)} - e^{-\sqrt{s/D}(R_{\rm esc}+r)}\right)\right] \tag{C.15}$$

Defining the diffusion length as $R_{\rm d}(t) = 2\sqrt{D(t-t_{\rm esc}(p))}$, the inverse Laplace transform of this function reads as

$$\begin{aligned}
\frac{u(r,t)}{f_{\rm conf}} =& M\left(1+{\rm Erf}\left[\frac{M-r}{R_{\rm d}}\right]\right) - \frac{1}{2}(r+R_{\rm esc}-2M)\left({\rm Erf}\left[\frac{|r+R_{\rm esc}-2M|}{R_{\rm d}}\right]-1\right)+\\
&-\frac{1}{2}(r+R_{\rm esc}-2M)\left(\frac{R_{\rm esc}}{|r+R_{\rm esc}-2M|}{\rm Erfc}\left[\frac{|r+R_{\rm esc}-2M|}{R_{\rm d}}\right]\right)+\\
&-\frac{1}{2}(r+R_{\rm esc}-2M)\left(\frac{R_{\rm d}}{\sqrt{\pi}}\frac{1}{|r+R_{\rm s}-2M|}\exp^{-\left(\frac{r+R_{\rm esc}-2M}{R_{\rm d}}\right)^2}\right)+\\
&+\frac{1}{2}\left(\frac{R_{\rm d}}{\sqrt{\pi}}\exp^{-\left(\frac{r+R_{\rm esc}}{R_{\rm d}}\right)^2} - r{\rm Erfc}\left[\frac{r+R_{\rm esc}}{R_{\rm d}}\right]\right)
\end{aligned} \tag{C.16}$$

This expression can be rearranged into

$$\begin{aligned}
\frac{u(r,t)}{f_{\rm conf}} =& \frac{R_{\rm d}}{2\sqrt{\pi}}\left(\exp^{-\left(\frac{r+R_{\rm esc}}{R_{\rm d}}\right)^2} - \exp^{-\left(\frac{r+R_{\rm esc}-2M}{R_{\rm d}}\right)^2}\right) - M{\rm Erf}\left[\frac{r-M}{R_{\rm d}}\right]+\\
&+\frac{r}{2}{\rm Erf}\left[\frac{r+R_{\rm esc}}{R_{\rm d}}\right] + \left(M-\frac{r}{2}\right){\rm Erf}\left[\frac{r+R_{\rm esc}-2M}{R_{\rm d}}\right]
\end{aligned} \tag{C.17}$$

or more compactly into Eq. (3.59).

C.2 Particle Acceleration Spectrum $\propto p^{-(4+1/3)}$

When considering an acceleration spectrum as steep as $p^{-(4+1/3)}$, the distribution function of confined particles grows linearly with the radial position, as shown in Eq. (3.67). Therefore, starting from Eq. (C.10), with $f_{\rm conf}(r) = k(t_{\rm esc})r\theta(R_{\rm esc}-r)$ and

$$k(t_{\rm esc}) = \frac{3\xi_{\rm CR}\rho_{\rm up}}{25\pi c(m_{\rm p}c)^{4-\alpha}\Lambda}\left(\frac{\xi_0 E_{\rm SN}}{\rho_{\rm up}}\right)^{1/5} t_{\rm esc}^{-8/5} \tag{C.18}$$

it holds that $u_0(r) = k(t_{\rm esc})r^2\theta(R_{\rm esc}-r)$ and therefore the Laplace transform solution of Eq. (C.5) reads as

$$
\begin{aligned}
\frac{\bar{u}(s,r,p)}{k(t_{esc})} &= \frac{1}{D} e^{-\omega r} \int_0^M dr' e^{2\omega r'} \int_{r'}^{R_{esc}} dr'' \, (r'')^2 e^{-\omega r''} = \\
&= \frac{1}{\omega^3 D} e^{-\omega r} \int_0^M dr' \Big[-e^{\omega(2r' - R_{esc})} (2 + 2\omega R_{esc} + \omega^2 R_{esc}^2) + \\
&\quad + e^{\omega r'} (2 + 2\omega r' + \omega^2 (r')^2) \Big] = \\
&= \frac{1}{\omega^3 D} e^{-\omega r} \left[-\frac{2}{\omega} + e^{\omega M} \left(\frac{2}{\omega} + \omega M^2 \right) + e^{-\omega R_{esc}} \left(\frac{1}{\omega} + \frac{\omega}{2} R_{esc}^2 + R_{esc} \right) + \right. \\
&\quad \left. + e^{\omega(2M - R_{esc})} \left(-\frac{1}{\omega} - \frac{\omega}{2} R_{esc}^2 - R_{esc} \right) \right]
\end{aligned}
\tag{C.19}
$$

where $\omega = \sqrt{s/D}$. By applying the anti-transform Laplace operator, the final solution is

$$
\begin{aligned}
\frac{u(t,r,p)}{k(t_{esc})} &= 2M^2 - 2Mr + \frac{R_d}{\sqrt{\pi}} \left[r e^{-\frac{r^2}{R_d^2}} + (M - r) e^{-\frac{(M-r)^2}{R_d^2}} \right] + \frac{R_d}{2\sqrt{\pi}} (R_{esc} - r) e^{-\frac{(r+R_{esc})^2}{R_d^2}} + \\
&\quad + \frac{R_d}{\sqrt{\pi}} (r - 2M + R_{esc}) e^{-\frac{(r-2M+R_{esc})^2}{R_d^2}} \left[\frac{1}{2} - \frac{R_{esc}}{|r - 2M + R_{esc}|} \right] + \\
&\quad + \left(r^2 + \frac{R_d^2}{2} \right) \mathrm{Erf} \left[\frac{r}{R_d} \right] + \left(2M^2 - 2Mr + r^2 + \frac{R_d^2}{2} \right) \mathrm{Erf} \left[\frac{M - r}{R_d} \right] + \\
&\quad + \frac{1}{2} \left(r^2 + \frac{R_d^2}{2} \right) \mathrm{Erfc} \left[\frac{r + R_{esc}}{R_d} \right] - \frac{1}{2(r - 2M + R_{esc})} \cdot \\
&\quad \cdot \left(4M^2 + r^2 + 2r R_{esc} + R_{esc}^2 - 4M(r + R_{esc}) + \frac{R_d^2}{2} \right) \cdot \\
&\quad \cdot \left(|2M - r - R_{esc}| - (2M - r - R_{esc}) \mathrm{Erf} \left[\frac{2M - r - R_{esc}}{R_d} \right] \right) + \\
&\quad + R_{esc} \mathrm{Erfc} \left[\frac{|r - 2M + R_{esc}|}{R_d} \right] (r - 2M + R_{esc}) \left(1 - \frac{R_{esc}}{2} \frac{1}{|r - 2M + R_{esc}|} \right)
\end{aligned}
\tag{C.20}
$$

From this expression, the non-confined particle distribution function can simply be compactly written as in Eq. (3.69).

C.3 The Shock Precursor Contribution

In order to account for particles escaping from the precursor, one should compute the Laplace transform of the precursor density function, by adopting the same approach described in Eq. (C.10). As already mentioned in Sect. 3.3.3, in order to simplify the computation, the particles contained in the precursor are schematically left free to expand from the escape radius (see Eq. (3.61)), without accounting for the exact radial shape of the precursor itself. Within such an approximation, it holds that

$$u_{\rm p,0}(r) = f_0(p,t)\frac{D_{\rm p}(p)}{v_{\rm s}(t_{\rm esc})}\delta(r - R_{\rm s})r \tag{C.21}$$

where $D_{\rm p}(p)$ is the diffusion coefficient at the shock transition region, and therefore the precursor Laplace transform reads as

$$\bar{u}_{\rm p}(r) = \frac{D_{\rm p}(p)}{D(p)}\frac{1}{v_{\rm s}(t_{\rm esc})}e^{-\omega r}\int_0^r dr' e^{2\omega r'}\int_{r'}^{\infty} dr'' f_0(p,t_{\rm esc})\, r'' e^{-\omega r''}\delta(r'' - R_{\rm esc}) \tag{C.22}$$

where instead $D(p)$ is the Kolmogorov-like diffusion coefficient in the ISM (which is also included in ω). The solution of this integral is

$$\begin{aligned} \bar{u}_{\rm p}(r) &= \frac{D_{\rm p}(p)}{D(p)}\frac{f_0(p,t_{\rm esc})}{v_{\rm s}(t_{\rm esc})}e^{-\omega r}\int_0^M dr' e^{2\omega r'} R_{\rm esc} e^{-\omega R_{\rm esc}} \\ &= f_0(p,t_{\rm esc})\frac{D_{\rm p}(p)}{D(p)}\frac{R_{\rm esc}}{v_{\rm s}(t_{\rm esc})}\frac{1}{2\omega}e^{-\omega(r+R_{\rm esc})}(e^{2\omega M} - 1) \end{aligned} \tag{C.23}$$

and finally, computing the inverse Laplace transform of such a function, one gets

$$u_{\rm p}(r,t) = \frac{f_0(p,t_{\rm esc})}{\sqrt{\pi}}\frac{R_{\rm esc}}{R_{\rm d}}\frac{D_{\rm p}(p)}{v_{\rm s}(t_{\rm esc})}\left[e^{-(\frac{r+R_{\rm esc}-2M}{R_{\rm d}})^2}{\rm Sign}(r + R_{\rm esc} - 2M) - e^{-(\frac{r+R_{\rm esc}}{R_{\rm d}})^2}\right] \tag{C.24}$$

References

1. Aab A et al (2017) Observation of a large-scale anisotropy in the arrival directions of cosmic rays above 8×10^{18} eV. Science 357:1266–1270. https://doi.org/10.1126/science.aan4338
2. Aartsen MG et al (2013a) Evidence for high-energy extraterrestrial neutrinos at the IceCube detector. Science 342:1242856. https://doi.org/10.1126/science.1242856
3. Aartsen MG et al (2013b) Search for time-independent neutrino emission from astrophysical sources with 3 yr of IceCube data. Astrophys J 779:132. https://doi.org/10.1088/0004-637X/779/2/132
4. Aartsen MG et al (2014) Searches for extended and point-like neutrino sources with four years of IceCube data. Astrophys J 796:109. https://doi.org/10.1088/0004-637X/796/2/109
5. Aartsen MG et al (2015) A Combined maximum-likelihood analysis of the high-energy astrophysical neutrino flux measured with IceCube. Astrophys J 809:98. https://doi.org/10.1088/0004-637X/809/1/98August
6. Aartsen MG et al (2016) Lowering IceCube's energy threshold for point source searches in the southern sky. Astrophys J Lett 824:L28. https://doi.org/10.3847/2041-8205/824/2/L28
7. Aartsen MG et al (2017a) All-sky search for time-integrated neutrino emission from astrophysical sources with 7 yr of IceCube data. Astrophys J 835(2):151. https://doi.org/10.3847/1538-4357/835/2/151
8. Aartsen MG et al (2017b) Measurement of the ν_μ energy spectrum with IceCube-79. Eur Phys J C 77:692. https://doi.org/10.1140/epjc/s10052-017-5261-3
9. Aartsen MG et al (2017c) IceCube-Gen2: a vision for the future of neutrino astronomy in Antarctica. PoS, FRAPWS2016:004
10. Aartsen MG et al (2018) Multimessenger observations of a flaring blazar coincident with high-energy neutrino IceCube-170922A. Science 361:eaat1378. https://doi.org/10.1126/science.aat1378
11. Abdalla H et al (2018a) Characterising the VHE diffuse emission in the central 200 parsecs of our galaxy with H.E.S.S. Astron Astrophys 612:A9. https://doi.org/10.1051/0004-6361/201730824
12. Abdalla H et al (2018b) H.E.S.S. observations of RX J1713.7-3946 with improved angular and spectral resolution: evidence for gamma-ray emission extending beyond the X-ray emitting shell. Astron Astrophys 612:A6. https://doi.org/10.1051/0004-6361/201629790
13. Abdo AA et al (2010) Observation of supernova remnant IC 443 with the fermi large area telescope. Astrophys J 712:459–468. https://doi.org/10.1088/0004-637X/712/1/459
14. Abdo AA et al (2009) Fermi LAT discovery of extended gamma-ray emission in the direction of supernova remnant W51C. Astrophys J Lett 706:L1–L6. https://doi.org/10.1088/0004-637X/706/1/L1
15. Abdo AA et al (2010) Fermi large area telescope observations of the supernova remnant W28 (G6.4-0.1). Astrophys J 718:348–356. https://doi.org/10.1088/0004-637X/718/1/348

S. Celli, *Gamma-ray and Neutrino Signatures of Galactic Cosmic-ray Accelerators*,
Springer Theses, https://doi.org/10.1007/978-3-030-33124-5

16. Abdo AA et al (2011) Observations of the young supernova remnant RX J1713.7-3946 with the fermi large area telescope. Astrophy J 734:28. https://doi.org/10.1088/0004-637X/734/1/28
17. Abeysekara AU et al (2013) Sensitivity of the high altitude water Cherenkov detector to sources of multi-TeV gamma rays. Astropart Phys 50–52:26–32. https://doi.org/10.1016/j.astropartphys.2013.08.002
18. Abeysekara AU et al (2017) The 2HWC HAWC observatory gamma ray catalog. ArXiv:1702.02992
19. Abramowski A et al (2016) Acceleration of petaelectronvolt protons in the galactic centre. Nature 531:476. https://doi.org/10.1038/nature17147
20. Acciari VA et al (2009) Observation of extended very high energy emission from the supernova remnant IC 443 with VERITAS. Astrophy J Lett 698:L133–L137. https://doi.org/10.1088/0004-637X/698/2/L133June
21. Acero F et al (2017) Prospects for Cherenkov telescope array observations of the young supernova remnant RX J1713.7-3946. Astrophys J 840:74. https://doi.org/10.3847/1538-4357/aa6d67
22. Acharya BS et al (2017) Science with the Cherenkov telescope array. ArXiv:1709.07997
23. Ackermann M et al (2013) Detection of the characteristic pion-decay signature in supernova remnants. Science 339:807–811. https://doi.org/10.1126/science.1231160
24. Actis M et al (2011) Design concepts for the Cherenkov telescope array CTA: an advanced facility for ground-based high-energy gamma-ray astronomy. Exp Astron 32:193–316. https://doi.org/10.1007/s10686-011-9247-0
25. Adrián-Martínez S et al (2012) Search for cosmic neutrino point sources with four years of data from the ANTARES telescope. Astrophys J 760:53. https://doi.org/10.1088/0004-637X/760/1/53
26. Adrian-Martinez S et al (2014) Searches for point-like and extended neutrino sources close to the galactic centre using the ANTARES neutrino telescope. Astrophys J 786:L5. https://doi.org/10.1088/2041-8205/786/1/L5
27. Adrian-Martinez S et al (2016a) Letter of intent for KM3NeT 2.0. J Phys G43(8):084001. https://doi.org/10.1088/0954-3899/43/8/084001
28. Adrian-Martinez S et al (2016b) Constraints on the neutrino emission from the galactic ridge with the ANTARES telescope. Phys Lett B 760:143–148. https://doi.org/10.1016/j.physletb.2016.06.051
29. Ageron M et al (2011) ANTARES: the first undersea neutrino telescope. Nucl Instrum Methods Phys A656:11–38. https://doi.org/10.1016/j.nima.2011.06.103
30. Aharonian F, Bergström L, Dermer C (2013) Astrophysics at very high energies. Astrophysics at very high energies: saas-fee advanced course 40. Swiss society for astrophysics and astronomy. Saas-fee advanced course, vol 40. Springer, Berlin. https://doi.org/10.1007/978-3-642-36134-0 (ISBN 978-3-642-36133-3)
31. Aharonian FA (2004) Very high energy cosmic gamma radiation: a crucial window on the extreme universe. World Scientific Publishing Co, Singapore. https://doi.org/10.1142/4657
32. Aharonian FA, Atoyan AM (1999) On the origin of TeV radiation of SN 1006. Astron Astrophys 351:330–340
33. Aharonian FA, Hofmann W, Konopelko AK, Völk HJ (1997) The potential of ground based arrays of imaging atmospheric Cherenkov telescopes. I. Determination of shower parameters. Astropart Phys 6:343–368. https://doi.org/10.1016/S0927-6505(96)00069-2
34. Aharonian FA, Buckley J, Kifune T, Sinnis G (2008) High energy astrophysics with ground-based gamma ray detectors. Rep Prog Phys 71:096901. https://doi.org/10.1088/0034-4885/71/9/096901
35. Aharonian F, Peron G, Yang R, Casanova S, Zanin R (2018) Probing the "sea" of galactic cosmic rays with fermi-LAT. arXiv:1811.12118
36. Aharonian F, Yang R, de Oña Wilhelmi E (2019) Massive stars as major factories of galactic cosmic rays. Nat Astron, p 232. https://doi.org/10.1038/s41550-019-0724-0

37. Aharonian F et al (2004) The Crab nebula and pulsar between 500 GeV and 80 TeV: observations with the HEGRA stereoscopic air Cerenkov telescopes. Astrophys J 614:897–913. https://doi.org/10.1086/423931
38. Aharonian FA et al (2005) Very high energy gamma rays from the composite SNR G 0.9+0.1. Astron Astrophys 432:L25–L29. https://doi.org/10.1051/0004-6361:200500022
39. Aharonian FA et al (2006) Discovery of very-high-energy gamma-rays from the galactic centre ridge. Nature 439:695–698. https://doi.org/10.1038/nature04467
40. Aharonian FA et al (2007) Primary particle acceleration above 100 TeV in the shell-type supernova remnant RX J1713.7-3946 with deep H.E.S.S. observations. Astron Astrophys 464:235–243. https://doi.org/10.1051/0004-6361:20066381
41. Aharonian FA et al (2008) Discovery of very high energy gamma-ray emission coincident with molecular clouds in the W 28 (G6.4-0.1) field. Astron Astrophys 481:401–410. https://doi.org/10.1051/0004-6361:20077765
42. Aharonian FA et al (2009) Spectrum and variability of the Galactic center VHE γ-ray source HESS J1745–290. Astron Astrophys 503:817–825. https://doi.org/10.1051/0004-6361/200811569
43. Ahlers M (2016) Multi-messenger aspects of cosmic neutrinos. In: European physical journal web of conferences, vol 116, p 11001. https://doi.org/10.1051/epjconf/201611611001
44. Ahlers M, Halzen F (2018) Opening a new window onto the universe with IceCube. Prog Part Nucl Phys 102:73–88. https://doi.org/10.1016/j.ppnp.2018.05.001
45. Aiello S et al (2018) Sensitivity of the KM3NeT/ARCA neutrino telescope to point-like neutrino sources
46. Albert A et al (2017a) New constraints on all flavor galactic diffuse neutrino emission with the ANTARES telescope. Phys Rev D 96(6):062001. https://doi.org/10.1103/PhysRevD.96.062001
47. Albert A et al (2017b) First all-flavor neutrino point-like source search with the ANTARES neutrino telescope. Phys Rev D 96(8):082001. https://doi.org/10.1103/PhysRevD.96.082001
48. Albert J et al (2007) Discovery of very high energy gamma radiation from IC 443 with the MAGIC telescope. Astrophys J Lett 664:L87–L90. https://doi.org/10.1086/520957
49. Aleksić J et al (2012) Morphological and spectral properties of the W51 region measured with the MAGIC telescopes. Astron Astrophys 541:A13. https://doi.org/10.1051/0004-6361/201218846
50. Aleksić J et al (2016) The major upgrade of the MAGIC telescopes, part II: A performance study using observations of the Crab Nebula. Astropart Phys 72:76–94. https://doi.org/10.1016/j.astropartphys.2015.02.005
51. Allison P et al (2018) Design and performance of an interferometric trigger array for radio detection of high-energy neutrinos. arXiv:1809.04573
52. Alvarez-Muniz J, Halzen F (2002) Possible high-energy neutrinos from the cosmic accelerator RX J1713.7-3946. Astrophys J 576:L33–L36. https://doi.org/10.1086/342978
53. Ambrogi L, De Oña Wilhelmi E, Aharonian FA (2016) On the potential of atmospheric Cherenkov telescope arrays for resolving TeV gamma-ray sources in the galactic plane. Astropart Phys 80:22–33. https://doi.org/10.1016/j.astropartphys.2016.03.004
54. Ambrogi L, Celli S, Aharonian F (2018) On the potential of Cherenkov telescope arrays and KM3 neutrino telescopes for the detection of extended sources. Astropart Phys 100:69–79. https://doi.org/10.1016/j.astropartphys.2018.03.001
55. Avrorin AD et al (2016) Baikal-GVD: results, status and plans. In: European physical journal web of conferences, vol 116, p 11005. https://doi.org/10.1051/epjconf/201611611005
56. Axford WI, Leer E, Skadron G (1977) The acceleration of cosmic rays by shock waves. International cosmic ray conference 11:132–137
57. Baade W, Zwicky F (1934) Cosmic rays from super-novae. Proc Natl Acad Sci 20:259–263. https://doi.org/10.1073/pnas.20.5.259
58. Bell AR (1978) The acceleration of cosmic rays in shock fronts. I. Mon Not R Astron Soc 182:147–156. https://doi.org/10.1093/mnras/182.2.147

59. Bell AR (2004) Turbulent amplification of magnetic field and diffusive shock acceleration of cosmic rays. Mon Not R Astron Soc 353:550–558. https://doi.org/10.1111/j.1365-2966.2004.08097.x
60. Bell AR, Schure KM, Reville B (2011) Cosmic ray acceleration at oblique shocks. Mon Not R Astron Soc 418:1208–1216. https://doi.org/10.1111/j.1365-2966.2011.19571.x
61. Bell AR, Schure KM, Reville B, Giacinti G (2013) Cosmic-ray acceleration and escape from supernova remnants. Mon Not R Astron Soc 431:415–429. https://doi.org/10.1093/mnras/stt179
62. Berezhko EG, Ksenofontov LT, Völk HJ (2013) The nature of gamma-ray emission of Tycho's supernova remnant. Astrophys J 763:14. https://doi.org/10.1088/0004-637X/763/1/14
63. Berezinskii VS, Ginzburg VL (1981) On high-energy neutrino radiation of quasars and active galactic nuclei. Mon Not R Astron Soc 194:3–14. https://doi.org/10.1093/mnras/194.1.3
64. Berezinskii VS, Bulanov SV, Dogiel VA, Ptuskin VS (1990) Astrophysics of cosmic rays
65. Berezinsky V (2014) Extragalactic cosmic rays and their signatures. Astropart Phys 53:120–129. https://doi.org/10.1016/j.astropartphys.2013.04.001
66. Bernlöhr K et al (2013) Monte Carlo design studies for the Cherenkov telescope array. Astropart Phys 43:171–188. https://doi.org/10.1016/j.astropartphys.2012.10.002
67. Black JH, Fazio GG (1973) Production of gamma radiation in dense interstellar clouds by cosmic-ray interactions. Astrophys J Lett 185:L7. https://doi.org/10.1086/181310
68. Blandford R, Eichler D (1987) Particle acceleration at astrophysical shocks: a theory of cosmic ray origin. Phys Rep 154:1–75. https://doi.org/10.1016/0370-1573(87)90134-7
69. Blandford RD, Cowie LL (1982) Radio emission from supernova remnants in a cloudy interstellar medium. Astrophys J 260:625–634. https://doi.org/10.1086/160284
70. Blandford RD, Ostriker JP (1978) Particle acceleration by astrophysical shocks. Astrophys J 221:L29–L32. https://doi.org/10.1086/182658
71. Blasi P (2013) The origin of galactic cosmic rays. Astron Astrophys Rev 21:70. https://doi.org/10.1007/s00159-013-0070-7
72. Breit G, Wheeler JA (1934) Collision of two light quanta. Phys Rev 46:1087–1091. https://doi.org/10.1103/PhysRev.46.1087
73. Caprioli D, Blasi P, Amato E (2009a) On the escape of particles from cosmic ray modified shocks. Mon Not R Astron Soc 396:2065–2073. https://doi.org/10.1111/j.1365-2966.2008.14298.x
74. Caprioli D, Blasi P, Amato E, Vietri M (2009b) Dynamical feedback of self-generated magnetic fields in cosmic ray modified shocks. Mon Not R Astron Soc 395:895–906. https://doi.org/10.1111/j.1365-2966.2009.14570.x
75. Cardillo M, Amato E, Blasi P (2015) On the cosmic ray spectrum from type II supernovae expanding in their red giant presupernova wind. Astropart Phys 69:1–10. https://doi.org/10.1016/j.astropartphys.2015.03.002
76. Casanova S, Jones DI, Aharonian FA, Fukui Y, Gabici S, Kawamura A, Onishi T, Rowell G, Sano H, Torii K, Yamamoto H (2010) Modeling the gamma-ray emission produced by runaway cosmic rays in the environment of RX J1713.7-3946. Publ Astron Soc Jpn 62:1127–1134. https://doi.org/10.1093/pasj/62.5.1127
77. Casse F, Lemoine M, Pelletier G (2002) Transport of cosmic rays in chaotic magnetic fields. Phys Rev D 65(2):023002. https://doi.org/10.1103/PhysRevD.65.023002
78. Casse M, Paul JA (1980) Local gamma rays and cosmic-ray acceleration by supersonic stellar winds. Astrophys J 237:236–243. https://doi.org/10.1086/157863
79. Celli S, Morlino G, Gabici S, Aharonian FA (2019a) Supernova remnants in clumpy media: particle propagation and gamma-ray emission. Mon Not R Astron Soc 487(3):3199–3213. https://doi.org/10.1093/mnras/stz1425
80. Celli S, Palladino A, Vissani F (2017) Neutrinos and γ-rays from the galactic center region after H.E.S.S. multi-TeV measurements. Eur Phys J C 77(2):66. https://doi.org/10.1140/epjc/s10052-017-4635-x
81. Celli S, Morlino G, Gabici S, Aharonian F (2019b) Exploring particle escape in supernova remnants through gamma rays. Mon Not R Astron Soc 490(3):4317–4333. https://doi.org/10.1093/mnras/stz2897

82. Cesarsky CJ, Montmerle T (1983) Gamma rays from active regions in the galaxy - the possible contribution of stellar winds. Space Sci Rev 36:173–193. https://doi.org/10.1007/BF00167503
83. Chalmé-Calvet R, Holler M, de Naurois M, Tavernet JP (2015) Exploiting the time of arrival of Cherenkov photons at the 28 m H.E.S.S. telescope for background rejection: methods and performance. ArXiv:1509.03544
84. Chevalier RA (1982) Self-similar solutions for the interaction of stellar ejecta with an external medium. Astrophys J 258:790–797. https://doi.org/10.1086/160126
85. Chevalier RA (1999) Supernova remnants in molecular clouds. Astrophys J 511:798–811. https://doi.org/10.1086/306710
86. Cirelli M, Panci P (2009) Inverse compton constraints on the dark matter e excesses. Nucl Phys B 821:399–416. https://doi.org/10.1016/j.nuclphysb.2009.06.034
87. Costantini ML, Vissani F (2005) Expected neutrino signal from supernova remnant RX J1713.7-3946 and flavor oscillations. Astropart Phys 23:477–485. https://doi.org/10.1016/j.astropartphys.2005.03.003
88. Costantini ML, Vissani F (2005) Neutrinos from supernovas and supernova remnants. In: Tricomi A, Albergo S, Chiorboli M (eds) IFAE 2005: 17th italian meeting on high energy physics. American institute of physics conference series, vol 794, pp 219–223. https://doi.org/10.1063/1.2125656
89. Cowie LL, McKee CF (1977) The evaporation of spherical clouds in a hot gas. I - classical and saturated mass loss rates. Astrophys J 211:135–146. https://doi.org/10.1086/154911
90. Cox P, Laureijs R (1989) Iras observations of the galactic center. Symposium - international astronomical union 136:121–128. https://doi.org/10.1017/S0074180900186401
91. Cristofari P, Gabici S, Casanova S, Terrier R, Parizot E (2013) Acceleration of cosmic rays and gamma-ray emission from supernova remnants in the galaxy. Mon Not R Astron Soc 434:2748–2760. https://doi.org/10.1093/mnras/stt1096
92. Crocker RM, Aharonian FA (2011) Fermi bubbles: giant, multibillion-year-old reservoirs of galactic center cosmic rays. Phys Rev Lett 106(10):101102. https://doi.org/10.1103/PhysRevLett.106.101102
93. D'Angelo M, Morlino G, Amato E, Blasi P (2018) Diffuse gamma-ray emission from self-confined cosmic rays around galactic sources. Mon Not R Astron Soc 474: https://doi.org/10.1093/mnras/stx2828
94. Dorman LI (1979) Pumping of energy from stellar winds to cosmic rays. In: International cosmic ray conference, vol 2, p 49
95. Draine BT, Roberge WG (1984) CO line emission from shock waves in molecular clouds. Astrophys J 282:491–507. https://doi.org/10.1086/162227
96. Drury L (1983a) On particle acceleration in supernova remnants. Space Sci Rev 36:57–60. https://doi.org/10.1007/BF00171901
97. Drury LO (1983b) An introduction to the theory of diffusive shock acceleration of energetic particles in tenuous plasmas. Rep Prog Phys 46:973–1027. https://doi.org/10.1088/0034-4885/46/8/002
98. Drury LO (2011) Escaping the accelerator: how, when and in what numbers do cosmic rays get out of supernova remnants? Mon Not R Astron Soc 415:1807–1814. https://doi.org/10.1111/j.1365-2966.2011.18824.x
99. Drury LO, Downes TP (2012) Turbulent magnetic field amplification driven by cosmic ray pressure gradients. Mon Not R Astron Soc 427:2308–2313. https://doi.org/10.1111/j.1365-2966.2012.22106.x
100. Drury LO, Duffy P, Kirk JG (1996) Limits on diffusive shock acceleration in dense and incompletely ionised media. Astron Astrophys 309:1002–1010
101. Dumas G, Vaupré S, Ceccarelli C, Hily-Blant P, Dubus G, Montmerle T, Gabici S (2014) Localized SiO emission triggered by the passage of the W51C supernova remnant shock. Astrophys J Lett 786:L24. https://doi.org/10.1088/2041-8205/786/2/L24
102. Dwarkadas VV (2007) The evolution of supernovae in circumstellar wind bubbles. II. case of a Wolf-Rayet star. Astrophys J 667:226–247. https://doi.org/10.1086/520670

103. Ellison DC, Patnaude DJ, Slane P, Raymond J (2010) Efficient cosmic ray acceleration, hydrodynamics, and self-consistent thermal X-ray emission applied to supernova remnant RX J1713.7-3946. Astrophys J 712:287–293. https://doi.org/10.1088/0004-637X/712/1/287
104. Enberg R, Reno MH, Sarcevic I (2008) Prompt neutrino fluxes from atmospheric charm. Phys Rev D 78:043005. https://doi.org/10.1103/PhysRevD.78.043005
105. Esmaili A, Serpico P (2015) Gamma-ray bounds from EAS detectors and heavy decaying dark matter constraints. J Cosmol Astropart Phys 10:014. https://doi.org/10.1088/1475-7516/2015/10/014
106. Evoli C, Linden T, Morlino G (2018) Self-generated cosmic-ray confinement in TeV halos: implications for TeV gamma-ray emission and the positron excess. Phys Rev D 98:063017. arXiv:1807.09263
107. Fermi E (1949) On the origin of the cosmic radiation. Phys Rev 75:1169–1174. https://doi.org/10.1103/PhysRev.75.1169
108. Fermi E (1954) Galactic magnetic fields and the origin of cosmic radiation. Astrophys J 119:1–6. https://doi.org/10.1086/145789
109. Fiasson A, Marandon V, Chaves RCG, Tibolla O (2009) Discovery of a VHE gamma-ray source in the W51 region. In: 31th International cosmic ray conference (ICRC2009)
110. Fraschetti F (2013) Turbulent amplification of a magnetic field driven by the dynamo effect at rippled shocks. Astrophys J 770:84. https://doi.org/10.1088/0004-637X/770/2/84
111. Freitag M, Amaro-Seoane P, Kalogera V (2006) Stellar remnants in galactic nuclei: mass segregation. Astrophys J 649:91–117. https://doi.org/10.1086/506193
112. Fujita Y, Kimura SS, Murase K (2015) Hadronic origin of multi-TeV gamma rays and neutrinos from low-luminosity active galactic nuclei: implications of past activities of the galactic center. Phys Rev D 92(2):023001. https://doi.org/10.1103/PhysRevD.92.023001
113. Fukuda Y et al (1998) Evidence for oscillation of atmospheric neutrinos. Phys Rev Lett 81:1562–1567. https://doi.org/10.1103/PhysRevLett.81.1562
114. Fukui Y, Moriguchi Y, Tamura K, Yamamoto H, Tawara Y, Mizuno N, Onishi T, Mizuno A, Uchiyama Y, Hiraga J, Takahashi T, Yamashita K, Ikeuchi S (2003) Discovery of interacting molecular gas toward the TeV gamma-ray peak of the SNR G 347.3-0.5. Publ Astron Soc Jpn 55:L61–L64. https://doi.org/10.1093/pasj/55.5.L61
115. Fukui Y, Sano H, Sato J, Torii K, Horachi H, Hayakawa T, McClure-Griffiths NM, Rowell G, Inoue T, Inutsuka S, Kawamura A, Yamamoto H, Okuda T, Mizuno N, Onishi T, Mizuno A, Ogawa H (2012) A study of the molecular and atomic gas toward the γ-ray supernova remnant RX J1713.7-3946: spatial TeV γ-ray and interstellar medium gas correspondence. Astrophys J 746:82. https://doi.org/10.1088/0004-637X/746/1/82
116. Gabici S (2011) Cosmic ray escape from supernova remnants. Mem Della Società Astron Ital 82:760
117. Gabici S (2013) Cosmic rays and molecular clouds. In: Torres DF, Reimer O (eds) Cosmic rays in star-forming environments. Astrophysics and space science proceedings, vol 34, p 221. https://doi.org/10.1007/978-3-642-35410-6_16
118. Gabici S, Aharonian FA (2014) Hadronic gamma-rays from RX J1713.7-3946? Mon Not R Astron Soc 445:L70–L73. https://doi.org/10.1093/mnrasl/slu132
119. Gabici S, Aharonian FA (2016) Gamma-ray emission from young supernova remnants: hadronic or leptonic? In: European physical journal web of conferences, vol 121, p 04001. https://doi.org/10.1051/epjconf/201612104001
120. Gabici S, Montmerle T (2015) On the connection of gamma rays from supernova remnants interacting with molecular clouds and cosmic ray ionization measured. In: 34th International cosmic ray conference (ICRC2015), vol 34, p 29
121. Gabici S, Aharonian FA, Casanova S (2009) Broad-band non-thermal emission from molecular clouds illuminated by cosmic rays from nearby supernova remnants. Mon Not R Astron Soc 396:1629–1639. https://doi.org/10.1111/j.1365-2966.2009.14832.x
122. Gabici S, Casanova S, Aharonian FA, Rowell G (2010) Constraints on the cosmic ray diffusion coefficient in the W28 region from gamma-ray observations. In: Boissier S, Heydari-Malayeri M, Samadi R, Valls-Gabaud D (eds) SF2A-2010: proceedings of the annual meeting of the French society of astronomy and astrophysics, p 313

123. Gaisser T, Albrecht K (2017) Neutrino astronomy: current status, future prospects. World Scientific Publishing Company, Singapore. https://books.google.it/books?id=meqtDgAAQBAJ. ISBN 9789814759427
124. Gaisser TK, Protheroe RJ, Stanev T (1998) Gamma-ray production in supernova remnants. Astrophys J 492:219–227. https://doi.org/10.1086/305011
125. Gaisser TK, Engel R, Resconi E (2016) Cosmic rays and particle physics, 2nd edn. Cambridge University Press, Cambridge. https://doi.org/10.1017/CBO9781139192194
126. Genzel R, Eisenhauer F, Gillessen S (2010) The galactic center massive black hole and nuclear star cluster. Rev Mod Phys 82:3121–3195. https://doi.org/10.1103/RevModPhys.82.3121
127. Giacalone J, Jokipii JR (2007) Magnetic field amplification by shocks in turbulent fluids. Astrophys J Lett 663:L41–L44. https://doi.org/10.1086/519994
128. Ginzburg VL, Syrovatsky SI (1961) Origin of cosmic rays. Prog Theor Phys Suppl 20:1–83. https://doi.org/10.1143/PTPS.20.1
129. Green DA (2017) Galactic SNRs: summary data
130. Greisen K (1966) End to the cosmic-ray spectrum? Phys Rev Lett 16:748–750. https://doi.org/10.1103/PhysRevLett.16.748
131. Gritschneder M, Naab T, Walch S, Burkert A, Heitsch F (2009) Driving turbulence and triggering star formation by ionizing radiation. Astrophys J Lett 694:L26–L30. https://doi.org/10.1088/0004-637X/694/1/L26
132. Gusdorf A, Cabrit S, Flower DR, Des Forêts GP (2008a) SiO line emission from C-type shock waves: interstellar jets and outflows. Astron Astrophys 482:809–829. https://doi.org/10.1051/0004-6361:20078900
133. Gusdorf A, Des Forêts GP, Cabrit S, Flower DR (2008b) SiO line emission from interstellar jets and outflows: silicon-containing mantles and non-stationary shock waves. Astron Astrophys 490:695–706. https://doi.org/10.1051/0004-6361:200810443
134. Hassan T, Arrabito L, Bernlör K, Bregeon J, Hinton J, Jogler T, Maier G, Moralejo A, Di Pierro F, Wood M (2016) Second large-scale Monte Carlo study for the Cherenkov telescope array. PoS, ICRC2015:971
135. Hennebelle P (2013) Theory of molecular cloud formation through colliding flows: successes and limits. In: Kawabe R, Kuno N, Yamamoto S (eds) New trends in radio astronomy in the ALMA era: the 30th anniversary of Nobeyama radio observatory. Astronomical society of the Pacific conference series, vol 476, p 115
136. Hewitt JW (2015) Resolving the hadronic accelerator IC 443: a joint study with fermi-LAT and VERITAS. https://fermi.gsfc.nasa.gov/science/mtgs/symposia/2015/program/wednesday/session9/JHewitt.pdf
137. Hillas AM (1984) The origin of ultrahigh-energy cosmic rays. Ann Rev Astron Astrophys 22:425–444. https://doi.org/10.1146/annurev.aa.22.090184.002233
138. Hinton JA, Hofmann W (2009) Teraelectronvolt astronomy. Ann Rev. Astron Astrophys 47:523–565. https://doi.org/10.1146/annurev-astro-082708-101816
139. Holder J (2017) Latest results from VERITAS: gamma 2016. AIP Conf. Proc. 1792(1):020013. https://doi.org/10.1063/1.4968898
140. Honda M (2016) Atmospheric neutrino flux calculation with NRLMSISE-00 atmosphere model and new cosmic ray observations. JPS Conf. Proc. 12:010008. https://doi.org/10.7566/JPSCP.12.010008
141. Hörandel JR (2008) Cosmic-ray composition and its relation to shock acceleration by supernova remnants. Adv Space Res 41:442–463. https://doi.org/10.1016/j.asr.2007.06.008
142. Humensky B et al (2015) The TeV morphology of the interacting supernova remnant IC 443. In: 34th International cosmic ray conference (ICRC2015)
143. Inoue T, Yamazaki R, Inutsuka S-I, Fukui Y (2012) Toward understanding the cosmic-ray acceleration at young supernova remnants interacting with interstellar clouds: possible applications to RX J1713.7-3946. Astrophys J 744:71. https://doi.org/10.1088/0004-637X/744/1/71
144. Jogler T, Funk S (2016) Revealing W51C as a cosmic ray source using fermi-LAT data. Astrophys J 816:100. https://doi.org/10.3847/0004-637X/816/2/100

145. Jones TW, Ryu D, Tregillis IL (1996) The magnetohydrodynamics of supersonic gas clouds: MHD cosmic bullets and wind-swept clumps. Astrophys J 473:365. https://doi.org/10.1086/178151
146. Kafexhiu E, Aharonian F, Taylor AM, Vila GS (2014) Parametrization of gamma-ray production cross sections for p p interactions in a broad proton energy range from the kinematic threshold to PeV energies. Phys Rev D 90(12):123014. https://doi.org/10.1103/PhysRevD.90.123014
147. Kahn FD (1975) Supernova remnants. In: International cosmic ray conference, vol 11, p 3566
148. Kappes A, Hinton J, Stegmann C, Aharonian FA (2007) Potential neutrino signals from galactic gamma-ray sources. Astrophys J 656:870–896. https://doi.org/10.1086/508936 [Erratum: Astrophys J 661:1348(2007)]
149. Katz B, Waxman E (2008) In which shell-type SNRs should we look for gamma-rays and neutrinos from P-P collisions? J Cosmol Astropart Phys 1:018. https://doi.org/10.1088/1475-7516/2008/01/018
150. Kelner SR, Aharonian FA, Bugayov VV (2006) Energy spectra of gamma rays, electrons, and neutrinos produced at proton-proton interactions in the very high energy regime. Phys Rev D 74(3):034018. https://doi.org/10.1103/PhysRevD.74.034018
151. Klein RI, McKee CF, Colella P (1994) On the hydrodynamic interaction of shock waves with interstellar clouds. 1: Nonradiative shocks in small clouds. Astrophys J 420:213–236. https://doi.org/10.1086/173554
152. Klepach EG, Ptuskin VS, Zirakashvili VN (2000) Cosmic ray acceleration by multiple spherical shocks. Astropart Phys 13:161–172. https://doi.org/10.1016/S0927-6505(99)00108-5
153. Kopper C (2017) Observation of astrophysical neutrinos in six years of IceCube data. Proc Sci 301:981. https://doi.org/10.22323/1.301.0981
154. Koyama K, Inui T, Matsumoto H, Tsuru TG (2008) A time-variable X-ray echo: indications of a past flare of the galactic-center black hole. Publ Astron Soc Jpn 60:S201–S206. https://doi.org/10.1093/pasj/60.sp1.S201
155. Krymskii GF (1977) A regular mechanism for the acceleration of charged particles on the front of a shock wave. Akademiia Nauk SSSR Doklady 234:1306–1308
156. Kulsrud R, Pearce WP (1969) The effect of wave-particle interactions on the propagation of cosmic rays. Astrophys J 156:445. https://doi.org/10.1086/149981
157. Kulsrud RM, Cesarsky CJ (1971) The effectiveness of instabilities for the confinement of high energy cosmic rays in the galactic disk. Astrophys J Lett 8:189
158. Lagage PO, Cesarsky CJ (1983) The maximum energy of cosmic rays accelerated by supernova shocks. Astron Astrophys 125:249–257
159. Lahmann R (2016) Acoustic detection of neutrinos: review and future potential. Nucl Part Phys Proc 273:406–413. https://doi.org/10.1016/j.nuclphysbps.2015.09.059
160. Lalakulich O, Mosel U (2015) GiBUU and shallow inelastic scattering. AIP Conf. Proc. 1663:040004. https://doi.org/10.1063/1.4919474
161. Landau LD, Lifshitz EM (1959) Fluid mechanics, 2nd edn. 1987. Pergamon, Oxford
162. Lee S (1998) Propagation of extragalactic high energy cosmic and γ rays. Phys Rev D 58(4):043004. https://doi.org/10.1103/PhysRevD.58.043004
163. Lozinskaya TA (1992) Supernovae and stellar wind in the interstellar medium New York: American Institute of Physics, 1992
164. Mac Low M-M, McKee CF, Klein RI, Stone JM, Norman ML (1994) Shock interactions with magnetized interstellar clouds. I. Steady shocks hitting nonradiative clouds. Astrophys J 433:757. https://doi.org/10.1086/174685
165. Malkov MA (1997) Analytic solution for nonlinear shock acceleration in the Bohm limit. Astrophys J 485:638–654. https://doi.org/10.1086/304471
166. Malkov MA, Drury LO (2001) Nonlinear theory of diffusive acceleration of particles by shock waves. Rep Prog Phys 64:429–481. https://doi.org/10.1088/0034-4885/64/4/201
167. Malkov MA, Diamond PH, Sagdeev RZ, Aharonian FA, Moskalenko IV (2013) Analytic solution for self-regulated collective escape of cosmic rays from their acceleration sites. Astrophys J 768:73. https://doi.org/10.1088/0004-637X/768/1/73

168. Markov MA, Zheleznykh IM (1961) On high energy neutrino physics in cosmic rays. Nucl Phys 27:385–394. https://doi.org/10.1016/0029-5582(61)90331-5
169. Maxted NI, Rowell GP, Dawson BR, Burton MG, Nicholas BP, Fukui Y, Walsh AJ, Kawamura A, Horachi H, Sano H (2012) 3 to 12 millimetre studies of dense gas towards the western rim of supernova remnant RX J1713.7-3946. Mon Not R Astron Soc 422:2230–2245. https://doi.org/10.1111/j.1365-2966.2012.20766.x
170. McKee CF, Cowie LL (1975) The interaction between the blast wave of a supernova remnant and interstellar clouds. Astrophys J 195:715–725. https://doi.org/10.1086/153373
171. McKee CF, Ostriker JP (1977) A theory of the interstellar medium - three components regulated by supernova explosions in an inhomogeneous substrate. Astrophys J 218:148–169. https://doi.org/10.1086/155667
172. McKenzie JF, Voelk HJ (1982) Non-linear theory of cosmic ray shocks including self-generated Alfven waves. Astron Astrophys 116:191–200
173. Mertsch P (2011) A new analytic solution for 2nd-order Fermi acceleration. J Cosmol Astropart Phys 12:010. https://doi.org/10.1088/1475-7516/2011/12/010
174. Mignone A, Bodo G, Massaglia S, Matsakos T, Tesileanu O, Zanni C, Ferrari A (2007) PLUTO: a numerical code for computational astrophysics. Astrophys J Suppl Ser 170:228–242. https://doi.org/10.1086/513316
175. Montmerle T (1979) On gamma-ray sources, supernova remnants, OB associations, and the origin of cosmic rays. Astrophys J 231:95–110. https://doi.org/10.1086/157166
176. Montmerle T, Cesarsky CJ (1981) Cosmic ray acceleration by stellar winds and self-confinement in giant HII regions. In: Colgate SA (ed) ESA special publication, vol 161, p 319
177. Moriguchi Y, Tamura K, Tawara Y, Sasago H, Yamaoka K, Onishi T, Fukui Y (2005) A detailed study of molecular clouds toward the TeV gamma-ray supernova remnant G347.3-0.5. Astrophys J 631:947–963. https://doi.org/10.1086/432653
178. Morlino G (2017) High-energy cosmic rays from supernovae, p 1711. https://doi.org/10.1007/978-3-319-21846-5_11
179. Morlino G, Blasi P (2016) Spectra of accelerated particles at supernova shocks in the presence of neutral hydrogen: the case of Tycho. Astron Astrophys 589:A7. https://doi.org/10.1051/0004-6361/201527761
180. Morlino G, Caprioli D (2012) Strong evidence for hadron acceleration in Tycho's supernova remnant. Astron Astrophys 538:A81. https://doi.org/10.1051/0004-6361/201117855
181. Morlino G, Amato E, Blasi P (2009) Gamma-ray emission from SNR RX J1713.7-3946 and the origin of galactic cosmic rays. Mon Not R Astron Soc 392:240–250. https://doi.org/10.1111/j.1365-2966.2008.14033.x
182. Morlino G, Blasi P, Amato E (2009) Gamma rays and neutrinos from SNR RX J1713.7-3946. Astropart Phys 31:376–382. https://doi.org/10.1016/j.astropartphys.2009.03.007
183. Moskalenko IV, Porter TA, Strong AW (2006) Attenuation of very high energy gamma rays by the milky way interstellar radiation field. Astrophys J Lett 640:L155–L158. https://doi.org/10.1086/503524
184. Murase K, Guetta D, Ahlers M (2016) Hidden cosmic-ray accelerators as an origin of TeV-PeV cosmic neutrinos. Phys Rev Lett 116(7):071101. https://doi.org/10.1103/PhysRevLett.116.071101
185. Murray SD, White SDM, Blondin JM, Lin DNC (1993) Dynamical instabilities in two-phase media and the minimum masses of stellar systems. Astrophys J 407:588–596. https://doi.org/10.1086/172540
186. Nava L, Gabici S (2013) Anisotropic cosmic ray diffusion and gamma-ray production close to supernova remnants, with an application to W28. Mon Not R Astron Soc 429:1643–1651. https://doi.org/10.1093/mnras/sts450
187. Nava L, Gabici S, Marcowith A, Morlino G, Ptuskin VS (2016) Non-linear diffusion of cosmic rays escaping from supernova remnants - I. The effect of neutrals. Mon Not R Astron Soc 461:3552–3562. https://doi.org/10.1093/mnras/stw1592

188. Neronov A, Semikoz DV (2016) Evidence the galactic contribution to the IceCube astrophysical neutrino flux. Astropart Phys 75:60–63. https://doi.org/10.1016/j.astropartphys.2015.11.002
189. Ohira Y, Murase K, Yamazaki R (2010) Escape-limited model of cosmic-ray acceleration revisited. Astron Astrophys 513:A17. https://doi.org/10.1051/0004-6361/200913495
190. Ohira Y, Murase K, Yamazaki R (2011) Gamma-rays from molecular clouds illuminated by cosmic rays escaping from interacting supernova remnants. Mon Not R Astron Soc 410:1577–1582. https://doi.org/10.1111/j.1365-2966.2010.17539.x
191. Orlando S, Peres G, Reale F, Bocchino F, Rosner R, Plewa T, Siegel A (2005) Crushing of interstellar gas clouds in supernova remnants. I. The role of thermal conduction and radiative losses. Astron Astrophys 444:505–519. https://doi.org/10.1051/0004-6361:20052896
192. Orlando S, Bocchino F, Reale F, Peres G, Pagano P (2008) The importance of magnetic-field-oriented thermal conduction in the interaction of SNR shocks with interstellar clouds. Astrophys J 678:274–286. https://doi.org/10.1086/529420
193. Ostriker JP, McKee CF (1988) Astrophysical blastwaves. Rev Mod Phys 60:1–68. https://doi.org/10.1103/RevModPhys.60.1
194. Padovani M, Galli D, Glassgold AE (2009) Cosmic-ray ionization of molecular clouds. Astron Astrophys 501:619–631. https://doi.org/10.1051/0004-6361/200911794
195. Pagliaroli G, Evoli C, Villante FL (2016) Expectations for high energy diffuse galactic neutrinos for different cosmic ray distributions. J Cosmol Astropart Phys 11:004. https://doi.org/10.1088/1475-7516/2016/11/004
196. Palladino A, Vissani F (2015) The natural parameterization of cosmic neutrino oscillations. Eur Phys J C 75:433. https://doi.org/10.1140/epjc/s10052-015-3664-6
197. Palladino A, Vissani F (2016) Extragalactic plus galactic model for IceCube neutrino events. Astrophys J 826(2):185. https://doi.org/10.3847/0004-637X/826/2/185
198. Palladino A, Winter W (2018) A multi-component model for the observed astrophysical neutrinos. Astron Astrophys 615(A168):17
199. Palladino A, Spurio M, Vissani F (2016) On the IceCube spectral anomaly. J Cosmol Astropart Phys 12:045. https://doi.org/10.1088/1475-7516/2016/12/045
200. Pohl M, Wilhelm A, Telezhinsky I (2015) Reacceleration of electrons in supernova remnants. Astron Astrophys 574:A43. https://doi.org/10.1051/0004-6361/201425027
201. Ponti G, Morris MR, Terrier R, Goldwurm A (2013) Traces of past activity in the galactic centre. In: Torres DF, Reimer O (eds) Cosmic rays in star-forming environments. Astrophysics and space science proceedings, vol 34, p 331. https://doi.org/10.1007/978-3-642-35410-6_26
202. Porter TA, Moskalenko IV, Strong AW, Orlando E, Bouchet L (2008) Inverse compton origin of the hard X-ray and soft gamma-ray emission from the galactic ridge. Astrophys J 682:400–407. https://doi.org/10.1086/589615
203. Press WH, Teukolsky SA, Vetterling WT, Flannery BP (1992) Numerical recipes in C. The art of scientific computing. Cambridge University Press, Cambridge
204. Ptuskin VS, Zirakashvili VN (2003) Limits on diffusive shock acceleration in supernova remnants in the presence of cosmic-ray streaming instability and wave dissipation. Astron Astrophys 403:1–10. https://doi.org/10.1051/0004-6361:20030323
205. Ptuskin VS, Zirakashvili VN (2005) On the spectrum of high-energy cosmic rays produced by supernova remnants in the presence of strong cosmic-ray streaming instability and wave dissipation. Astron Astrophys 429:755–765. https://doi.org/10.1051/0004-6361:20041517
206. Rybicki GB, Lightman AP (1986) Radiative processes in astrophysics pp. 400. ISBN 0-471-82759-2. Wiley-VCH , June 1986
207. Sano H, Sato J, Horachi H, Moribe N, Yamamoto H, Hayakawa T, Torii K, Kawamura A, Okuda T, Mizuno N, Onishi T, Maezawa H, Inoue T, Inutsuka S, Tanaka T, Matsumoto H, Mizuno A, Ogawa H, Stutzki J, Bertoldi F, Anderl S, Bronfman L, Koo B-C, Burton MG, Benz AO, Fukui Y (2010) Star-forming dense cloud cores in the TeV gamma-ray SNR RX J1713.7-3946. Astrophys J 724:59–68. https://doi.org/10.1088/0004-637X/724/1/59
208. Sano H, Fukuda T, Yoshiike S, Sato J, Horachi H, Kuwahara T, Torii K, Hayakawa T, Tanaka T, Matsumoto H, Inoue T, Yamazaki R, Inutsuka S, Kawamura A, Yamamoto H, Okuda T,

Tachihara K, Mizuno N, Onishi T, Mizuno A, Acero F, Fukui Y (2015) A detailed study of non-thermal X-ray properties and interstellar gas toward the γ-ray supernova remnant RX J1713.7-3946. Astrophys J 799:175. https://doi.org/10.1088/0004-637X/799/2/175

209. Sano T, Nishihara K, Matsuoka C, Inoue T (2012) Magnetic field amplification associated with the Richtmyer-Meshkov instability. Astrophys J 758:126. https://doi.org/10.1088/0004-637X/758/2/126
210. Schmid J (2014) Searches for high-energy neutrinos from gamma-ray bursts with the Antares neutrino telescope
211. Schonert S, Gaisser TK, Resconi E, Schulz O (2009) Vetoing atmospheric neutrinos in a high energy neutrino telescope. Phys Rev D 79:043009. https://doi.org/10.1103/PhysRevD.79.043009
212. Schure KM, Bell AR (2014) From cosmic ray source to the galactic pool. Mon Not R Astron Soc 437:2802–2805. https://doi.org/10.1093/mnras/stt2089
213. Schure KM, Bell AR, O'C Drury L, Bykov AM, (2012) Diffusive shock acceleration and magnetic field amplification. Space Sci Rev 173:491–519. https://doi.org/10.1007/s11214-012-9871-7
214. Sedov LI (1959) Similarity and dimensional methods in mechanics. New York: Academic Press
215. Senno N, Murase K, Mészáros P (2016) Choked jets and low-luminosity gamma-ray bursts as hidden neutrino sources. Phys Rev D 93(8):083003. https://doi.org/10.1103/PhysRevD.93.083003
216. Seo J, Kang H, Ryu D (2018) The contribution of stellar winds to cosmic ray production. J Korean Astron Soc 51:37–48. https://doi.org/10.5303/JKAS.2018.51.2.37
217. Skilling J (1971) Cosmic rays in the galaxy: convection or diffusion? Astrophys J 170:265. https://doi.org/10.1086/151210
218. Skilling J (1975) Cosmic ray streaming. I - effect of Alfven waves on particles. Mon Not R Astron Soc 172:557–566. https://doi.org/10.1093/mnras/172.3.557
219. Slane P, Gaensler BM, Dame TM, Hughes JP, Plucinsky PP, Green A (1999) Nonthermal X-ray emission from the shell-type supernova remnant G347.3-0.5. Astrophys J 525:357–367. https://doi.org/10.1086/307893
220. Springer RW (2016) The high altitude water Cherenkov (HAWC) observatory. Nucl Part Phys Proc 279-281:87–94. Proceedings of the 9th cosmic ray international seminar. https://doi.org/10.1016/j.nuclphysbps.2016.10.013, http://www.sciencedirect.com/science/article/pii/S2405601416301948. ISSN 2405-6014
221. Spurio M (2014) Constraints to a galactic component of the Ice Cube cosmic neutrino flux from ANTARES. Phys Rev D 90(10):103004. https://doi.org/10.1103/PhysRevD.90.103004
222. Stecker FW (1974) Cosmic gamma rays. NASA Special Publication, Washington, p 249
223. Su M, Slatyer TR, Finkbeiner DP (2010) Giant gamma-ray bubbles from fermi-LAT: active galactic nucleus activity or bipolar galactic wind? Astrophys J 724:1044–1082. https://doi.org/10.1088/0004-637X/724/2/1044
224. Szanecki M, Sobczyńska D, Niedźwiecki A, Sitarek J, Bednarek W (2015) Monte Carlo simulations of alternative sky observation modes with the Cherenkov telescope array. Astropart Phys 67:33–46. https://doi.org/10.1016/j.astropartphys.2015.01.008
225. Tanaka T, Uchiyama Y, Aharonian FA, Takahashi T, Bamba A, Hiraga JS, Kataoka J, Kishishita T, Kokubun M, Mori K, Nakazawa K, Petre R, Tajima H, Watanabe S (2008) Study of nonthermal emission from SNR RX J1713.7-3946 with suzaku. Astrophys J 685:988–1004. https://doi.org/10.1086/591020
226. Tavani M et al (2010) Direct evidence for hadronic cosmic-ray acceleration in the supernova remnant IC 443. Astrophys J Lett 710:L151–L155. https://doi.org/10.1088/2041-8205/710/2/L151
227. Taylor G (1950) The formation of a blast wave by a very intense explosion. I. Theoretical discussion. Proc R Soc Lond Ser A 201:159–174. https://doi.org/10.1098/rspa.1950.0049
228. Thomas LH (1949) Elliptic problems in linear differential equations over a network. Watson Science Computer Laboratory Report, Columbia University, New York

229. Tiffenberg J (2009) Limits on the diffuse flux of ultra high energy neutrinos set using the Pierre Auger Observatory. In: 31th International cosmic ray conference (ICRC2009)
230. Truelove JK, McKee CF (1999) Evolution of nonradiative supernova remnants. Astrophys J Suppl Ser 120:299–326. https://doi.org/10.1086/313176
231. Uchiyama Y, Aharonian FA (2008) Fast variability of nonthermal X-ray emission in Cassiopeia a: probing electron acceleration in reverse-shocked ejecta. Astrophys J Lett 677:L105. https://doi.org/10.1086/588190
232. Uchiyama Y, Aharonian FA, Tanaka T, Takahashi T, Maeda Y (2007) Extremely fast acceleration of cosmic rays in a supernova remnant. Nature 449:576–578. https://doi.org/10.1038/nature06210
233. Uchiyama Y, Funk S, Katagiri H, Katsuta J, Lemoine-Goumard M, Tajima H, Tanaka T, Torres DF (2012) Fermi large area telescope discovery of GeV gamma-ray emission from the vicinity of SNR W44. Astrophys J Lett 749:L35. https://doi.org/10.1088/2041-8205/749/2/L35
234. Unger M, Farrar GR, Anchordoqui LA (2015) Origin of the ankle in the ultrahigh energy cosmic ray spectrum, and of the extragalactic protons below it. Phys Rev D 92(12):123001. https://doi.org/10.1103/PhysRevD.92.123001
235. Vannoni G, Gabici S, Aharonian FA (2009) Diffusive shock acceleration in radiation-dominated environments. Astron Astrophys 497:17–26. https://doi.org/10.1051/0004-6361/200809744
236. Vercellone S (2014) The next generation Cherenkov telescope array observatory: CTA. Nucl Instrum Methods Phys Res A 766:73–77. https://doi.org/10.1016/j.nima.2014.04.015
237. Vietri M (2008) Foundations of high-energy astrophysics. Theoretical astrophysics. Chicago University Press, Chicago. http://cds.cern.ch/record/1109401
238. Villante FL, Vissani F (2008) How precisely neutrino emission from supernova remnants can be constrained by gamma ray observations? Phys Rev D 78:103007. https://doi.org/10.1103/PhysRevD.78.103007
239. Vink J (2012) Supernova remnants: the X-ray perspective. Astron Astrophys Rev 20:49. https://doi.org/10.1007/s00159-011-0049-1
240. Vissani F (2006) Neutrinos from galactic sources of cosmic rays with known gamma-ray spectra. Astropart Phys 26:310–313. https://doi.org/10.1016/j.astropartphys.2006.07.005
241. Vissani F, Aharonian FA (2012) Galactic sources of high-energy neutrinos: highlights. Nucl Instrum Methods Phys A692:5–12. https://doi.org/10.1016/j.nima.2011.12.079
242. Vissani F, Aharonian FA, Sahakyan N (2011) On the detectability of high-energy galactic neutrino sources. Astropart Phys 34:778–783. https://doi.org/10.1016/j.astropartphys.2011.01.011
243. Völk HJ, Bernlöhr K (2009) Imaging very high energy gamma-ray telescopes. Exp Astron 25:173–191. https://doi.org/10.1007/s10686-009-9151-z
244. Völk HJ, Berezhko EG, Ksenofontov LT (2005) Magnetic field amplification in Tycho and other shell-type supernova remnants. Astron Astrophys 433:229–240. https://doi.org/10.1051/0004-6361:20042015
245. Wang ZR, Qu Q-Y, Chen Y (1997) Is RX J1713.7-3946 the remnant of the AD393 guest star? Astron Astrophys 318:L59–L61
246. Warren JS, Hughes JP, Badenes C, Ghavamian P, McKee CF, Moffett D, Plucinsky PP, Rakowski C, Reynoso E, Slane P (2005) Cosmic-ray acceleration at the forward shock in Tycho's supernova remnant: evidence from chandra X-ray observations. Astrophys J 634:376–389. https://doi.org/10.1086/496941
247. Weaver R, McCray R, Castor J, Shapiro P, Moore R (1977) Interstellar bubbles. II - structure and evolution. Astrophys J 218:377–395. https://doi.org/10.1086/155692
248. Woltjer L (1972) Supernova remnants. Annu Rev. Astron Astrophys 10:129. https://doi.org/10.1146/annurev.aa.10.090172.001021
249. Yang RZ, Jones DI, Aharonian F (2015) Fermi-LAT observations of the Sagittarius B complex. Astron Astrophys 580:A90. https://doi.org/10.1051/0004-6361/201425233

250. Yang RZ, Kafexhiu E, Aharonian FA (2018) Exploring the shape of the γ-ray spectrum around the "π^0-bump". Astron Astrophys 615:A108. https://doi.org/10.1051/0004-6361/201730908
251. Zatsepin GT, Kuz'min VA (1966) Upper limit of the spectrum of cosmic rays. ZhETF Pisma Redaktsiiu 4:114
252. Zheleznykh I (2006) Early years of high-energy neutrino physics in cosmic rays and neutrino astronomy (1957–1962). Int J Mod Phys A 21:1–11. https://doi.org/10.1142/S0217751X06033271
253. Zirakashvili VN, Aharonian F (2007) Analytical solutions for energy spectra of electrons accelerated by nonrelativistic shock-waves in shell type supernova remnants. Astron Astrophys 465:695–702. https://doi.org/10.1051/0004-6361:20066494
254. Zirakashvili VN, Aharonian FA (2010) Nonthermal radiation of young supernova remnants: the case of RX J1713.7-3946. Astrophys J 708:965–980. https://doi.org/10.1088/0004-637X/708/2/965
255. Zweibel EG, Shull JM (1982) Confinement of cosmic rays in molecular clouds. Astrophys J 259:859–868. https://doi.org/10.1086/160220

Index

S. Celli, *Gamma-ray and Neutrino Signatures of Galactic Cosmic-ray Accelerators*, Springer Theses, https://doi.org/10.1007/978-3-030-33124-5

Printed in the United States
By Bookmasters